THE EDGE
OF AN
UNFAMILIAR
WORLD

THE EDGE
OF AN
UNFAMILIAR
WORLD

A History of Oceanography

SUSAN SCHLEE

E. P. DUTTON & CO., INC. | NEW YORK | 1973

Published simultaneously in Canada
by Clarke, Irwin & Company Limited, Toronto and Vancouver

SBN: 0–525–09673–6
Library of Congress Catalog Card Number: 72–82705

To Scottland,
Lynn Forbes, and John

ACKNOWLEDGMENTS

From the time that Herman Mauskopf, my history of science professor at Duke Unversity, encouraged me to undertake a book on the history of oceanography, till the time the last proofs were read, a most generous group of people have helped me with this history.

There has been, for example, C. Lloyd Claff, who introduced me to the old and venerable Marine Biological Laboratory at Woods Hole, Massachusetts, and to H. Burr Steinbach, who was then its director. With their help I was given a place to work at the Laboratory. With still more help from Jane Fessenden, Virginia Brandenburg, Doris Ricker, and Lenora Joseph, MBL's patient librarians, I was shown how to "turn over half a library to write one book."

During the course of the next several years, I regularly left the stacks and walked along Water Street at the edge of Great Harbor to visit, interview, badger and question the scientists who work at the Woods Hole Oceanographic Institution. Mary Sears' office was my most frequent stop and Mary my most constant advisor. Next door, K. O. Emery gave me references on every imaginable subject, and on the floor below, Dave Ross read several of my chapters. (His textbook on oceanography was a valuable reference.)

7

As the book progressed, Allyn Vine and Dean Bumpus guided me through the endless documents pertinent to World War II, Tom Kelley let me read the log books he kept as First Officer aboard the *Atlantis*, and Frank Manheim at the U.S. Geological Survey, Roland Wigley at the National Marine Fisheries Service, and John Milliman, Dick Backus, and Bill von Arx at the Oceanographic all gave me advice in their special fields.

Some days I walked farther along Water Street, crossed School Street, and went up Maury Lane to visit Alfred Redfield, Senior Scientist Emeritus at WHOI. Before writing almost every chapter I tried out my ideas on Dr. Redfield, who listened, advised, and encouraged.

Behind Dr. Redfield's house sits Blake Building and there, until recently, the Oceanographic's archives were stored. Jeff Allen knows his way through those files and boxes better than anyone and he and Stanley Reeves took time to unearth the letters, reports, and photographs that I wished to see.

I am indebted also to persons and facilities far from Woods Hole—to the Museum of Comparative Zoology at Harvard University, to the British Museum of Natural History in London, and to my friends Daniel Merriman at Yale University, Orrin Pilkey at Duke University, and Robert Gerard at the Lamont-Doherty Geological Observatory.

Help of still another sort came from Jane Peterson, who typed and retyped the manuscript; from Frank Medeiros, who struggled to photograph many old and wrinkled maps and cruise reports; and from Ann Martin, who prepared the index.

Lastly, a most special thanks must go to Lynn Forbes, and to my husband, son, and parents who have lived with this book and who have welcomed its presence for so long a time.

Susan Schlee
Woods Hole, Massachusetts
March, 1972

CONTENTS

INTRODUCTION

A few years ago, oceanography in the United States enjoyed a popularity that was unprecedented in its intensity and extent. As a key that could unlock federal funds and private moneys, "Oceanography" was invoked by congressmen, investors, educators, and scientists, until, at the height of its popularity, such disparate topics as fishmeal, tuna-tagging, submarines, oil wells, beach erosion, sewage disposal, marine law, college curricula, scuba diving, summer camps, magazines, and endless committees and commissions were all considered to be in some way oceanographic.

That time is past now, and oceanography—and oceanographers—have returned to the relative obscurity of the scientific community, and there they are doing very well indeed. In its *Directory of Oceanographers* for 1969, the National Academy of Sciences listed some 1,500 in the United States alone who were working in the broad fields of biological, chemical, geological, physical, and geophysical oceanography. Most of these persons are employed by federal agencies, private institutions, and universities, and today they are concerned with such problems as whether the sea floor spreads, how ocean waters circulate, how salts, minerals, and gases are distributed throughout the sea, and where marine animals live. Also what causes the great fluctuations in the world's fisheries, and how were the deep ocean basins formed?

The money to pay for these expensive studies, which includes funds for the operation of the nation's 500 or so research vessels, comes for the most part from the federal government, whose annual budget for oceanographic work of all kinds is now nearly half a billion dollars. Many millions more come from private sources, and together public and private funds support a substantial undertaking. These examples and statistics could be multiplied many times over by considering the oceanographic programs supported by Russia, Japan, and western European countries which involve thousands of ships and men and millions of dollars. Yet a hundred years ago no one used the word "oceanography,"* and the exploration of the sea in its own right had hardly begun. What happened? Why has oceanography become so important and how has it grown so fast?

To ascribe the beginnings of oceanography to a particular time in history it is first necessary to choose one of a dozen definitions of the word. One of the clearest and most inclusive was given by the American oceanographer Henry Bigelow in 1928.

> Oceanography . . . is the study of the world beneath the surface of the sea; it should include the contact zone between the sea and the atmosphere. . . . It has to do with all the characteristics of the bottoms and margins of the sea, of the sea water, and of the inhabitants of the latter. It is thus widely inclusive, combining Geophysics, Geochemistry and Biology. Inclusiveness is, of course, characteristic of any "young" science, and modern Oceanography is in its youth. But in this case it is not so much youth that is responsible for the fact that these several sub-sciences are still grouped together, but rather the realization that the Physics and Chemistry and Biology of the sea water are not only important

---

* In French the word "océanographie" was used in the late 1500's, but did not survive. It reappeared in 1878 (in a supplementary volume of the *Grand Dictionnaire Universel*) and was defined as "*la déscription de l'océan.*" At about the same time, the Germans were using the word "*Thalassographie*" to mean the study of gulfs and seas. In the early 1880's the German chemist William Dittmar was among the first to use the word "oceanography" in English and by the close of the century, it was in common use.

per se, but that in most of the basic problems of the sea all three of these subdivisions have a part. And with every advance in our knowledge of the sea making this interdependence more and more apparent, it is not likely that we shall soon see any general abandonment of this concept of Oceanography as a mother science, the branches of which, though necessarily attacked by different disciplines, are intertwined too closely to be torn apart.Ψ*

Using this definition, the search for the science's beginnings becomes an attempt to recognize the time when enough had been learned of the sea to make the interdependence of its phenomena apparent and when this realization inspired the organization of oceanic investigations broad enough to consider at least several of the interrelated aspects of the sea.

Of course, much earlier examples of marine science are well known, and oceanography, as a new combination and reorganization of older disciplines, was built upon the information gathered from the sea by generations of inspired polymaths, natural historians, and physicians. In seventeenth-century Britain, for example, Robert Boyle recorded his "Observations and Experiments on the Saltness of the Sea," Sir Isaac Newton advanced an explanation of the tides, and Robert Hooke delivered lectures on methods that might be used for deep-sea research. Like the early studies of marine life, these inquiries were concerned with one aspect of the sea at a time, and usually with such a one as could be poured into a jar and studied in the laboratory without further reference to the rest of the ocean. Nor did the adventurers of the following century study the sea with an eye for its interdependencies, although some had the advantage of government support and could range more extensively over the seas.

In the eighteenth century, European sea powers explored the world to locate new sources of raw materials and to open markets for an increasing abundance of finished goods. To aid the traders who, it was hoped, would venture into the newly opened territories, many nations founded surveying bureaus and hydrographic offices to provide charts, maps, and other navigational aids for the seafarers. These organizations sponsored their own exploratory cruises, and by the early 1800's Napoleon Bonaparte was sending *Le Géographe* and *Le Naturaliste* around the world, Russians and Danes were active on the seas, and the

---

* Throughout the text Ψ refers to notes on sources and other matters, which will be found on pages 363-374.

British, following their own auspicious example set by Captain James Cook, were launching numerous expeditions of an economic, political, and scientific nature. Within this growing tradition of geographical exploration were opportunities for scientific study, although such investigations were generally considered to be a fringe benefit accruing to more important tasks. The scientists on board surveying vessels were often naval surgeons (a profession whose studies included a great deal more zoology and natural history than at present) or guests (like Charles Darwin) invited by the captain to join the expedition. In either case the scientists had no control over the equipment provided, the ports of call, or the time allotted to their labors. Under these conditions, isolated measurements of the sea were made—some deep water sounded, some animals dredged up—but the activities were too unorganized to inspire a widespread interest in the sea.

In the same period, scientists working without the help of government surveying bureaus were studying the less exotic gulfs and bays near home. Scholars, interested in what would later be called marine biology or zoology, spent their summers wading in and rowing over shallow coastal waters. In the 1770's the Danish scientist Otto Müller spent four summers on the southern coast of Norway and after the first could report that "Much more is required to search out the creations of the sea . . . [it] abounds with expense and many forms of danger."⍦

In spite of the difficulties, natural historians were increasingly drawn to the sea, and summer collecting trips, in the manner of the Reverend William Spiers' *Rambles and Reveries of a Naturalist*, enjoyed tremendous popularity. More exacting, and less popular, were the chemical and physical studies of the sea. The French chemist Antoine Lavoisier published an analysis of sea water in 1776; in Britain Charles Blagden related the variable freezing point of water to the concentration of dissolved substances it might contain; and by 1820 a retired London physician, Alexander Marcet, had observed that sea water from different parts of the ocean contained the same ingredients and that these existed in nearly the same proportions relative to each other.

In physics, the French mathematician Marquis Pierre Simon de Laplace elaborated on the Newtonian explanation of the tides; Franz Joseph von Gerstner published the first theory of surface waves in deep water in 1802; James Rennell's studies of currents in the Atlantic Ocean appeared posthumously in 1832; and Gaspard Gustave de

Coriolis described the force named after him which became an important ingredient in the calculations made by physical oceanographers.

In addition to the scientific interest in the sea, there developed more and more practical reasons for studying the ocean. Governments that could afford the expense increased their efforts to have compass deviations charted, prevailing winds plotted, anti-fouling sheathings and compounds tested, and as before, more accurate maps constructed. A very different reason for studying the sea arose from the desire and ability to lay telegraph cables along the bottom of gulfs and channels. By mid-century, more ambitious plans were being considered, and cable companies needed help from government surveying offices to sound and sample prospective cable routes that crossed entire oceans. These surveys gathered more information from the sea than was intended or expected, and their discoveries of deep-sea animals and planktonic oozes added to both the knowledge of and the interest in the sea.

The situation in the middle of the nineteenth century, then, was such that a growing number of scientists knew enough about the sea to suspect that the few pieces of information available were parts of a fascinating and complexly interrelated system; better instruments and new machines (such as the important steam-powered winch) were available; and governments, having at hand the means to launch extensive explorations, were becoming accustomed to paying for what they considered to be useful information. From this combination of interest, technical ability, and financial support there arose two distinctive styles of oceanography.

In the United States, the physical geography of the sea, as early oceanographic studies were sometimes called, was a kind of seaward extension of that country's rapid territorial expansion and exploration. The Coast Survey (now the National Ocean Survey) was established to chart the nation's coastline and survey her inshore waters, while the high seas were the province of the Navy and the concern of the Depot of Charts and Instruments (now the Naval Oceanographic Office). Since the improvement of navigation was the motivation and guiding force behind these bureaus' investigations, it is not surprising that their scientists and surveyors studied the surface condition of the sea—its tide, currents, and winds—rather than its abyssal depths or inhabitants.

In Britain, oceanographic activities began in response to different needs and interests, and rather than evolving as a branch of hydrography, were closely allied to marine biology. The early investigations of the 1860's and 1870's were undertaken to open a whole new collecting ground for natural historians and to gather information believed to be pertinent to the controversial theory of evolution. The summer cruises of the 1860's led to the organization of the well-known *Challenger* expedition, which, circling the globe in three and one half years, collected thousands of animals and sea-floor deposits and inspired numerous other voyages. Its dates, 1872 to 1876, are often used to mark the beginning of an era of oceanographic exploration in which scientists embarked upon long cruises to collect the information necessary for a general description of the sea.

At the turn of the century a new style of oceanographic investigation evolved in response to scientists' desire to understand the ocean as a constantly changing system. These dynamic studies, as they were called, were started primarily by a group of Scandinavian physicists and chemists. Their work in physical oceanography received considerable support from northern European ministries of agriculture and fisheries when it was found that the behavior of certain food fish was apparently influenced by the physical properties of the sea. With support from ailing fisheries, scientists also undertook marine biological studies and the International Council for the Exploration of the Sea, which was established in 1902 to give fisheries the benefit of scientific research, fostered and coordinated a great portion of European oceanography in the first decades of the twentieth century. Some of the studies thus organized were temporarily suspended during World War I (and in a few instances replaced with acoustical studies), but the war did not alter the familiar patterns of research and these continued to be dominated by an interest in marine biology.

World War II radically changed both the structure and the scale of oceanographic research. Acoustical studies, begun to meet the need for submarine detection, led to new studies of the physical properties of the sea, and investigations of waves and tides, made in anticipation of amphibious landings, placed a new emphasis on dynamic studies which had changed very little since their promising start at the turn of the century. Geological and geophysical investigations expanded after the war and benefited considerably from the new instruments and techniques that had been developed. Biological studies, on the other hand, were supported at a low level during the war and subsequently declined in importance relative to physical and geophysical oceanography.

In the years immediately following the war, European nations did not have the financial resources to invest in the expensive pursuit of oceanography, and although the United States government was not sure whether it wanted, or needed, to support the field at its expanded wartime level, the leaders and leadership in the field crossed to America. Dependence on federal funds continued, and oceanography was linked, as it had been during the war, to the needs of the nation as understood by the draftsmen of the national budget.

Since World War II, and more especially since 1960, when the decision was made to build and maintain a fleet of Polaris submarines, the most expensive national need involving oceanography has been that of security, and the Navy in particular is committed to learning all it can of the undersea environment. Other needs have emerged that may be met in part by the sea's resources. Offshore oil and natural gas have been tapped since the 1920's and each year provide a greater percentage of the world's fuel. More and more food is taken from the sea as an exploding population rapidly outgrows the crops produced on land. More seaside recreation areas are desired, more industries want waterfront sites, and more dumping areas for incredible amounts of refuse are needed.

These primary needs have created secondary ones, and the past decades' explosive increase in oceanographic activity has required expanded educational programs, a clearer formulation of the law of the sea, large data management systems, and the establishment of numerous administrative bodies.

By the 1960's the costly advantages of ocean exploration—the military superiority supposedly to be gained and the materials to be acquired—were no longer considered options, but necessities. The questions remaining were not *whether* men would use the sea, but how, when, and with what foresight.

In discussing the history of oceanography, boundaries for the breadth of the topic must be established, as well as chronological limits. Considering the latter limitation first, I have chosen to begin my study in the mid-1800's with the emergence on both sides of the Atlantic of relatively large-scale, government-sponsored activities. I have continued to trace the growth of oceanography through World War II, and have concluded with a chapter on contemporary geophysical studies which I believe describes the kind of work being done today.

As to breadth or scope, I have chosen to focus on the scientific

content of oceanography and to include its instrumentation and application only insofar as they affect the direction of scientific thought. As
in other sciences, the progress of oceanography in its largest sense has
not been determined by technical ability but by ideas. The more basic
question to answer is not when did men design this or that piece of
deep-sea equipment, but what scientific question did they hope to
answer by designing it? Why, in other words, have men wanted to
study the sea?

The selection of a framework within which to arrange the myriad of
persons, thoughts, inventions, and events which have combined in so
many ways to form a part of the study of the sea, is bound to distort
the subject—to neglect some aspects while artificially emphasizing
others. I have not attempted to present an exhaustive picture of
oceanographic activity during the last century and a quarter but have
tried instead to trace the various motivations which have led scientists
to study the sea and to describe the irregular progress that they have
made toward understanding a vast and mysterious region.

Mermaids and tow net, the first illustration in the *Challenger* Report.
(*From the* Challenger *Report, Narrative, Vol. 1, First Part*)

Strange and beautiful things were brought to us from time to time, which seemed to give us a glimpse of the edge of some unfamiliar world.

—Sir Charles Wyville Thomson,
"The *Challenger* Expedition," 1876.

THE EDGE
OF AN
UNFAMILIAR
WORLD

The U.S. surveying brig *Washington* was caught off Cape Hatteras in the hurricane of September, 1846, just after completing one of the earliest surveys of the Gulf Stream. (*Courtesy Nantucket Historical Association*)

CHAPTER I

Oceanography in Nineteenth-Century America

U.S. brig *Washington*, off Cape Hatteras on a survey of the Gulf Stream for U.S. Coast Survey. September 8, 1846. Pilot's statement:

Blowing a perfect hurricane;—never knew it to blow so; sea making breach over vessel; white foam in every direction. . . . Sea curled over and tumbled upon vessel. Tremendous sea—crushing deck cabin and sweeping deck of everything. . . . Water and everything in hold. Tore up berth deck, beams and all; all but those lashed were washed over;—stooping over to saw [anchor] chain [I] felt crash, and found [myself] underwater—came up under lee of vessel; . . . swam and got hold of starboard anchor. When getting on deck saw Capt. Bache and several men overboard astern, apparently as if stunned or hurt . . . tried to get some moveable article to bend on to end of line and throw overboard . . . another sea came and was almost suffocated. This was the last I saw of the captain. Pieces of wreck were buried in water under a crushing sea . . . looked, but saw him no more.

Early in the 1800's a cautious and deliberate United States Congress agreed to support those studies of the sea which promised to extend the life expectancy of the country's seafarers and to minimize the sacrifice of cargo that was spread each year on poorly charted waters. To attain these goals, a task which required much more surveying and mapping than it did oceanographic investigation, the government established two agencies—the United States Coast Survey in 1807, and the Depot of Charts and Instruments within the United States Navy in 1830. From these two came the required navigational aids and, in addition, the nation's first sustained programs of marine science. Neither the Survey nor the Depot became immediately involved in oceanographic studies, but by the 1840's both were measuring the depth of the sea and were examining its currents, temperatures, salinities, and bottom deposits. From this time until the last quarter of the century, when private institutions began supporting marine science and when the government organized its Fish Commission, the Coast Survey and the Depot—and for a brief time the infamous United States Exploring Expedition—were the only forces in the country to advance the very shaky young science of oceanography. An account of nineteenth-century oceanography in the United States must begin, then, with an account of the Coast Survey, the Depot, and the Exploring Expedition.

In 1807, when the future first superintendent of the Coast Survey asked President Thomas Jefferson for a job in the government that would make use of his talents as a surveyor, it was generally agreed that the Constitution prohibited the federal government from supporting scientific activities.

"It is their opinion," the House of Representatives had stated, referring to its members who were denying an amateur astronomer's request for funds, "that application to Congress for pecuniary encouragement of important discoveries, or of useful arts, cannot be complied with, as the Constitution . . . appears to have limited the powers of Congress to granting patents only."✢

This openly negative attitude toward science was not as intransigent a position as it appeared, however, for many of the discoveries and useful arts which could not rightfully be supported as scientific activities could be tailored to fit a less controversial category of government activity—namely, the support of commerce. If commerce could be given as the reason for the government's support of the Lewis and Clark Expedition (and it had been), then commerce could serve equally well as the reason for establishing a surveying bureau or launching a great exploring expedition. And serve it did.

By the time Jefferson had been in office some half-dozen years, he had been persuaded by a powerful and articulate group of East Coast merchants that demands for a marine survey were both reasonable and constitutionally correct. In 1807, therefore, the Coast Survey was established as a civilian agency within the Treasury Department. It was given the handsome budget of $50,000 and the ambitious task of charting the country's harbors, ports, and coastal waters. Jefferson appointed a Swiss, Ferdinand Rudolph Hassler, as the first superintendent, and the organization was off to a slow and indecisive start.

Hassler's failure to circumvent the multitude of problems that, for twenty-five years, blocked the establishment of a viable survey, was not due to any lack of personal merit (although his manner with Congress was less than engaging), but resulted instead from the government's lack of experience as a patron of science.

Should the government support any scientific activities at all? Congress asked. And for those members who answered "yes," the next decision to make was, what kind of science? Should only practical investigations which promised immediate benefits to seamen or fishermen be undertaken? Or could more basic studies which might only someday be useful be justified as well? Hesitantly, inconsistently, the government began to answer, or at least to face, some of these questions as it dealt with the Coast Survey.

In spite of the troubles besetting the Survey, Hassler and the men he had trained finally embarked upon an uninterrupted period of mapping and charting in the summer of 1832.*

Work was begun in the vicinity of New York harbor, and while some men surveyed the shoreline, others, with boats borrowed from the Revenue Cutter Service (precursor of the Coast Guard), began to chart the waters. Although a few haphazard surveys of the country's territorial waters had been made by foreign navies, private firms, and the U.S. Navy, none of the waters which lapped and occasionally pounded the eastern seaboard had been carefully or systematically examined. Consequently, Hassler's men made frequent discoveries of unsuspected rocks, shoals, sinks, and channels. The discovery of an alternate deep-water channel into New York harbor, for example, convinced many of the Survey's worth, but Hassler's more usual policy of *not* publishing the findings he made each season soon outweighed the excitement generated by a new channel.

By the close of the 1842 field season, the last before Hassler's death, the Survey had gathered enough data to chart coastal waters from Rhode Island to the Chesapeake Bay, but few of these charts had been published, and Congress was tired of waiting for results. Hassler died in 1843 and was succeeded by Alexander Dallas Bache, who knew very well how to get results. In the 24 years that Bache led the Survey, he

---

* Actually, Hassler had begun surveying in 1816 and 1817, but his work had been stopped almost immediately when Congress passed a provision which required all Survey employees to be military or naval officers. Out went Hassler and with him all ideas of how to organize a survey. Commercial pressure did not reactivate the Survey until 1832, at which time Congress reversed its previous decision.

made that organization into the largest, most powerful scientific agency within the government.

While Hassler had confined his work to surveying, and had not seen the pursuit of oceanographic investigations as part of the Survey's work as Bache certainly did, he had successfully organized a new agency whose work was in some degree scientific and whose constitutionality was, for a time at least, unquestioned.

The second public agency to find its way into the emerging field of oceanography was the Depot of Charts and Instruments and, like the Survey, the Depot got off to a slow start as far as marine science was concerned. In 1830, a Lieutenant L. M. Goldsborough, perturbed by the Navy's lack of concern for the care and storage of its navigational instruments, took his complaints to the Secretary of the Navy. Goldsborough claimed that "such instruments as are not on ship-board are dispersed about among the naval stores of our yards, in charge of individuals perfectly unacquainted with such matters, and corroding and becoming ruined for want of proper attention."◆

And that was not all. He reported that there was no system for testing or comparing the instruments that the Navy used and there was no central storage facility for either instruments or charts, both of which were needed on every cruise.

The Secretary of the Navy agreed that the situation was not good and in 1830 authorized the establishment of a Depot of Charts and Instruments, which was patterned to some extent after the French Dépôt des Cartes, Plans, Journaux et Mémoires Relatifs à la Navigation and the British Hydrographic Office. Lieutenant Goldsborough became superintendent of the new Depot, and at first it seemed that he had in fact prepared the ground for an astronomical observatory rather than for an agency concerned with marine science. Even within Goldsborough's short term as superintendent—he was succeeded in 1833— the astronomical observations necessary for testing ships' chronometers were begun, and the second superintendent, Lieutenant Charles Wilkes, built a small observatory on Capitol Hill. The astronomical aspects of the Depot's work were still further expanded by the third superintendent.

Then, in 1842, a year before Bache lit a fire under the modestly advancing Survey, Lieutenant Matthew Fontaine Maury became superintendent of the Depot and abruptly replaced his predecessors' emphasis on astronomy with his own interest in marine meteorology and in what he called the physical geography of the sea.* Like Bache, Maury inherited a small, relatively young agency which he built into a much larger, more active organization. Under Maury, the Depot planned and carried out some of the earliest investigations of the deep oceans.

But just before the Depot and the Survey embarked upon their respective adventures, Congress extended a significantly different kind of support to marine science. In 1838 it sent the United States Exploring Expedition straggling around the world to collect all the information it could from lands and seas alike.

There were many reasons for organizing such an expensive venture as an exploring expedition, and the project's supporters leaped agilely from one to another according to the interests of their listeners. Statesmen and scientists were encouraged to see an exploring and surveying expedition as an important mark of political and scientific stature. Britain had already sent such famous explorers as Captain James Cook, Sir John Ross, and Robert Fitzroy (with Charles Darwin) all over the world, and France had sponsored the daring explorations of Dumont d'Urville, the Chevalier de Bougainville, and the Comte de la Pérouse. It was time, some people thought, that America showed that she could do the same.

To others, with a more economic turn of mind, the expedition was presented as an opportunity to enhance the country's commercial prospects. There were new areas to be opened for trade, better routes to be found for New England whalers, and at least a few new islands to be discovered for the insatiable fur-seal industry. These commercial reasons had great appeal to Congress, while men of science did not have to be reminded that "countries have looked beyond the mere discovery of new lands, new commercial resources, and territorial aggrandizement. Their efforts have been directed towards an increase of knowledge in every branch of science. . . ."⸸

With all these benefits to be gained, Congress authorized the President to send out a surveying and exploring expedition to the Pacific Ocean and the South Seas and in 1836, $300,000 was appropriated for the project.

As originally conceived, the scientific portion of the expedition was

---

* In 1866 the two interests within the Depot were finally separated into the Naval Hydrographic Office and the Naval Observatory.

to be of especial benefit to the wide field of natural history and studies
were to be made in geology, mineralogy, botany, vegetable chemistry,
zoology, ichthyology, ornithology, and ethnology. Some physical
studies such as meteorology and astronomy were also included, and all
in all, at least 20 civilian scientists were considered necessary to carry
out the broad and comprehensive program. As for the Navy's part in
the undertaking, its men were expected to provide the transportation
and perhaps to do a little "naval science" if the opportunity arose.

Naturally such a division of labor, and of glory, did not appeal to the
Navy, and especially not to the Secretary of the Navy, Mahlon
Dickerson. If an expedition were to be sent out at all, and the project
seemed one of dubious merit to Dickerson, then the Navy should
control the entire operation. The problem that he immediately encountered was the absence of scientists within the Navy. Had there
been a naval academy analogous to the military one at West Point,
there might have been some naval scientists, but this was idle speculation. The Navy had neither an academy nor the scientists that such an
institution might have produced. This deficiency did not, however,
deter Dickerson, and he insisted that the Navy control the expedition.
From 1836 to 1838 plans were dragged through endless complications
and embarrassments. Personnel was a particular problem, and it seemed
that as soon as scientists and enlisted men were chosen, the command
of the expedition was changed or its goals altered and consequently,
those who didn't like the new conditions resigned and a new round of
appointments was begun. Lieutenant Maury, for example, who was
then the superintendent of the U.S. Gold Mine, was chosen to accompany the expedition as an astronomer, but when command of the
expedition changed, Maury resigned. This kind of folderol continued
until 1838. Then, as the congressional appropriation was about to run
out with nothing but scandal to show for $300,000, Secretary Dickerson
got sick. With that impediment temporarily removed, ships, men, and
equipment were hastily assembled, and the expedition scrambled out to
sea under the command of Lieutenant Wilkes.

This officer, who in 20 years of naval service had earned for himself
the confusing reputation of being scientifically able, but proud and
overbearing, had his own ideas as to how an exploring expedition
ought to be organized, and in the last minutes' hustle and confusion he
had largely got his way. Wilkes clearly preferred physical studies over
natural ones, and naval officers over civilian scientists.

Rear Admiral Charles Wilkes who, in 1838, as a lieutenant, was given command of the United States Exploring Expedition. (*Courtesy U.S. Naval Oceanographic Office*)

"All the duties appertaining to Astronomy, Surveying, Hydrography, Geography, Geodesy, Magnetism, Meteorology, and Physics [are] generally to be exclusively confined to the Navy officers," he had directed, and "these [subjects] are deemed the great objects of this expedition. . . .

"The other scientific Depts. consisting of Zoology, Geology and Mineralogy, Botany and Conchology, it is proposed to fill up as far as can be from among the medical corps [also naval personnel] that will be attached to the Expedition."⅄

There were only nine positions which Wilkes did not fill with Navy men, and these berths were given to civilians. But civilians they hardly were, for under Wilkes' authoritarian rule the scientists were bound by naval regulations, and this amounted to the censorship of all communications "referring to discoveries or any circumstances connected with the progress of your enterprise."⅄

This was control of the worst sort as far as scientists were concerned, and with their admittedly limited experience in dealing with the government, they assumed that such control was an integral part of all military projects. These problems, however, belonged to the future. What was important in August of 1838 was that the United States Exploring Expedition was finally under way.

From Norfolk, Virginia, the warships *Vincennes, Peacock,* and *Porpoise,* the store-ship *Relief,* and the tenders *Sea Gull* and *Flying Fish* sailed slowly across the Atlantic to the Portuguese island of Madeira, recrossed the ocean to Rio de Janeiro, then sailed south to round Cape Horn and enter the Pacific Ocean, where their most important work was to be done. By the time the ships had sailed up the west coast of South America as far as Callao, Peru, however, storms had put half their number out of commission, and much of the work that was to follow was done by vessels working alone or in pairs. What remained of the expedition crossed the Pacific, and while "the scientific gentlemen were occupied making observations and collections in New Holland and New Zealand," two ships, the *Vincennes* and the *Porpoise,* sailed south into the cold Antarctic Sea. Wilkes was looking for a land of great extent that was rumored to lie beyond a drifting barrier of ice.

In the winter of 1839–40, the Antarctic's summer season, Wilkes sailed his ships south until blocked by the edge of the pack ice, then turned west and cruised along the icebound coast of Antarctica. The ice that year was less extensive than usual, and Wilkes was able to sail so close to the unexplored continent that he could see land at intervals for 1,500 miles. At one point his vessels were able to come within a mile of the coast of "Termination Land," as he named it, and collections were made of large masses of rocks which had been rafted toward the open sea on ice floes. This was the most interesting part of the expedition as far as Wilkes was concerned, and his alleged discovery of Antarctica—a claim contested by the British explorer Sir James Clark Ross—was the only well-known achievement of the entire expedition.

There was, however, more to the venture than a polar cruise, and after the civilian scientists had been retrieved from New Zealand, the ships sailed on through many parts of the South Seas. The expedition gave its name to a cluster of islands in the Fiji group, the Exploring Isles, but as the natives in that region were known to be cannibals, no very thorough investigation was made of their territory.

The brig *Porpoise,* one of the ships used for the United States Exploring Expedition. (*Courtesy U.S. Navy Department*)

The expedition worked slowly westward, charting and collecting as it went, until in the spring of 1842 what vessels were left rounded the Cape of Good Hope, sailed one last time across the Atlantic, and finally made landfall off New York harbor.

"America has at last taken her part in the labors of exploration," proclaimed an unusually enthusiastic editorial in the *American Journal of Science and Arts.* "An Exploring Expedition has been sent out, and has returned. . . . The duties have been extremely laborious, beyond the conception of the comfortable house-dweller at home. The loss of one schooner with all hands, including two officers; the total wreck of another vessel—the sloop of war *Peacock*—stripping the crew of everything but their lives; the massacre of two officers by the savages of the Feejee Islands, and of a sailor by the treacherous Kingsmill Islanders, are the only fatal disasters; but they are a few only of its perils. Indeed there were dangers every where, by land as well as by sea. The personal adventures in the course of the cruise, told as simple tales, without exaggeration, would make a volume full of startling incidents, and replete with interest."[¥]

Regardless of the expedition's adventures, personal and scientific, its return aroused little interest. Wilkes and his men had been gone for three years and ten months and had sailed between 80 and 90 thousand miles, yet they were received without ceremony and their exploits regarded with indifference. Little had appeared in newspapers or magazines to sustain public interest, scientists were appalled by the fact that much of the expedition's vast collection had been lost or mishandled, and many members of Congress, when they considered the hundreds of thousands of extra dollars already spent on the endless project, were unabashedly "tired of all this thing called Science here."ᵠ It was hardly the time to ask for additional funds for sorting the collection and for "printing pictures of bugs," yet this was what Wilkes had to do.

Prior to the expedition's return, the newly organized National Institute (later absorbed by the Smithsonian) had been given responsibility for the care of the expedition's collection, and as boxes and crates dribbled in from intermediate ports of call, the specimens were added to the "Extraordinary Curiosities of the National Institute" and stacked in the Great Hall of the Patent Office. Here, the Institute's curator, Henry King, ineptly unpacked the collection, parts of which he inadvertently ruined and parts of which he proudly gave away to his admiring friends.

Some of the collection—which pound for pound was the largest scientific collection in the country—never made it into King's destructive hands. Many jars and boxes had gone down with the two wrecked ships, and other specimens were still on their long and devious way from foreign ports to the Patent Office.

"The *Oregon and California* plants were shipped in 1841 from the Sandwich Isl. [Hawaiian Islands] direct for the United States," wrote a botanist a year after the expedition's return. "The vessel as well as I have been able to make out, touched at Valparaiso and thence proceeded on a voyage to China! then to Europe, where she was sold! and subsequently going on a voyage to the West Indies, finally dropped our plants at Havanna! They have been written for by the Department, and we are in hopes of some day seeing them."ᵠ

The usual procedure for handling such a large collection—usual, that is, for the European scientific societies that generally oversaw the distribution of their countries' collected treasures—was to divide each expedition's booty among specialists of many nationalities. These men would examine whatever plants, animals, bones, or shells fell within their area of expertise, write a report on what they saw, and return their work to a central office for publication. It was understood by all involved that the quicker an expedition could publish and distribute its results, the more influence it would have on the scientific thinking of the day.

Unfortunately, Wilkes could not handle the Exploring Expedition's collection in the same way. First, he was not closely associated with the few weakly organized scientific societies that existed in the United States in his day and therefore could not hope to know all the men to whom the collection should be distributed. Secondly, Congress was reluctant to advance sufficient funds for the publication of the reports; and finally, Congress had insisted that all reports be written by American scholars, which narrowed Wilkes' already limited choice of specialists still further.

In spite of these difficulties, which almost guaranteed the inadequacy of the expedition's report, Wilkes succeeded in browbeating Congress into paying for the preparation of 24 volumes. Twenty of these were eventually published, but it was 32 years before the last of them appeared. Equally damning for any expedition's report was the decision that only 100 copies of each volume were to be published.

"It would be hard to contrive a more effectual plan for defeating the very object of the publication,"* wrote the botanist Asa Gray, whose second volume on plants, written for the expedition, was never even published.

Of the meager number of volumes finally issued, the naturalists, and not the Navy's amateur physical scientists, had written the large majority. Ten of the expedition's 20 volumes concerned natural history, another three dealt with ethnology (the study of the races of man), five were cruise narratives, and only two—one each on hydrography and meteorology—were geophysical.

For the encouragement of marine science the expedition produced four volumes on the natural history of marine animals, notably shallow-water molluscs and crustaceans, and Wilkes' volume on hydrography. Three of the natural-history volumes were written by the expedition's geologist, James Dwight Dana, and all four volumes, with atlases, were published within ten years of the expedition's return, which was a fine record. The reports on marine animals encouraged other scientists to study starfish, clams, sea urchins, or whatever they could find in their locale, and also gave Dana and Augustus Gould, author of *Mollusca and Shells,* an unparalleled opportunity to devote themselves to marine biological studies. Dana, for example, developed his explanation for the formation of coral reefs from the data he'd collected in the Fijis.*

Wilkes' volume on hydrography, which was not distributed until 1873, was less influential than Dana's work; in fact, it apparently had no effect at all on subsequent studies of the seas. By the time his volume appeared, new methods and better equipment were already being used to make much more accurate measurements of depth, temperature, salinity, and density than Wilkes had been able to do. Wilkes had intended to supplement the volume on hydrography with one on physics which was to include a chapter "On the Circulation of the Oceans." Unfortunately, that volume, like several others, was never written.

Although his publications were unfinished and unpublished, there was nothing to stop Wilkes from taking his enthusiastic reports directly to scientists who might be inspired to study the sea. Wilkes toured the eastern states and urged these men to consider "the vast space of our globe occupied by the great ocean, it cannot but strike every one what a wide field is open for investigation and experiment."

To geographers and geologists he proclaimed that sufficient soundings had been made to prove that the topography of the sea floor was, in some places, more mountainous and more remarkable than that of the land and, when large blank spaces on his charts were noticed, he justified the lack of deep-sea soundings with an explanation of the difficulties involved.

> Though in theory the depth is easily to be arrived at, yet to obtain it practically is exceedingly troublesome, requiring much time as well as favorable opportunities.
>   . . . Few are aware that it requires from two to three hours for a well appointed vessel to make a sounding to a depth of 1500 or 2000 fathoms, for which opportunities seldom occur: calms or light winds, and a smooth sea are requisite.
>   . . . It will thus be seen that it is out of the power of an ordinary vessel to make experiments; in order that this interesting enquiry may advance . . ., [continued Wilkes in a prophetic vein] it becomes necessary that some new mode of sounding be

* Dana's part in the controversy over coral reefs is briefly described in Chapter IV.

adopted whereby both the time may be lessened and the opportunities multiplied. It has been suggested to obtain an echo from the bed of the ocean by the explosion of a shell just beneath the surface, the depth to be measured through the propagation of reflected sound.*♅

In 1874, Congress had finally seen enough of the United States Exploring Expedition, and funds for its incomplete report were cut off. Thirty-eight years after its inception, the expedition was over.

On the positive side of any balance intended to weigh the venture's effect on science there must be placed, in addition to collected curiosities, the rather loose precedent which the expedition set in using government funds for the support of a scientific voyage. More than a dozen comparatively small naval expeditions were sent out in the 20 years following the Exploring Expedition's return, and although these were not as apparently interested in science as Wilkes' expedition had been, they did use many of the same channels of funding and communication that Wilkes had so laboriously dredged·across Capitol Hill.

On the negative side were the bitter feelings that the expedition had left with almost everyone involved. The continual confusion that attended the expedition's every phase, the costs that had risen to three times the original estimate, and the scholarly reports that when finally

---

* Wilkes may have been referring to an experiment "with the view of determining the depth of the sea by the echo"♅ which was carried out in 1838 by Professor C. Bonnycastle and R. M. Patterson with the help of the Coast Survey. The Survey provided a brig and over the vessel's stern Bonnycastle suspended a loaded petard which was to be detonated at the proper moment by an officer on board. Bonnycastle then rowed away in a dory to set up a listening station which consisted of a floating platform from which was suspended a long hearing tube made from a stovepipe. An earpiece protruded through a hole in the platform and Bonnycastle hoped that with this half-submerged ear trumpet he would be able to hear both the original explosion and its echo as it came bouncing up from the sea floor. The interval between the two sounds he planned to measure, and since earlier investigators had determined the velocity of sound in water, the length of time between the two noises could be used to determine the depth of the water.

The experiment was performed and two sounds, a third of a second apart, were received. Unfortunately—both for Bonnycastle and the future of oceanography—both sounds came directly from the explosion; the second was not an echo. This became apparent when the experiment was repeated in water of varying depths but the interval between the two sounds remained constant. In spite of the disappointing results, Bonnycastle was still enthusiastic about the principle of his experiment, and blamed his listening device for the temporary failure. He warned others who might follow him not to underestimate the extent to which "the ruffled surface of the ocean acted on sounds proceeding from its depths."♅ (Practical methods of echo sounding were not developed until the 1920's.)

published were unintelligible to most senators and representatives, made the question of the government's promotion of science in any form an extremely delicate and emotional issue. Congress was less certain than ever that it wanted to support science. It seemed an immensely expensive headache.

Another bad effect was the animosity that the expedition generated between civilian and naval scientists. For civilian scientists, many of whom were thoroughly disgusted by the handling of the whole expedition, the experience had taught them what they felt they should have known—science and military support were incompatible. How, they asked, could officers or congressmen be expected to support "work, the nature and scope of which not one . . . in 20 understands?"

For military scientists, on the other hand, the lesson to be learned was to keep the snobbish savants away from their projects. Without their professorial insistence on Latin names and their time-consuming exposition of theories, a most practical sort of science could be undertaken and its benefits easily explained to a willing Congress.

"I was determined to ask no advice or instruction from the savants [in setting up the Depot's scientific program]," wrote Maury, "but to let it be out and out a Navy work."⸸

The quarrel which the Exploring Expedition intensified was as much between two kinds of scientists as it was between two kinds of support. Men like Maury, practical-minded, self-taught military men, had little in common with college-educated scholars and specialists like Bache, and any argument that pitted one against the other further widened the rift that already existed between them.

In 1842, the year the Exploring Expedition returned, Lieutenant Matthew Fontaine Maury was put in charge of the Depot of Charts and Instruments. Almost immediately he embarked upon a series of self-appointed tasks and for the first of these, which was the compilation of wind and current charts, he had a special reason.

When Maury was 25 years old, with several years' experience in the Navy, he had been given the job of sailing master or navigator aboard the *Falmouth* and was, in that capacity, expected to choose the best route for the ship to travel between the country's Atlantic and Pacific coasts. Maury had wanted the trip to be a fast one, a credit to himself as navigator, and so had tried to find wind and current charts which would help him select a route with favoring winds and currents.

Finding that such charts did not exist, Maury had then asked the masters of merchant ships to advise him of the best routes, but again had been disappointed, and surprised as well, to find that navigating on the high seas depended so much more on luck and on jealously guarded personal experience than it did on any sort of cooperative pooling of knowledge.

The cruise aboard the *Falmouth* had been a long one, but the experience convinced Maury that a way ought to be found to consolidate mariners' knowledge of the sea.

When Maury returned to the East Coast he was not assigned immediately to the Depot but was given several jobs which took him away from the sea. In 1839 he was again assigned to sea duty but on his way to the ship suffered a stagecoach accident which permanently crippled his right leg. With broken bones and a badly dislocated knee Maury withdrew to an inn in Somerset, Ohio, and while there, eased

Matthew Fontaine Maury, director of the U.S. Navy's Depot of Charts and Instruments from 1842 to 1861. (*Courtesy U.S. Naval Oceanographic Office*)

the pain and boredom of a long convalescence by writing anonymous articles on the inequities and inefficiencies of the Navy and on the need for immediate reform. In spite of these articles (or, who knows, perhaps because of them), Maury was at last reassigned to active duty and given the superintendentship of the Depot, a position which very much suited his tastes. Once settled in Washington he started to compile the wind and current charts which he had needed so badly aboard the *Falmouth*.

To make such charts, Maury began to employ a method which had been used by a meteorologist, William Redfield, to chart the paths of hurricanes.*

Like Redfield, Maury gathered the information he needed from ships' logs. First he divided certain well-traveled portions of the sea into squares, five degrees on a side, and then he proceeded to list all the wind and current entries for each square that he could find in the logs of naval ships. Maury wished to collect similar information from private ship owners and, to do this, he set up a mutually advantageous system whereby he promised to supply sea captains with copies of the wind and current charts (as soon as they should appear) if they, in turn, would complete special forms called abstract logs, which Maury gave them. His system of barter worked well.

"Never before has such a corps of observers been enlisted in the cause of any department of physical science . . . ," he wrote, "and never before have men felt such an interest with regard to this knowledge."⍦

With the observations thus culled from thousands of voyages, Maury was soon able to suggest the most "probable force and direction of the winds" and, to a lesser extent, of the currents that might be encountered in each of the five-degree squares. The first of his Wind and Current Charts was published in 1847, and with it Maury showed seamen the best routes to choose for voyages to the West Indies, South America, or to the increasingly popular ports of California's golden coast. To seamen who were used to setting a fairly direct

---

* In the 1820's Redfield had begun to reconstruct the paths of hurricanes from information he was able to extract from ships' logs. Putting many accounts of a single storm together, he had been able to trace both the hurricane's course and the varying directions of its winds. In one instance he had used 70 sources to track a storm from St. Thomas in the Virgin Islands, all the way to Nova Scotia. From these studies he had arrived at the unexpected conclusion that "most storms, if not all . . . blow in the form of extensive eddies or whirlwinds."⍦

course from port to port regardless of intervening doldrums or contrary currents, many of Maury's new routes seemed unnecessarily long and devious. In spite of their complexities, however, a few of the bolder masters dared to try them, and, as Maury was not loath to admit, the results were good and, on occasion, spectacular.*

Maury next started work on a whale chart, which, like the wind and current ones, utilized an extensive collection of observations.

"I hope to present our whalers with a chart which will show them the routes and stopping places which the whales make in their annual migrations,"Ψ wrote Maury, and soon whalers were sending him abstract logs which included remarks on every whale they sighted. With these Maury compiled a whale chart which was published in 1851, but while greatly appreciated by whalers, the chart could not by itself do much to halt the steady decline of that industry.

The third, and in some ways the most ambitious, of Maury's collecting schemes was his encouragement of an International Maritime Meteorological Conference that met in 1853 to establish a worldwide system of meteorological observations for the sea as well as for the land. Maury proposed that his system of distributing abstract logs be extended to the ships of many nations and the results used for the benefit of all. Ten nations adopted his plan, and nine others joined the program later. "Thus the sea," said Maury, "has been brought regularly within the domains of philosophical research, and crowded with observers."Ψ

Several observers, however, were notably absent from the conference. Maury had asked the director of the Smithsonian Institution, Joseph Henry, and Coast Survey Superintendent Bache to participate. Neither accepted the invitation. Henry had already organized a network of land-based weather stations under the jurisdiction of the Smithsonian and wished to keep them there, and Bache, although not particularly concerned with marine meteorology, could see that Maury's other interests overlapped his own. He did not like to see the Depot's director becoming too popular with the political hands which fed them both.

Maury's Wind and Current Charts had scarcely begun to appear in

---

* The 15,000-mile passage from New York to California was "the great racecourse of the ocean," and with Maury's charts and Sailing Directions clipper ships could sail from New York to San Francisco in a little over three months. (The record was 89 days.) Some vessels saved as much as a month's time.

the late 1840's, and the work on his whale chart was just beginning, when it became clear to him that a study of the sea, such as he wished to make, required more than the compilation of three or four huge collections of observations. What the Depot needed, or at least what Maury wanted, was a research vessel. With it Maury believed he could undertake a series of scientific explorations similar to those already in progress at the Coast Survey.

Although the superintendents of the Depot and the Survey did not like each other, and certainly the scientific philosophy of one was anathema to the other, still, each kept careful track of what the other was doing and was quick to adopt methods of research which promised good results. The Survey's explorations, begun in 1844, were yielding good results, and consequently, when in 1848 Maury got a ship which he could use for research, he embarked upon a series of explorations which to a considerable extent were patterned on the work begun by the Survey five years before.

In 1843, Alexander Dallas Bache, great-grandson of Benjamin Franklin, became the second superintendent of the Coast Survey. An ambitious man, Bache proudly considered himself one of the country's new professional scientists, and professional too was his ability to interest Congress in his projects. During his reign the Survey's annual budget rose to $500,000, a tremendous sum for the support of activities even secondarily scientific, and the Survey had the added distinction of being the largest employer of mathematicians, astronomers, and physicists in antebellum America.

Bache, who had graduated first in his class from West Point at the age of 19, with interests in physics and chemistry, had subsequently taught these subjects at the University of Pennsylvania. From teaching, his interests had moved toward more general problems of education, and he had gone to Europe to study the organization and philosophies of the educational institutions there. Shortly after his return to the United States, he accepted the position of superintendent of the Survey.

The new "Chief," as he was called by members of an active Washington-Cambridge scientific clique, marked out five areas of activity for the Survey, and one of these was hydrography. Within this category were included the practical duties of sounding, marking shoals and rocks, measuring the direction and velocity of inshore

Alexander Dallas Bache, director of the Coast Survey from 1843 to 1867. (*Courtesy National Ocean Survey*)

currents, and maintaining tide gauge stations. It also could, and soon did, include much of the "pure science" which for Bache was the icing on an otherwise plain cake.

Within a year of his appointment as superintendent, Bache had expanded the Survey's hydrographic activities to include sounding in offshore waters (within 60 miles of the coast), the routine collection and preservation of sediments from the sea floor, and, most important, the first sustained study of the powerful Gulf Stream current.*

To members of Congress and influential friends, Bache proclaimed that the Gulf Stream "is the great sea mark of the coast of the United States, both Gulf and Atlantic, and its qualities as hindrances and aids to navigation require that the navigator should be well informed in regard to it."* This was certainly no exaggeration, but if Bache had been concerned only with navigation he might have been content to

---

* Bache and Maury each claimed the study of the Gulf Stream as the rightful prerogative of his agency. The Survey's responsibility for charting coastal waters put most of the Stream within its province while the Depot's concern for deep water put offshore and even some inshore portions of the Stream within its realm.

repeat a few of his great-grandfather's temperature measurements, update the map of the Gulf Stream that Franklin had had made in the 1770's, and leave it at that. But Bache was motivated at least as much by scientific curiosity as by the expectation of useful results and between 1844 and 1860, the Survey, under his direction, completed 14 transects or lines of study across the Gulf Stream at points between Florida and New Jersey. These investigations, the most sophisticated to be undertaken in the United States, set the pattern for hydrographic research in the western Atlantic until the closing decades of the century.

The data that Bache wished to collect from the Gulf Stream area included soundings, samples of bottom sediments, measurements of the direction and speed of the current itself, and temperature readings. To gather this information Bache brought together a variety of techniques and devices which had been developed by independent scientists and curious mariners over the previous hundred years.

For temperature measurements, which were made at the sea surface and at fairly standard intervals all the way to the bottom, British maximum-minimum thermometers were employed and were usually attached at four, five, or even 10 points along a sounding line. Each instrument registered the warmest and the coldest temperatures that it encountered, and it was assumed that the coldest readings had been made near the bottom and the warmest on top. (This is usually, but not always, the case.) For deep casts, the thermometers were enclosed in glass globes which protected them from the water's pressure. A second way of measuring the temperature of deep water—one which for reasons both physical and mechanical did not work very well—was the use of tubs or insulated water bottles. With these, a sample of deep water was brought to the surface and its temperature measured on board the ship. It was hoped that the water's temperature upon arrival was approximately the same as it had been below. (Maury's officers at the Depot relied on this second method, for the models of thermometers they chose to use were crushed when sent much below 500 fathoms.)*

Using several kinds of thermometers, the officers of the Coast

---

* At 500 fathoms, an instrument is subjected to a pressure of approximately 440 pounds per square inch. Maury tried to protect his thermometers by enclosing them in metal cylinders, but even so they rarely returned undamaged from below 400 or 500 fathoms.

Survey measured water temperatures as they sailed across the Stream, and in the summer of 1846 one of them, Lieutenant George Bache, the superintendent's brother, discovered that "the Gulf Stream is divided into alternate bands of hot, or warm and cool or cold water, the most distinct of which is that containing the axis of the Gulf Stream."[*] Prior to this time it had been thought that the Stream was an unbroken band of warm water.

When the curious temperature readings gathered by Lieutenant Bache were shown to the superintendent, the latter theorized that the alternate bands of warm and cool water were permanent features of the current and were caused by the submarine hills and valleys that were being discovered along the path of the Stream. Others before him had advanced the idea that bottom topography could affect the character of the overlying waters, and Bache extended their hypotheses to the Gulf Stream.

"The correspondence of these [submarine] features with the bands of temperature is plainly marked,"[*] he stated. (When further studies of the Gulf Stream were made by the Survey after the Civil War, Bache's interpretation was shown to be incorrect. It was found that the sea floor beneath the Stream, although irregular, was not corrugated in a pattern of hills and valleys, nor did the position of the warm and cool bands of water remain fixed.)

A study of ocean temperatures gave members of the Survey a fairly good idea of where the Gulf Stream flowed—its waters were so noticeably warmer than the surrounding sea—but to investigate the incredibly swift flow of waters within the Stream itself, other information besides temperatures was needed.

To measure the speed and direction of the current, surveyors released drift bottles which they hoped would be picked up and returned to them, or calculated the current's progress by the drift of their ship. Using the latter method, accurate measurements could be made only if the drift of the ship were plotted in relation to some fixed point. When out of sight of land, Survey officers sometimes tried mooring buoys in midstream to serve as their reference points.

Soundings were an important part of the Gulf Stream study too, and a variety of procedures and devices were used to plumb the depths. The simplest, of course, was a hand-held line with a sounding lead attached, but in water more than several hundred fathoms deep, and especially in areas where a current was known to flow, it was not possible to use this method, for a sailor could not feel the line slacken when the sounding lead hit the bottom. The reason he could not was that the weight of the line already in the water or the drag of the current kept pulling out the line even after the lead had reached bottom. To overcome this difficulty, several new methods of sounding were devised. Some involved gadgets with rotating propellers and mechanical counters, but the one most commonly employed by the Survey made use of a regular sounding line which was marked at intervals with strips of cloth or leather. As this line was pulled off a reel by the descending lead these marks could be timed and when they rolled off the reel more slowly, the weight was assumed to have reached the bottom. Although this method seemed fairly accurate, and had the added advantage of bringing up a little bottom mud on the end of the lead, it was terribly time-consuming. The chore of hauling in a heavy line and weight after a very deep cast could require several hours. This was not a particular problem for the officers of the Coast Survey who rarely worked in more than 1,000 fathoms, but it was for Maury who wished to sound in mid-ocean. Consequently, he devised another method of deep-sea sounding. In 1850, he circulated a pamphlet in which he advocated the use of sounding twine, which was a line so thin that it could not be used to retrieve the lead. After a weight had pulled the marked twine to the bottom, the line was cut and the depth determined by measuring the length of twine that remained on a spool of known length. A great deal of time was saved using this method, but the disadvantage was that no sample of the bottom was retrieved. In an effort to combine the best of both methods sounding "machines" began to be designed in the 1850's, many of which used sounding weights which were automatically detached from the line when they hit bottom. Only a thin line and a small cup or tube, used to collect bottom samples, were hauled back to the surface.

Regardless of how they were made, the Survey's soundings were added to coastal charts, and Bache made sure that these were published at a rate pleasing to Congress. To further assuage the anxieties of those who might worry at the Survey's rapid expansion, Bache began sending surveying and hydrographic teams into several different areas during each field season so that congressmen from the Gulf Coast states, Middle Atlantic ones, and New England were all kept reasonably happy.

Still another part of the Survey's Gulf Stream study was the collection and classification of sea-floor sediments. Bache gave credit to his brother, Lieutenant George Bache, a commander of one of the Survey's ships, for instituting these investigations, and in 1844, presumably at the latter's suggestion, the superintendent ordered all bottom samples retained.*

The first of the Survey's samples to arouse the interest of the academic community were ones collected off the coast of New Jersey in 1848. Sixteen of these samples were sent to Jacob W. Bailey, a professor of chemistry, mineralogy, and geology at Bache's old alma mater, the U.S. Military Academy. In examining them, Bailey expected to find that the common sands and muds known to lie near shore extended into deeper waters without much change in character, and from a casual look at the samples the progression certainly seemed to go from sand to mud or clay.

Bailey was not giving them just a casual look, however, and he studied each of the samples under a microscope. "Nothing of interest," he reported upon examining the shallow sediments taken from depths of 10 to 50 fathoms. In these he saw only familiar particles of sand and mud. But when Bailey looked at the deeper sediments, taken from 50 to 100 fathoms, he was startled to find "a truly wonderful development of minute organic forms. . . ."Ψ

He was looking at samples composed entirely of animal plankton, or rather of the remains of such plankton, and through his microscope he saw thousands of small, delicate shells which had been built by simple animals called Foraminifera. Bailey correctly surmised that these animals had lived near the surface of the sea, and when dead, had sunk to

---

* The classification of sea-floor oozes may have been the Survey's innovation, but their collection, at least their temporary collection, was not. For years seamen had examined the mud on their sounding leads—had even armed their leads with tallow to help pick up some sand or ooze—for a sample of the sediment could help an experienced seaman determine his location or spot a good fishing ground. According to an old story, there was a skipper from Nantucket who claimed he could tell exactly where his ship was just by the color and the taste of the lead. As a joke, his mate, Jotham Marden, brought dirt from a Nantucket parsnip bed which he smeared on the lead as the ship was approaching New York. He woke the captain from a sound sleep, asked him "please to taste," and, in the words of James Thomas Fields,

> The skipper stormed and tore his hair,
> Hauled on his boots, and roared to Marden,
> "Nantucket's sunk, and here we are
> Right over old Marm Hackett's garden!"

the sea floor. Fascinated by his discovery, Bailey willingly studied many more of the Coast Survey's sediment samples and, in the mid-1850's, some of the Depot's deep-sea samples as well.

While the different kinds of investigations, which together made up the Survey's Gulf Stream study, were often treated separately—the sediments sent to Bailey, the temperatures studied by Bache—all were made together on each expedition in whatever sequence the weather allowed.

The work and results of one cruise, which in most respects was typical, were described by Lieutenant Bache in a letter sent to his brother. The letter was written while the *Washington* was in the Gulf Stream and was given to a passing whaler to deliver.

U.S. Surveying Brig *Washington*
The Gulf Stream, August 5, 1846

My Dear Brother:
I will first give you some of the incidents of the cruise, and then display its results, which are highly interesting. . . .
We are now in the Gulf Stream. Observations growing more and more interesting. On the 31st, water the warmest and warm water the deepest; got a fine set of temperatures by the globe, from 600 fathoms up to five fathoms; current setting to the eastward. Here threw overboard a spar 28 feet long, with a gilt ball on top, and a weight at the heel, having at its upper end a brass case . . . marked with the latitude and longitude, and containing a request that the position of the spar, when picked up, should be made known to the superintendent of the coast survey, and that the spar should be set adrift again. Threw drift bottles over also. We are now closing up with our first observations, and on the same line. During the night experienced a strong easterly current. The next day (August 1st) took up position No. 13, and found the old system of curves again, the water being shallow. Took temperatures here from 1,500 fathoms up to 5; the intermediate ones being at 500 fathoms, 300, 200, 100, *etc.* The globes stood the pressure very well, and brought up a temperature of [*sic*] 1,500 fathoms. This cast was taken under particularly favorable circumstances; the first opportunity I had of getting so deep a cast; and still continuing the observations at lesser depths. My note book says: "Line commenced running out at 10h. 57m.; lead weighs 80 lbs.; line out to 1,500 fathoms, at 11h. 36m.; a good up and down cast. Brig under close-reefed maintopsail and foretopmast staysail; wind light from southward; course W.SW., 6/10 of a mile per hour, which I had found, on trial previously, to counteract exactly the current.
(Sent the crew to dinner.)

At 12h. 27m. commenced hauling in the line, 30 men employed; this number includes our cooks and waiters. Globe came up at 2h. 56m.; register 37°, index 40°. . . ."

I would like to be with you when you look at and admire this section, as admire it you must, and speculate on it together. Here on the left we have the main current of the stream turned to the eastward, by Hatteras, and butting up against a bank of cold water, which it overflows, and on the right mingling with a vast reservoir of warm water, which is probably brought there by the eddies of the stream itself. How beautifully the line is defined to the left, or westward, and how well the observations of the 2nd of August come in to verify the others! I hit upon this by having the marine thermometer going at the depth of 30 fathoms; and as soon as it brought up warm water, hove to and made the observations. The tracing of the cold wall from Hatteras up will be highly inter-esting, and will lead to useful practical results, if it is permanent, (and can it be otherwise?). . . . [The term "cold wall," coined by Bache and still in use today, refers to the sudden drop in tem-perature occurring where the Gulf Stream water meets the cooler coastal water. The cold wall marks the landward edge of the Stream but is not stationary as was then imagined.]

The brig has been improved very much in sailing, and all other respects by taking in the ballast and the armament, and is much more comfortable and safer than she was before. Still she leaks very much in her upper works and decks, and continues to wet everyone below, and we feel that we have to be very careful of her in heavy weather. A good hard gulf sea would rack her very much. By the time we make Sandy Hook, New Jersey we will have sailed 1,327 miles. We have taken 532 surface temperatures, 149 temperatures at from 5 fathoms to 1,500 fathoms, and col-lected 17 specimens of bottoms on soundings.

Your affectionate brother,

G. M. Bache[Ψ]

Unfortunately, the leaky *Washington* encountered more than a hard gulf sea before coming into port. After completing the last of the three transects of the Gulf Stream that were to be made that summer, the brig was caught in a violent hurricane off Cape Hatteras and wrecked.

"Of that loss I cannot trust myself to speak . . ." wrote the super-intendent. "He was returning from his crowning work upon the survey . . . when overtaken by the hurricane of the 8th of Septem-ber. Securing by his own exposure the safety of others, he was swept from the deck of the vessel which he commanded, and perished off that dangerous coast, the perils of which to others it was his object to diminish."[Ψ]

The information that the Survey gathered from the sea, sometimes at such terrible expense, was published in an annual report, and Bache further publicized the information in scientific and popular journals. In almost all cases, he presented the Survey's data without interpretation, for the superintendent wished to abstain from "mingling doubtful speculation upon the causes."⁺ That, in his opinion, was Maury's sin.

The Coast Survey's rare but noteworthy excursions into natural history date from 1847, when Bache invited the Swiss naturalist Louis Agassiz to accompany a surveying party then working in Massachusetts Bay. Agassiz willingly joined them and was able to dredge for animals and sediments in shallow waters of the bay. He collected enough bottom material from the rows of submerged shoals that lay beneath the water to theorize that these hilly structures were related to similar features on land. He believed correctly that both were formed of glacial deposits that had been dropped as the ice sheet had retreated thousands of years before.

In 1851 Agassiz was again invited to accompany the Survey, this time for a two-month study of the Florida reefs and keys.

"This examination was imperiously called for," wrote Bache, "by the contradictory statements in regard to the character of the reef . . . being by some represented as composed of living and growing coral—by others of boulder masses of dead coral; sustaining, in the two cases, altogether a different relation to navigation, and to the questions of sites for light-houses and sea-marks."⁺

The excursion to Florida, and other cruises which followed, helped Agassiz as well as the Survey. The naturalist was allowed to keep the specimens he collected, and these he used in his new position as professor of natural history at Harvard's school of science. For its part, the Survey received Agassiz' scientific reports, which, although rarely of any practical importance, did give the Survey a certain prestige within the academic community and inspired others within the organization to undertake scientific investigations.

Shortly after Agassiz became involved with the Survey's program, his close friend and countryman Louis François de Pourtalès, joined the Survey and was encouraged by Bache to follow his interests in marine zoology and geology. Although Pourtalès had been formally educated as an engineer, he had learned a great deal about natural history from Agassiz, and by 1848 had begun to study the muds and

Louis Agassiz, adviser to the Coast Survey, professor of natural history at Harvard University, and founder of the Museum of Comparative Zoology. (*Courtesy Museum of Comparative Zoology, Harvard University*)

oozes brought up on the Survey's sounding leads. The examination of sediments was not Pourtalès' main responsibility at the Survey, however (he was officially head of the tidal division), and the marine sediments were still sent to West Point, where they were studied and reported upon by Professor Bailey. Only after Bailey's death, in 1857, was Pourtalès given charge of the thousands of samples collected by the Survey, and a dozen or so years later he used these samples to construct a map which showed the distribution of sediments off the East Coast of the United States. (See Chapter IV.)

The study of corals and echinoderms was another of Pourtalès' specialties, and he collected many of these animals from the shallow waters in which the Survey did most of its work. After the Civil War, when the organization extended its limited dredging operations into deeper water, Pourtalès was given a chance to explore the sea floor at depths greater than one or two hundred fathoms. But before 1860, Pourtalès, like other American and European naturalists, was content to let the deep sea lie, for he believed it devoid of life.

While the Coast Survey sent out ships to study the Gulf Stream and the Florida reefs and after each surveying season presented new and interesting observations, Maury, in the Depot, was busy arranging for a ship of his own. Finally, in 1849, an Act of Congress assigned three small ships to the Depot, and Maury was not particularly surprised or disappointed to receive only one of these, the schooner *Taney*.

The apparent reason for assigning the *Taney* to the Depot was to meet a demand posed by Boston shipowners, but suggested by Maury, that the "new routes . . . made by Lieutenant Maury in the course of his investigations of the winds and currents of the ocean"* be given an official trial.

Maury, of course, had much more than routes in mind. In an amazingly short time he had the *Taney* refitted, reprovisioned, and reequipped to serve as a research vessel. In October of 1849 the unstable 74-foot schooner, so heavily laden with reels, lines, winches, and diverse gadgets that she wet everyone both above and below decks, sailed sluggishly from New York harbor bound for the Canary Islands. The ship, under the command of Lieutenant Joseph C. Walsh, encountered heavy weather in the North Atlantic, but by November conditions were more favorable and the men had settled down to a cautious routine. Every 30 miles or so the *Taney* was brought into the wind so that the temperature of the surface water could be measured and other thermometers could be let down to depths of 50 and 100 fathoms. Soundings, which took a much longer time, were supposed to be made every 200 miles.

In mid-November there finally dawned a day of almost flat calm and it was decided that the *Taney*'s unique sounding machine could at last be tried. Up from the hold came a hand-cranked winch and when this was bolted onto the deck, up came a huge reel carrying mile upon mile of galvanized steel wire. The reel was cradled on the winch frame and when all was in readiness a weight was attached to the end of the wire, the brake was released, and down into the sea zipped the wire with dizzying speed. On and on it went, minute after minute, until, with nearly six miles out, it broke. The fantastic depth of 5,700 fathoms with "no bottom" was recorded for the spot. (The *Taney* was actually riding in about half that depth of water, but the weight of the wire had continued to strip more off the reel.) The sounding machine was tried some dozen times during the remainder of the voyage, but an accurate sounding was never obtained.

"Reeling up from great depths is a long and great labor," wrote Walsh to Maury. "It takes half a day to put together and rig the machine . . . and it is a dangerous business with any motion [of the ship]."⚥ On the last point, Walsh was probably referring to the time that the reel broke away from sailors who were trying to unship it from its frame at the stern of the ship.

"The heavy reel surged violently three or four times from side to side [and] Mr. Marcy, the [Sailing] Master . . . was struck and knocked down . . . his right collar bone was fractured and the hammock nettings on both sides [of the ship] were shattered. . . ."⚥

The temperature stations, less time-consuming and less dangerous, were also disappointing. The day before the sounding machine was first tried, Walsh ordered a thermometer and a cylindrical wooden water sampler sent down to 800 fathoms. The thermometer burst, the sampler split, and after a series of such accidents, the *Taney* arrived in the Canary Islands with only one thermometer, no working samplers, and a precarious supply of short sounding lines.

From the Canaries the leaky schooner proceeded south to the Cape Verde Islands, and there, upon a thorough inspection, she was found to be "a floating coffin."⚥ After makeshift repairs she was ordered to sail straight for St. Thomas in the Virgin Islands in the company of the brig *Porpoise*. This crossing was not meant to be slow or scientific and the *Taney* had to abandon most of her investigations.*

The *Taney* was laid up in St. Thomas for three months before she was declared fit enough to start upon the last leg of her unusual voyage. Early in May she put out to sea, and Walsh set a devious zigzag course for New York harbor in order to cover as much of the

---

* But not all. Since the *Taney*, a schooner, could outsail the *Porpoise*, Walsh took it upon himself to crowd on all sail each morning and pull 8 or 10 miles ahead of his escort. Then, "the *Porpoise* being so far astern as to give us time to sound for temperatures,"⚥ Walsh managed to squeeze in a few of the tasks that Maury had set him.

sea as he could in the time allotted for his return. In addition to resuming both temperature measurements and unsuccessful soundings, he decided to investigate an undercurrent which he believed was flowing beneath his ship. Walsh was led to suspect the existence of such a countercurrent by the temperature readings his men had made that indicated a seven-degree difference between the surface water and the water flowing only 100 fathoms below. This was a much greater difference than usual, and Walsh decided to investigate.

To test this hypothesis, Walsh used a weighted drogue (a kind of modified sea anchor) that he had attached to a small float. The large drogue, which was weighted to sink to a depth of 120 fathoms, offered much more resistance to any flow of water than did the small surface float linked to it and so the whole contraption was bound to drift with whatever current might exist at 120 fathoms. By watching the surface float, it was easy to see if the drogue's movement agreed or disagreed with the drift of the ship.

On four consecutive days Walsh had the drogue heaved over the side, and each time the surface buoy showed that it was moving contrary to the drift of the ship. By changing the weights on the drogue Walsh found he could determine the approximate boundary of the surface and subsurface currents.*

The final days of the cruise were spent sailing back and forth across the eastern side of the Gulf Stream (an area which Maury thought Bache had neglected) and making temperature measurements from the surface to 500 fathoms. Finally, on June 4 the *Taney* put into New York harbor and was promptly condemned as "unfit for any service." Walsh retired temporarily from sea duty to write the *Taney*'s cruise report.

A year after the *Taney*'s return, Maury received the second of his would-be research vessels. This was the *Dolphin*, a much larger ship, and Maury provided her with thermometers, water samplers, and mile upon mile of thin sounding twine which was to take the place of Walsh's winch and wire. The *Dolphin* made three cruises in the Atlantic, and on the first two brought back the same kinds of information

---

* Actually, Walsh was probably measuring two layers of a single current in which the surface layer moved faster than the subsurface layer but both were flowing in the same direction. Since the *Taney* was not anchored and was not within sight of land, Walsh had no way of judging the absolute direction of the drogue's drift, only its movement relative to the ship's progress.

that the *Taney* had obtained. On the third, however, she was equipped with a new sounding device designed by Passed Midshipman John Mercer Brooke, which was intended not only to measure the great depths of the open ocean but to bring back samples of the unknown sediments as well.

"What is the use of knowing how deep it is," asked Maury rhetorically, "unless we know what is at the bottom of it and where was the mechanical skill that would contrive for us the means of bringing up from miles below the surface the feathers from old ocean's bed, be it ooze, or mud, or rock, or sand? . . ."[Ψ]

The Brooke Patent Sounding Lead was Maury's answer. The device consisted of a detachable weight, usually a cannonball with a hole drilled through it, which was suspended on a sounding line just above a sampling tube. The weight and tube were dropped to the bottom, and their momentum drove the tube into the sea floor. When the sounding line slackened, the lead was automatically detached and the sailors above had only to heave in the relatively light line with the sampling tube still attached.*

Maury was in a frenzy of excitement to have Brooke's machine tested, and he carefully instructed the *Dolphin*'s master, Lieutenant Otway Berryman, on its proper use. The device was first tried in deep water on July 7, 1853. "Recovered line with specimen of bottom," states the log entry. "Yellowish-white chalky clay, 2,800 fathoms."[Ψ]

This was indeed exciting. Not only did Brooke's machine make more accurate soundings than either of the methods commonly used before, but it also was capable of procuring materials that no man had ever seen. Eight of the chalky samples were promptly sent to Professor Bailey.

"They are exactly what I have wanted to get hold of," he wrote happily to Maury. "The bottom of the ocean at a depth of *more than*

---

* Brooke's device was not the first of the detachable-weight machines. Although he did not know it, a wild young professor at the Lycée in Algiers, Georges Aimé, had built a "*sondeur à plomb perdu*" some 10 years before. This device utilized a sizable weight which took a line quickly to the bottom and, once the depth had been marked on the line, a messenger (a small sliding weight) was sent down the line to trip a trigger and detach the weight. The line and a bottom sampling device were then hauled to the surface.

The remarkable Aimé also designed a water bottle, a reversing thermometer, and a deep-sea current meter. He was killed at the age of thirty-five in a horseback riding accident.

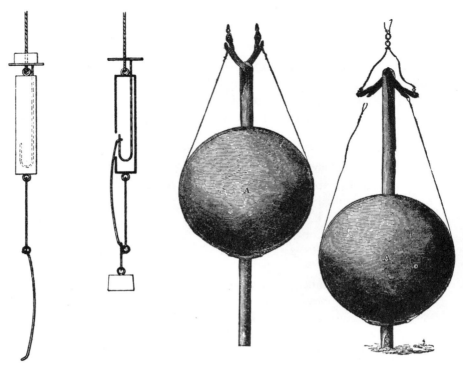

Sounding devices with detachable weights. On the left is Georges Aimé's *sondeur à plomb perdu*, designed about 1843. On the right is a later model designed by Passed Midshipman John Brooke and used by Maury's officers in their deep-sea explorations. (*From* Annales de Chimie et de Physique, *Vol. 7, and from Matthew Fontaine Maury*, The Physical Geography of the Sea)

two miles I hardly hoped ever to have a chance of examining; yet thanks to Brooke's contrivance, we have it . . . so that it can be put under a microscope. I was greatly delighted to find that all these deep soundings are filled with microscopic shells; not a particle of sand or gravel exists in them. They are chiefly made up of perfect little calcareous shells (Foraminiferae), and contain also a small number of silicious shells (Diatomaceae)."⁺

Bailey described these new and unexpected wonders in articles which were distributed in America and in Europe, where similar discoveries were beginning to be made. Aside from such occasional reports, however, the information gathered on Maury's experimental cruises largely disappeared. The official report of the *Taney's* cruise was tucked away among the appendices of the *U.S. National Observatory Astronomical Observations*, while the reports of the *Dolphin's* exploits were swallowed up in Senate Executive Documents. Neither was readily available to scientists, especially to those working in Europe, and it is not surprising that the scientists of Maury's time did not mention or even know much about the cruises of the *Taney* and the *Dolphin*.

While the information gathered on Maury's cruises and on the Navy's other, less specifically scientific voyages might sink prematurely into dusty archives, Maury's own light on the subject was not to be hid. Using bits and pieces of the data his officers had gathered, Maury first produced a dramatic profile of the Atlantic sea floor (between the Yucatán Peninsula and the Cape Verde Islands), then constructed a *Bathymetrical Map of the North Atlantic Basin with Contour Lines Drawn in at 1,000, 2,000, 3,000, and 4,000 Fathoms.* This map, published in 1854, was the first contour map to be drawn of an entire ocean basin. Some 200 deep soundings had been available for its preparation, and considering the small amount of data and the vastness of the ocean, the map was as accurate as could be expected. Maury distrusted some of the soundings he had received, especially Walsh's, which delineated an imaginary depression six to seven miles deep south of the Grand Banks, but he was unable to have the work redone, and so entered the depths on his chart and followed them with question marks. The soundings made in the middle of the Atlantic by the *Dolphin* were more accurate, and the rise or swell in the ocean floor which they showed, and which Maury named the "Dolphin Rise," was the first indication of the unimaginably extensive Mid-Atlantic Ridge.

Among the men most interested in Lieutenant Maury's map were members of the Atlantic Telegraph Company, who were proposing to lay a transatlantic submarine telegraph cable between Newfoundland and Ireland. While shorter lengths of cable had been laid across gulfs and straits, no one had yet attempted to drop 2,000 miles of cable through two miles of sea water. The success of the venture depended in part on the nature of the sea floor. Was it rough and rocky? Or, as was hoped, flat and silty?

Although the Telegraph Company was not organized until 1856, its founder, Cyrus Field, wrote Maury late in 1853 to ask him what the sea floor was like along the proposed route. Since the *Dolphin* had spent some time investigating that portion of the North Atlantic, Maury was able to say that "the bottom of the sea between the two

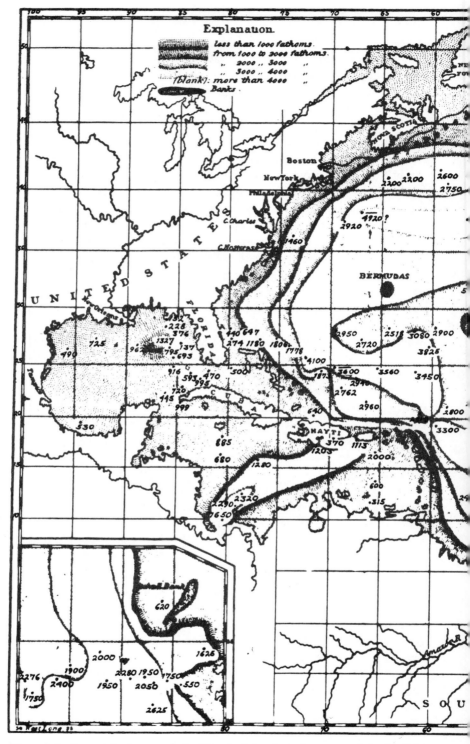

Maury's bathymetrical map of the North Atlantic as it appeared in the 8th edition of his book (1861). Many of the soundings which appear on the map were made on exploratory cruises that Maury organized aboard the

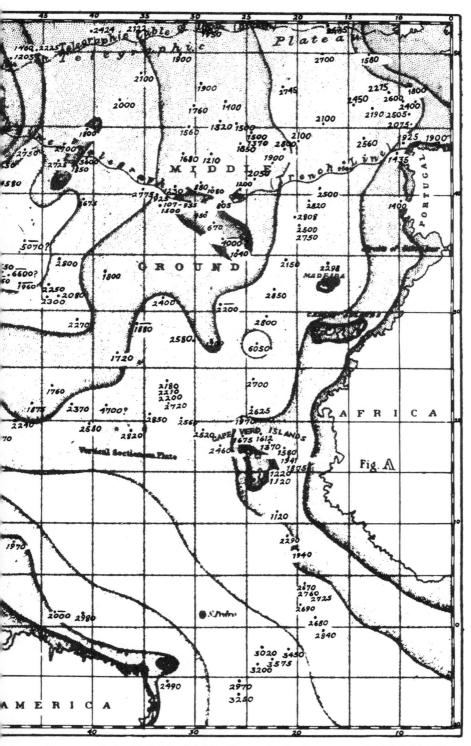

Taney and the *Dolphin*. Middle Ground, also known as the Dolphin Rise, is actually a part of the Mid-Atlantic Ridge. (*From* The Physical Geography of the Sea)

places is a plateau which seems to have been placed there for the purpose of holding the wires of a submarine telegraph. . . ."⚇ This divinely inspired arrangement he inferred from about 30 soundings that had been made along the 2,000-mile route. The science of the sea, Maury pointed out, had innumerable practical applications.*

In 1855, just a year after the contour map appeared, Maury's book, *The Physical Geography of the Sea*, was published. It was an immediate and controversial success. With chapters on navigation, currents, winds, weather, soundings, red fogs and sea dusts, and with plenty of imaginative theories and explanations, the book had great appeal for many kinds of readers. And Maury had an elegant style.

"There is a river in the ocean," his book began. "In the severest droughts it never fails, and in the mightiest floods it never overflows. Its banks and its bottom are of cold water, while its current is of warm. The Gulf of Mexico is its fountain, and its mouth is in the Arctic Seas. It is the Gulf Stream."⚇

Readers were captivated. *The Physical Geography of the Sea* went through six editions in its first four years and was translated into German, French, Italian, Norwegian, Dutch, and Spanish. In spite of its popularity, however, the book had some serious shortcomings, and these did not go unnoticed.

"While the work contains much instruction," wrote a cautious editor in the *American Journal of Science and Arts*, "we cannot adopt some of its theories, believing them unsustained by facts."⚇

Maury's reputation is created "by those writing in favor of works they did not understand,"⚇ complained a less inhibited reviewer.

What bothered scientists about Maury's work was the oversimplified and often contradictory explanations which he insisted upon advancing for all the data that fell into his hands. While energy and ambition

---

* The Atlantic Telegraph Company and its heirs only slowly learned that their chosen route did not cross a "Telegraphic Plateau" as Maury had called it, but spanned mid-ocean canyons and climbed ridges as well. In spite of these unsuspected barriers, the first transatlantic cable was laid in 1858 by the U.S.S. *Niagara* and the H.M.S. *Agamemnon*. Unfortunately, the connection failed within three months, and it was not until 1866 that the two sides of the Atlantic were successfully and permanently linked. The accomplishment encouraged other companies to lay great lengths of cables, and as each project was preceded by a sounding survey, a great deal was learned of the sea floor, at least in certain spots. Cable projects were also responsible for many improvements in the design of sounding machines, and these in turn were used by scientists who wished to investigate the sea for less immediately practical purposes.

went a long way toward making up for his lack of formal education, these admirable qualities were not enough to initiate Maury into the intricacies of geophysics and fluid dynamics. Yet these were the areas which fascinated him, and he formulated theories concerning the most complex systems of the sea and supported his contentions with a few observations, a vast number of assumptions, and quotations from the Bible.

Maury's characteristic overextension of his powers, which so infuriated Bache, Louis Agassiz, Joseph Henry, and others, was most obvious in the theories that he advanced on the causes of the Gulf Stream. His chapter on the Stream reopened a controversy of long standing between those who believed ocean currents to be generated by the winds and those who believed them to be set in motion by differences in water density.

There were in Maury's time—as there have almost always been throughout the short history of oceanography—two major theories advanced to explain the steady currents of the open ocean. One states that water is propelled by the winds which blow across its surface, the other that it is moved by the more subtle forces of heating, cooling, evaporation, and precipitation. The relative importance of each of these mechanisms is still questioned, even though the modern concept of how the ocean waters are made to move has become much more sophisticated and complicated. In Maury's time, however, the problem appeared simpler.

Proponents of the wind-driven theory explained each current as the result of a strong and steady system of winds such as the Trades or the Westerlies. They felt no need to explain how winds might stir deeper ocean waters for these were not thought to move.

The other idea on ocean currents, which became known as the thermo-haline (temperature and salinity) theory, stated that masses of water moved both vertically and horizontally because of their different densities. It was understood that these differences in density were caused by the sun's unequal heating which made the water in tropical regions warm and therefore relatively light, while polar waters remained cold and heavy. Evaporation and precipitation could also help create masses of water that were saltier and therefore heavier or brackish and lighter. Freezing too could remove fresh water from the sea, in the form of ice, and leave heavy, salty water.

Taking the effects of temperature and salinity together, the proponents of the thermo-haline theory believed that cold and briny waters were continuously sinking in the polar seas and to replace them, warm, light water from the tropics flowed poleward. The Gulf Stream, thought Maury, was just such a flow of warm water from the equatorial Atlantic to the Arctic Sea. As for vertical circulation, the heavy polar water was imagined to be creeping toward the equator, but with literally immeasurable slowness.

As oversimplified as the nineteenth century's formulation of these two theories appears now, scientists even in Maury's time understood that both winds and differences in water density affected the flow of currents, and both forces were, and still are, used in combination to explain the ocean's circulation. Arguments arose, however, whenever a scientist insisted that one of these mechanisms was much more important than the other.

In the 1780's, for example, Benjamin Franklin explained how the Gulf Stream was caused by the Trade Winds. Remembering times when he had seen a strong wind drive lake water before it with such force that the waters at the end of the lake stood several feet above their normal level, Franklin reasoned that the Trade Winds that blew steadily across the mid-Atlantic from east to west could pile huge quantities of water against the tropical and subtropical coasts of America. He believed that the westward-flowing water would be deflected northward when pushed against the land, and this northward flow of water was the Gulf Stream.

In 1832, a similar theory appeared which had been worked out in greater detail by British scientist James Rennell. He stated that the Stream was a downhill flow of water, the top of the hill being the pile of water pushed by the Trade Winds into the Gulf of Mexico and along the American coast. From these high points the water flowed northward toward the normal level of the sea.

Rennell's theory was contested by the French physicist François Arago in 1836. The latter believed that the Gulf Stream could not possibly be maintained by the winds, for a leveling survey made across Florida had shown that the waters piled by the Trades in the Gulf of Mexico on Florida's west coast stood only about seven inches above those on the state's east coast. Arago did not believe that this seven-inch slope could generate a current as powerful as the Gulf Stream, and so he supported the other possibility—a difference in water densities.

Maury agreed with Arago, and in his book he stated that currents were driven by density differences, which were in turn caused both by unequal solar heating—which made polar waters colder and heavier— and by excessive equatorial evaporation—which made the tropical waters saltier and heavier.

"Either the one theory or the other may be true or neither," wrote British oceanographer C. Wyville Thomson with justifiable pique, "but it is logically impossible that both of these can, for one simple reason that the waters of the equator cannot at the same time be both lighter and heavier than the water at the poles. . . . Maury's theory of ocean current . . . is really chiefly remarkable for its ambiguity."✢

Luckily for those favoring the density theory, Maury did not have the last word, and more logical explanations of the mechanism he supported were put forward by other scientists.*

Another problem that stumped Maury was the effect of the earth's rotation on ocean currents. He was aware that the French mathematician Gaspard Gustave de Coriolis had sought to explain how motion on a spinning body would differ from motion in a void, but in spite of confident allusions to the "Coriolis effect," Maury did not really understand the forces involved.

But there were those who did. William Ferrel, a self-taught mathematician and geophysicist, was a deductive thinker of considerable talent whose interest lay in formulating general laws of atmospheric and oceanic circulation. From the scientific seclusion of Nashville, Tennessee, Ferrel wrote an "Essay on Winds and Currents of the

---

* In 1870 a British scientist, William B. Carpenter, investigated the circulation of the Mediterranean Sea and used his findings to support the view that differences in densities are responsible for the oceans' circulation. He had set out aboard the *Porcupine* to determine the Mediterranean Sea's physical and biological relations to the Atlantic, with "special attention given to the Gibraltar Current." Carpenter believed that a deep countercurrent ran out of the Mediterranean underneath the easily observable surface inflow. He further believed that this system of current and countercurrent was generated and maintained by differences in water density and that such a system was a sort of small-scale model of the circulation in the Atlantic Ocean. He understood that while excessive evaporation made the Mediterranean's waters very salty and dense, the differences in Atlantic waters resulted more from unequal temperatures than from unequal salinities.

In much the same way that Lieutenant Walsh aboard the *Taney* had investigated a current beneath the Gulf Stream, Carpenter used a drogue attached to a small boat to ascertain the existence of the Gibraltar undercurrent. His careful observations settled the question of the Mediterranean's invisible outlet and was a victory too for proponents of the density-generated current theory. Scientists who disagreed with Carpenter allowed that the mechanism he described was essentially correct but not nearly powerful enough to produce a current like the Gulf Stream. Carpenter, of course, said the same of the winds.

Ocean," which was published in 1856 in the *Nashville Journal of Medicine,* and read by few outside the state of Tennessee. The journal's editor, who had read Maury's book and knew that Ferrel disagreed with most of it, had urged the latter to "pitch into" Maury. With characteristic reserve, Ferrel had refused to skirmish, but did agree to write a paper in which he "noticed Maury's views a little in an incidental way."^Ψ Ferrel's essay included the first general discussion of the effect of the earth's rotation on ocean currents. His views, however, did almost nothing to modify Maury's mistaken ideas, and Ferrel's corrections were ignored in subsequent editions of *The Physical Geography of the Sea.*

From 1855 to 1861, Maury's last years at the Depot, the superintendent spent much of his time expanding and improving both his Wind and Current Charts and his book. He collected a considerable amount of new data pertinent to each by persuading the leaders of four Arctic expeditions (in the field between 1850 and 1861) to make observations for him. An idea which Maury found particularly fascinating, and which he finally believed was true, was the popular conviction that high in the Arctic, beyond a ring of pack ice, lay an open polar sea. To support this hypothesis, he asked members of the Rodgers North Pacific Expedition to measure water temperatures, densities, and currents as they passed through the Bering Sea and Straits. Maury received the data he requested and subsequently added several new sections on Arctic oceanography and the open polar sea to the last editions of *The Physical Geography of the Sea.* (The belief he energetically supported hung on almost to the turn of the century before being disproved.) The investigation of the Arctic Sea was one of Maury's last projects at the Depot, for when civil war broke out his career as a marine scientist ended.

In 1861 Maury's home state of Virginia seceded, and Maury went with her. He spent most of the war years in England, and then, after a brief sojourn in Maximilian's Mexico, returned to Virginia but not to the practice of marine science.

Once Maury had left the Depot, the contingencies of a civil war and the bitterness of his scientific peers combined to pull down much of the work that he had begun. In 1863, the newly organized National Academy of Sciences, headed by Bache, was requested to report on the advisability of continuing Maury's publications. His charts were said to "embrace much which is unsound in philosophy and little that is practically useful . . . therefore these publications ought no longer to be issued in their present form."*Ψ

This judgment, while admittedly harsh, did reflect more than personal animosity. Maury's amateurish approach to science, his reckless generalizations and careless contradictions, drew similarly negative evaluations from British and American scientists who were not connected with Bache, his friends, or the National Academy. It is unfortunate that civilian scientists working for the Coast Survey and for other institutions were so rarely able to cooperate with Maury and his officers. The animosity between the two groups did not help the Navy to develop a special scientific corps or even a productive scientific program, and the poor relations deprived civilian scientists of a source of funds, big ships, and equipment that they might well have used for cooperative projects.

With Maury gone, the Depot shed its time-consuming concern for the bales of abstract logs which he had continued to collect, but gone too was his stubborn and passionate interest in the sea and his ability to channel funds toward its exploration. The Depot's next directors, astronomer James Gillis (whom Maury had replaced in 1842) and Rear Admiral Charles Henry Davis, were both scientists of energy and reputation, but neither had the desire or the financial resources to reinspire the Depot's hydrographic work. In 1866, the Depot, renamed the Hydrographic Office, was officially separated from the Naval Observatory. In the following years the Office was led by men whose interest in science was less than passionate, and Maury's efforts to chart large portions of the sea gave way to an irregular series of minor surveys. In 1872, a last gasp of ambition generated an appropriation of $50,000 "for surveying in the Pacific," but when that meager sum was spent, no further support for the project was obtained. The Hydrographic Office had abandoned even the dream of comprehensive surveys, and its directors took no more than a passing interest and an occasional part in the advancement of oceanography until the onset of World War II.

The Coast Survey underwent a similar though less rapid decline from research to routine, and the germ of this unfortunate transition

---

* The publication of Maury's charts was suspended until 1883, 10 years after his death, at which time the Wind and Current Charts were reissued as Pilot Charts. These have been kept up to date and are still being published.

lay partly within the organization of the Survey itself. When, in 1832, Congress had revived the Survey, that parsimonious body had been careful to state that the Survey was a temporary organization established to accomplish a specific task. Once the country's coasts were surveyed, it was understood that the Survey would disband. Under this assumption long-range plans were difficult to justify, especially as the sections of the nation's coastline still to be charted became shorter and shorter. Another difficulty for the Survey was its dependence upon naval personnel for the operation and command of all surveying vessels. This was particularly troublesome during the Civil War, when it seemed certain that the Survey's work would be stranded by the withdrawal of naval support. Bache, however, had been able to translate all these threats into opportunities. He had extended the Survey's activities from the east to the west coast as Americans moved westward, and he had interested Congress in supporting studies of the Gulf Stream and of marine sediments which had only the most tenuous connection to surveying. As for the Survey's role during the war, Bache did not wait for naval personnel to be transferred out from under his control, but campaigned to have the Survey's work reassigned to harbor and near-shore waters for the benefit of naval operations.

After the war, a further reorganization of the Survey's aims and activities became necessary if the agency were to receive continued support from an impecunious Congress, but in this Bache, with his consummate skill in foreseeing and handling such crises, was no longer the guide. Toward the end of the war he had become seriously ill, and remained so until his death in 1867.

The Survey did not immediately lighten its load of scientific activities. Bache's successor, Harvard mathematician Benjamin Peirce, postponed the time when the Survey might complete its mission by starting a program of cross-country triangulation—which he felt was necessary to accurately connect the Survey's East and West Coast charts—and by allowing Louis Agassiz to convince him, as he had Bache, of the value of undertaking marine zoological and geological studies.*

---

* These cooperative investigations were possible because of the close friendships that existed among the relatively small group of American East Coast scientists. With few exceptions, each man knew what opportunities or materials his friends were looking for and, barring obvious conflicts of interest, would willingly supply

Under Peirce, the Survey's work on land became increasingly important, so much so that in 1878 the bureau's name was changed to the Coast and Geodetic Survey. But the agency's scientific interests on land or at sea were not shared by Congress, and it became more and more difficult for Peirce and his successors to gain approval for the research they wished to pursue.

The Coast and Geodetic Survey drifted slowly away from activities which might be called oceanographic until 1885. And then it was given a shove, but in the wrong direction. In that year Cleveland became President, and newly reinstated Democrats, willing to replace Republicans in well-paid positions, turned critical eyes on government agencies and their expenses. Both the Hydrographic Office and the Survey were investigated, and the condition of the latter was reported to be "one of demoralization," and its operations were described as "inefficient, unjust, and to some degree disreputable."* The removal of the old and ailing superintendent, J. E. Hilgard, was recommended. He was succeeded by F. M. Thorn, head of the investigating committee—clearly a victory for the forces fighting corruption.

"A year or two more such as the last will leave nothing worth preserving of an organization which was once the pride of American applied science,"* warned an editorial in *Science*, and for serious scientists, interested in more than routine surveying, this prophecy seemed almost correct. But only almost correct, because in the 1880's Naval Lieutenant John Elliot Pillsbury undertook a unique study of the Gulf Stream under the auspices of the Survey. His work is notable, not only because it was the last major oceanographic project to be conducted by the beleaguered Survey for many years, but also, in fact primarily, because it was the first time that the Gulf Stream had been measured with a current meter used from an anchored ship. Both metering and anchoring presented considerable problems, but the results promised to be more precise than previous measurements of the current, which had all been made from drifting ships.

In a lengthy appendix to the Survey's Annual Report for 1890, Pillsbury described the current meter (which he had devised in 1876) and explained the ship's complicated deep-sea anchoring gear. The

---

these if he could. Thus Louis Agassiz was offered a circumnavigation of South America when the Survey vessel *Hassler* was being sent from the East to the West Coast in 1871, and later, his son, Alexander Agassiz, accepted several invitations to use the Survey steamer *Blake*.

latter, a system of booms, spars, reels, and wire rope, was used to hold the 139-foot steamer *Blake* in a single spot for as long as 166 hours in water as deep as 2,180 fathoms.

From 1884 to 1890 the *Blake* spent a part of each year moored along six sections or lines which crossed the Gulf Stream at points between Cape Hatteras and the southern tip of Florida. At stations set along each of these lines the temperature of the water at various depths was recorded and the direction and speed of the current were measured. Observations were also made in many parts of the Caribbean, and all in all, the *Blake* made "at least 164 anchorages . . . at some 90 stations deeper than 350m [200 fathoms] between Cape Hatteras and the southernmost passages of the Lesser Antilles. One station in the Florida Current [located off the southern tip of Florida] was occupied 22 times during several years for a total of more than 600 hours of current observations."ᵂ

"In a vessel floating on the Gulf Stream," wrote Pillsbury after making all these observations, "one sees nothing of the current and knows nothing but what experience tells him; but to be anchored in its depths far out of the sight of land, and to see the mighty torrent rushing past at a speed of miles per hour, day after day and day after day, one begins to think that all the wonders of the earth combined can not equal this one river in the ocean."ᵂ

Yet to the river in the ocean Pillsbury ascribed some unexpected and most unriverlike characteristics. To the surprise of those who still considered major currents to be stable flows of water through the sea, Pillsbury's investigations showed that the Gulf Stream was constantly changing its position, its speed, its temperature, and even its direction. His report suggested to scientists on both sides of the Atlantic, and on both sides of the wind versus thermo-haline controversy, that currents were more complex than had, at first, been suspected.

As to the cause of the Gulf Stream, Pillsbury's observations did not yield any novel theories. "At this day, after so many thinking men have investigated and have written on the subject of ocean physics, particularly on the Gulf Stream, and have advanced and advocated every possible theory as to the causes of currents, the only question for a new explorer to decide is toward which of the ranks will his own research lead him to incline . . . .

"I place myself with those who advocate the wind theory . . . as the chief cause of most ocean currents . . . ."ᵂ

Unfortunately, Pillsbury's remarkably accurate investigations were made just as the Coast and Geodetic Survey's programs of marine research were being cut back and discouraged. Pillsbury's reports did not become well known either in the United States or in Europe, and no one seriously attempted to continue or to duplicate the promising program he had begun.*

After the Civil War, at a time when both the Survey and the Depot were struggling to retain a portion of their former power and prestige, a third government agency, one interested in a very different aspect of marine science, began to enjoy its initial and most vigorous period of growth. This was the United States Fish Commission.

The idea of establishing a fish commission, and subsequently its organization and direction, all came from a single man, Spencer Fullerton Baird. Baird was a competent naturalist of broad interests who spent most of his professional life filling the Smithsonian Institution with innumerable collections of birds, bats, bugs, bones, rocks, fish, and fossils. He had joined the Institution as a young man, and in 1850 had become its assistant secretary under Joseph Henry.

In 1870, when Baird was 47, he took on the additional responsibility of establishing a fish commission, and in doing so he was prompted by several motivations. The official one, which Congress intoned as it resolved to create the new organization in 1871, was that a commission was needed to investigate the nation's fish stocks "with the view of ascertaining whether any . . . diminution in the number of the food-fishes of the coast and the lakes of the United States has taken place; and, if so, to what causes the same is due; and also whether any and what protective, prohibitory, or precautionary measures should be adopted. . . ."†‡

The unofficial reason for establishing a commission, and the one which most appealed to Baird, was to initiate a sustained ecological

---

* Pillsbury's current meter was used aboard several British surveying vessels in the 1890's for making occasional measurements.

† The possibility that the government might have to impose fishing regulations was a problem that was very much on congressional minds in 1870 and 1871. The increasing use of fish traps and weirs, which conventional fishermen (using boats and lines) hated, and blamed for a disastrous reduction in their annual catch, had led to an argument of such intensity that both trappers and liners were calling on the government to protect their incompatible interests. Congress was not unhappy to be offered the chance to pass this emotional problem on to Baird and his new commission.

Spencer Fullerton Baird, left, pushes a small collecting boat into Little Harbor at Woods Hole. (*From the collection of Norman T. Allen, courtesy National Marine Fisheries Service*)

study of North American waters. Europeans, Baird knew, had already begun to explore their coastal waters and were finding a fascinating variety of marine plants and animals. But in the United States only the Coast Survey's short cruises with Louis Agassiz and his pupils and the Navy's occasional multipurposed expeditions had produced any marine zoological collections, and these were at best haphazard. Baird wanted to create a permanent bureau which could provide marine biologists with the support and equipment they needed to systematically study the sea over a long period of time.

Congress, acting on the official reason, established the United States Fish Commission, and President Grant appointed Baird its unsalaried commissioner. (Baird requested this arrangement in hopes that the directorship would remain a nonpolitical appointment.)

For the first summer's work Baird set up temporary quarters in a vacant Light House Board building in Woods Hole, Massachusetts. He chose this small, seaside town, located at the elbow of Cape Cod, because he had vacationed there in previous summers and knew that the waters along its shores were rich in both fish and bottom-dwelling animals. Baird brought a few assistants with him that summer, and chief among them was Addison E. Verrill, a professor of natural history at Yale and a former student of Louis Agassiz. Verrill was particularly interested in invertebrate marine animals such as starfish and sea anemones, and Baird put him in charge of surveying Martha's Vineyard Sound to determine what animals lived on the sound floor and how they were distributed.

Baird provided Verrill and his other assistants with boats borrowed from the Revenue Cutter Service and with the trawls, buckets, thermometers, preserving jars, and spirits that they needed for their investigations. While Verrill studied bottom animals, Baird spent the first summer collecting fish, and many of the 106 different species he gathered were obtained from the fykes and weirs which he examined as he toured up and down the Massachusetts coast. However they had been caught, the fish were preserved for the National Museum's collection (part of the Smithsonian), and duplicates were saved to trade with other institutions. Baird also prepared the life histories of two important food fish, the scup and bluefish. These studies were the last purely scientific contributions that Baird had time to make. The administration of the Commission and of the Smithsonian, which he directed after 1878, took increasing amounts of his time.

During the next several summers Baird and his peripatetic commission moved from Maine to Connecticut, repeating and expanding upon the work they had begun off Woods Hole in 1871. Baird wanted to examine as many different kinds of environments as he could—sandy shores and rocky headlands—and he also wanted to find the most suitable location for a permanent marine laboratory. By 1881 he had decided that Woods Hole offered the most advantages, and plans were made to settle the organization's field work in the unsuspecting village.

Baird consciously patterned his plans for a laboratory on the successful and well-known Stazione Zoologica in Naples, Italy. The Stazione, which included both laboratories and a public aquarium, had been started in 1872 by the German zoologist Anton Dohrn. Dohrn had adopted a plan of having "tables" or work areas, in the laboratory which a government, university, or scientific society might rent for the use of their scientists. The income from the tables was used to equip the laboratories and also to operate collecting boats, which set out twice a day to gather the starfish, worms, clams, seaweeds, and whatever else the scientists were studying.

Baird hoped that such a plan would work equally well in Woods Hole. (Eventually tables were set up at the new laboratory and rented to Johns Hopkins, Harvard, Princeton, and Williams.) But in 1881, before his plans had even been introduced to Congress, Baird needed more money than could be raised by renting a few unbuilt tables, and he wanted to obtain the first of it from private sources. The reason for this was that Baird, respecting the government's wish to keep clear of basic scientific investigations "except in an incidental way," knew that his request for a laboratory which was to be used at least partially for such basic studies would have a better chance if some of the money came from private sources. Baird effectively petitioned the "friends of science" and soon had obtained gifts of both land and money. As he had hoped, Congress was impressed by so tangible an expression of interest in marine zoology and responded with appropriations totaling $117,000 for a laboratory. In 1884 and 1885, two large buildings were erected, the first a residence and mess hall for visiting summer scientists, the second a laboratory with a public aquarium and a large fish hatchery.*

At the same time, or actually a few years earlier, Baird began working to expand the Commission's capabilities in a different direction—seaward. He asked members of Congress to pay for the country's first specially built research vessel, and with their approval the 234-foot steamer *Albatross* was built and delivered to the Commission in 1882. With two shipboard laboratories, a dark room, a powerful winch for dredging, and literally tons of oceanographic paraphernalia, the *Albatross* was fully equipped to carry out Baird's expanding program of

---

* In Baird's lifetime, and for a great many years afterward, fish hatching was the Commission's most important practical function. In 1872 the American Fish Culturists Association had successfully urged the government to enter the popular field of fish raising and stocking, and the time-consuming task fell naturally to the Fish Commission. By 1878, 75 percent of the Commission's budget went for hatching such fish as shad, salmon, whitefish, German carp, and ornamental goldfish (these last were for private fish ponds on Capitol Hill). The second of the Commission's practical tasks was the collection of statistics on major American fisheries.

The Fish Commission's laboratory, left, and residence, right, at Woods Hole, about 1890. The Commission's research vessel *Albatross* is moored at the left. (*Courtesy National Marine Fisheries Service*)

deep-sea studies. In the instructions, issued to the captain of the *Albatross* prior to each cruise, Baird enumerated what these studies were to include. At a time when most marine biologists were content to collect specimens for dissection in the laboratory, Baird's concern for the effects of temperature, salinity, currents, plankton, and bottom animals on the presence and behavior of fish was remarkably progressive.

Sir [he addressed Captain Zera Tanner]:

As soon as you can be ready . . . you will go to sea for the purpose of investigating the conditions which govern the movements of the mackerel, menhaden, bluefish, and other migratory species along the coast of the United States in the spring, commencing your investigations off Hatteras, or in the region where these fish usually make their first appearance, and following up the schools in their movements.

The special work to be performed will be to determine the rate of progress of the fish along the coast, their comparative abundance and condition, the places where they first show themselves, the physical condition of their surroundings as to temperature and currents of the water, its chemical and biological peculiarities, etc.

You will endeavor to ascertain whether the apppearance of the fish at or near the surface depends upon the condition of temperature, wind or sky, and also, by the use of the apparatus at your command, what character of food in the water [e.g., plankton or bait fish] seems to determine their movements. You will cause examination to be made of the stomachs of such of these fish as you can capture and carefully preserve a portion at least of the contents of the stomach for immediate or future examination.

Should you deem it expedient you will cruise off the coast a sufficient distance to determine the outward line of motion of the fish, and you will communicate to such fishing vessels as you may meet any information that may enable them the more successfully to prosecute their labors. The time of this work is left to your discretion. You will whenever you touch at any port of the United States send a telegram to me and await instructions as to further operations, if there be nothing to detain you.

You will give to the naturalist of the expedition all possible facilities for collecting and preserving such specimens as you may meet during the cruise.

Very respectfully,
Spencer F. Baird, Commissioner

P.S.—The operations of dredging and trawling should be carried on as frequently as opportunity offers; and if no suitable bait can be had, the trawling line should be used for the purpose of determining the currents of desirable fishing grounds.⁺

As in the laboratory at Woods Hole, the scope and success of the scientific programs undertaken aboard the *Albatross* were in large part determined by the visiting scientists who made use of the exceptional vessel. Again like the laboratory, the ship was for years the only one of its kind available to American marine biologists.

The *Albatross* was not intended for use as a collecting vessel and was sent instead on long voyages in both the Atlantic and Pacific oceans. In her 39 years' service, probably her best-known cruises were those directed by Alexander Agassiz at the turn of the century. Agassiz twice took the *Albatross* into the Pacific (he paid for the coal and for outfitting the ship) and returned with remarkable collections of animals and corals which were deposited in Harvard's Museum of Comparative Zoology. Although the younger Agassiz preferred to publish the results of his cruises in the museum's bulletin, the Fish Commission had an annual report and various bulletins of its own for the same purpose. For scientists less wealthy than Agassiz the opportunity to publish their reports at government instead of personal expense was an added reason for working with the Fish Commission.

Each summer, while the *Albatross* cruised in the waters off Hatteras or moved north to dredge and trawl off Cape Cod, a band of scientists and their graduate students came from universities throughout the East to occupy the Commission's new laboratory and dormitory at Woods Hole. There they spent the summer collecting shallow-water animals, studying the life cycles of fish or the diseases and parasites that affected them, and examining the embryological development of sea urchins or of other simple animals which was often easier than attempting the same study on a land animal. This was the sustained ecological study of the North American waters which Baird had hoped to see. The results of the Woods Hole studies, some of which were a help to fishermen, were published in the Commission's reports, and the specimens themselves were either sent to join the National Museum's valuable collection or were kept by the scientists in part payment for their work at the laboratory.

Baird, who arrived from Washington with his family every June or July, watched over all the projects. He visited the laboratory each morning to encourage, provide, or arbitrate as the situation required, and in the evenings he sat on the veranda of the residence hall talking with his friends and looking out past the buoys, tipped by the tide, that marked the narrow passage between Buzzards Bay and Vineyard Sound.

In the summer of 1886 Baird's personal routine changed, for while more scientists than ever crowded into the end of town, Baird was kept to his rooms in the residence hall by frequent headaches and a laboring heart. The following winter was hard on him and the next summer he arrived in Woods Hole in a wheelchair. In August he died.

"I remember the day and the hour," wrote one of his friends. "It was afternoon, and the tide was low. I recall a picture of a red sun hanging over Long Neck and reflected in the still waters of Great Harbor, of sodden masses of seaweed on the dripping piles and on the boulder-strewn shore; and there rises again the thought that kept recurring then, that the sea is very ancient, that it ebbed and flowed before man appeared on the planet, and will ebb and flow after he and his works have disappeared; and in a singular, indefinite impression, as if something had passed that was, in some fashion, great, and mysterious, and ancient, like the sea itself."⁴

In the summer following Baird's death the plans he had had for establishing a "summer university" at Woods Hole, plans which he had not been able to effect, were taken up by his friends, and the famous Marine Biological Laboratory (MBL) was begun. Baird had hoped to see a group of universities use the Fish Commission's facilities for education as well as for research. Undergraduates, he thought, could come to Woods Hole for the summer, hear a series of informal lectures given by the laboratory's specialists, and learn something of seashore ecology by roaming over the region's rich collecting grounds. At first he imagined that the Fish Commission could organize and coordinate such a project, but when the suspicious Cleveland administration came to power and began to probe the Depot and Coast Survey for "irregularities," and finally looked hard at the Commission as well, Baird disappointedly withdrew his plans for a summer university.

Baird was not the first in the country to envision such a plan. Louis Agassiz, who was both professor of natural history at Harvard and founder and director of that university's Museum of Comparative Zoology (MCZ, established in 1859), had also attempted to start a summer school for science. Agassiz was an old man, in fact he had less than a year to live, when his friends persuaded him that a science school for teachers and other young adults could be established somewhere in the vicinity of Woods Hole, if he would throw his weight behind the plan. This he agreed to do, and with funds donated by a wealthy gentleman (John Anderson), the Anderson School of Natural History was opened in 1873 on Penikese Island in Buzzards Bay after a scant three months' preparation.

"Forty students are enrolled," announced the New York *Tribune* that summer, "16 of them ladies."

There were almost as many courses as students. Pourtalès, who was leaving the Coast Survey to become keeper of the MCZ, arrived to lecture on "the animals and plants living in deep waters, and the peculiar conditions of their existence";[1] Spencer Baird was to come over from Woods Hole to talk on fish; Louis' son, Alexander, agreed to teach embryology, and the famous professor himself undertook the task of explaining what he could about reptiles, birds, eggs, fossils, marine animals, and microscopes. It was a hectic summer. But the following season was worse than hectic. It was disorganized. Professor Agassiz had died during the winter, and none of his followers wanted the responsibility of operating a school on an uninhabited island that was at best only marginally accessible. (Penikese was subsequently bought by the Commonwealth of Massachusetts and used as a leper colony.) The school struggled through its second season and then "was quietly and very wisely dropped."

"The spasmodic descent upon the sea-coast in a summer vacation . . . is a very delightful thing, and may sometimes lead to the collection of a few new species of one group or another," wrote the British critic who had felt it was wise to drop the school, "but it is not in this way that the zoology of to-day can be forwarded."⸸

The criticism was valid, although more so for Britain than for the scientifically less developed United States. If general marine biology were going to expand into marine ecology and eventually into biological oceanography, or, on the other hand, if it were going to be funneled into special niches such as invertebrate embryology and algalogy, then it was time that *The Rambles and Reveries of a Naturalist* were replaced by more systematic forms of investigation. This the new Marine Biological Laboratory at Woods Hole tried to do.

Just as scientists and students had been coming to the Fish Commission's laboratory each summer, and as they would come in the future to the Woods Hole Oceanographic Institution, so they came in the summer of 1888 to the unweathered shingle building put up by MBL. As Baird had envisioned, there were courses, field trips, and research projects. In addition, the privately run MBL promised its scientists a kind of continuity and security which a government agency could not offer and this atmosphere encouraged investigators to embark upon projects requiring five or 10 summers to complete.

In 1890 the Laboratory's tradition of having a special lecture each Friday evening on one of the "unsettled problems of the day" was begun, and soon the series became so popular "that the Friday-evening streets around the Laboratory have ever since resembled those steel prints of early New England Sabbaths with thronging worshipers on their way to church."⸸

The first lectures that those worshipers gathered to hear were on such diverse topics as "The Naturalist's Occupation," "Some Problems of Animal Morphology" (the study of structure and form), "The Relationships of the Sea Spiders," "The Ear of Man . . . ," and "The Study of Ocean Temperatures and Currents." It appeared that MBL was developing a broad program of studies which might eventually involve the lab in some aspect of oceanography. In the following years, however, a movement in the opposite direction took place, and the general topics of the 1890's were replaced by the special and very popular interests in experimental biology and embryology. In spite of this change, and the Laboratory's subsequent adherence to a program of specialization, the Marine Biological Laboratory did encourage the development of oceanography in a second-hand sort of way both by giving a boost to the founding of the Woods Hole Oceanographic Institution in 1930, and by providing that institution's oceanographers, and those working elsewhere, with one of the country's most complete libraries in the natural sciences. As for its own work, only in the 1950's did the Laboratory's board of directors decide to reemphasize the broader field of marine ecology, and only in 1962 did this lead to a program of year-round studies in that field.

On the West Coast, marine laboratories were started 10 or 15 years later than in the East, and in the West they tended to develop as a part of a college or university rather than as an independent institution. Although not the first permanent marine laboratory to be set up on the West Coast, the Scripps Institution for Biological Research (now the Scripps Institution of Oceanography) at La Jolla, California, grew to be the largest and best known. Like several other marine biological stations established at the turn of the century (but unlike MBL), Scripps changed from a marine laboratory into a fully oceanographic institution.

In 1892 Scripps was begun by William Ritter as a portable laboratory-in-a-tent. Ritter, who was then an instructor of biology at the University of California at San Diego, had studied at Harvard in the late 1880's just at the time when the Marine Biological Laboratory was being set up in Woods Hole and when Alexander Agassiz was inviting a few Harvard students to use his private laboratory in Newport, Rhode Island. After receiving his Ph.D., Ritter returned to California to teach and to spend his summers collecting marine animals from the rocky tidal pools along the coast. Like Baird, who had searched for the best place to establish his Fish Commission's facilities, Ritter moved up and down the coast looking for a good location for a marine laboratory.

He drifted from place to place without sufficient funds to settle down, until, while spending a summer in San Diego and using a hotel boathouse for a laboratory, he was introduced to a wealthy and philanthropic newspaper owner, E. W. Scripps, and to his sister, of

William Ritter, founder and first director of the Scripps Institution for Biological Research. (*Courtesy Scripps Institution of Oceanography, University of California, San Diego*)

like character and resources, Ellen. Between the two of them, the Scrippses donated a great deal of money to the young laboratory and helped it through a long and arduous adolescence. With the Scrippses' money, for example, a site was acquired on La Jolla Cove 14 miles north of San Diego. In 1905 a few frame buildings were erected on this site, and the Scrippses' personal yacht *Loma* was converted into the laboratory's first research vessel. Help in equipping the lab came both from Scrippses and from Alexander Agassiz, who visited La Jolla on his way back to Boston after a lengthy cruise aboard the *Albatross* in the Pacific.

But the very next year, 1906, the *Loma* ran aground and was wrecked before she could be pulled off the reef, and the Town Council of La Jolla granted permission for sewers to be built into the cove. This meant that the institution had no boat, and no guarantee of clean water for its laboratories. Luckily, the Scripps family came through with an unexpectedly large gift, and the next year a permanent site for the laboratory was purchased north of La Jolla and a new boat, the *Alexander Agassiz,* was built, launched, and pulled off the sand bar where her first brisk slide into the water had landed her.

After 1911 the laboratory was kept open all through the year, and in 1912 it became a department of the University of California. At this time too it began expanding into fields other than biology, and in this process it was guided both by Ritter's intuition and by the examples set by several European institutions which were already undergoing a similar expansion. The link between California and Europe was made in part by Charles Kofoid, a marine biologist at Scripps, who spent part of a year in Europe visiting the Stazione in Naples, several British and German laboratories, and the Bergen Museum and Biological Station in Norway. From each of these he brought back new ideas, new methods, and often new equipment as well. When Ritter retired in 1923, Scripps had programs in marine physics and chemistry as well as in marine biology, and was well on its way toward becoming an oceanographic institution.

Unlike Scripps, neither the Marine Biological Laboratory nor the Fish Commission developed into organizations onto which the new interests of oceanography could be grafted. MBL chose to enter very specialized fields, and the Fish Commission, although given a fine start under Baird, soon took on the characteristics of an enormous fish hatchery rather than of a laboratory engaged in basic research. One reason for this unfortunate change was the frequent appointment of a politician rather than a scientist as commissioner. Baird's insistence that the position of commissioner be unsalaried was impractical and was soon disregarded. Within a decade after Baird's death the position was given by President McKinley to G. M. Bowers in appreciation for the latter's political support in West Virginia. (Happily, Bowers made a good commissioner, but other political appointees were less adept.)

A second problem, which like the first resulted from combining politics with science, was the limbo in which summer scientists were left to drift as each administration and each Congress reworked the Commission's priorities and budget. Since the Fish Commission—renamed the Bureau of Fisheries in 1903—did not hire any permanent scientific staff for its laboratories until after World War II, summer scientists were the *only* scientists working for the Commission, and these men, instead of being assured a regular salary for their two or three months' work, had to rely on each year's uncertain appropriation for their pay. Occasionally scientists arrived in Woods Hole in June only to find out that the Commission's appropriation for the next fiscal year—which began that July—had been cut, which meant that there would be almost no money for salaries or even equipment. In such an atmosphere of uncertainty, with the scope of the Commission's program determined in large part by the political mood of the moment, it was difficult to attract competent and enthusiastic scientists or to encourage them to undertake long-term projects when they did come.

The final difficulty facing the Commission was the government's increasing insistence on practical results. In earlier years, Baird had been able to channel a sizable portion of the Commission's budget, and some private funds as well, into the research he knew was necessary for an understanding of fisheries' problems. In spite of his foresight, which was shared by others, the Commission's routine responsibilities grew faster than its budget, and the organization was forced to turn too soon and too completely to the application of the little knowledge that it had gained.

"The Fish Commission is hardly in condition to do more than attend to the problems that they have in hand,"[*] wrote Alexander Agassiz at the turn of the century, and his statement was equally applicable to the Depot and the Survey as well.

For marine science, a half-century of active if not sympathetic government support was over. In the next 40 years, those before the beginning of World War II, oceanography in the United States was largely supported by private institutions.

CHAPTER II

British Oceanography Before 1870

Hurrah for the dredge, with its iron edge,
  And its mystical triangle.
And its hided net with meshes set
  Odd fishes to entangle!
The ship may move thro' the waves above
  'Mid scenes exciting wonder,
But braver sights the dredge delights
  As it roves the waters under.
          —Edward Forbes,
            "The Song of the Dredge"

While Americans were exploring their coastal waters and ascertaining the limits of their own resources, European scholars were gazing at the sea with very different thoughts in mind. For British, French, German, and Scandinavian naturalists, who by the mid-nineteenth century had already collected and catalogued many shallow-water animals, the most exciting prospect held out by marine science was the possibility of studying the animals of the deep, abyssal sea. This impractical fascination led Europeans to organize their first truly oceanographic expeditions and this they could do because, unlike Americans, who had to promise useful results in return for government support, European investigators worked with the help of respected, well-organized scientific societies, and they could therefore elicit government support for more esoteric reasons. Especially in Britain, the prospect of capturing deep-sea animals—creatures which some thought to be living fossils—was reason enough for launching costly expeditions.

Surprisingly, this happy conjunction of interest and support arose quite suddenly, for while Europe's scientific establishment had been capable of commanding government funds and attention for many years, there had not been much interest in turning this support to a study of the deep sea until the mid-1800's. Before that time, the common and eminently sensible belief had been that the cold and black abyss, which stretched for unmeasured miles below the reach of fishermen's nets, could not possibly harbor any sort of life. With no specimens to be gathered, no expeditions were launched for their capture.

This abysmal concept of the deep sea resisted change until the late 1850's, and then, within an incredibly short span of years, the deep sea's imagined poverty was replaced by speculations of faunal profusion and geological wealth. By the late 1860's "the land of promise for the naturalist . . . was the bottom of the deep sea."* Predictably, it was not long before oceanographic expeditions set sail for the promised land.

In the early years of the nineteenth century, the generally accepted picture of the sea was one of a familiar, if unpredictable, surface layer underlain by a vast and mysterious abyss. The most obvious reason for this inequality of understanding was that a knowledge of the surface of the sea—its currents, waves, and roving inhabitants—had been needed by navigators and fishermen for centuries while no one's livelihood had depended on an understanding of what went on below. A second reason was that it was a great deal easier to examine surface or shallow waters than it was to sound or dredge through several miles of ocean. These, very generally, were the economic and, one might say, the mechanical or technical reasons why the deep sea had not been explored by the early 1800's, but neither explains why so few men were interested in the deep sea and why so few attempts had been made to learn something about it. To understand this lack of motivation it must be realized that while the abyss was mysterious and unplumbed, it was not a blank sheet in men's minds waiting to receive whatever information might someday be dredged from its depths. A lack of information, in other words, did not mean a lack of beliefs and speculations.

Long before oceanographic expeditions were launched, the laws of nature—those beliefs of man based on observations made in the more accessible regions of the world—had been somewhat recklessly extended to the deep sea and had been used to form a cohesive framework of "facts" which accounted for the conditions which were imagined to exist there. These mistaken beliefs were of both a physical and a biological nature and, being consistent among themselves, formed a structure that was difficult to break apart.

Among the more bizarre misunderstandings which continued well into the nineteenth century was the old belief that water is much more compressible than it actually is and that it would, under great pressure, become so dense a medium as to support the remains of wrecked ships and drowned sailors.

In 1844, for example, the anonymous author of *The Ocean, a Description of the Wonders and Important Products of the Sea*, insisted that "Heavy bodies, which will sink rapidly from the surface, do at length apparently cease to descend long before they have reached the bottom; the pressure of the water being such as to cause them to remain at certain depths, varying in proportion to their weights. Thus it is that the plumb line will not act beyond a certain length, and we have no means, of course, of extending our inquiries deeper."Ψ

Although this fallacy persisted with extraordinary tenacity into the second half of the century—some people were sure that the transatlantic telegraph cable would hang suspended above the bottom—and even into the twentieth century—relatives of passengers aboard the *Titanic* were worried that drowned bodies were floating through the abyss—it was not one which significantly affected the scientific exploration of the sea. At least by the late 1700's scientists had learned that the density of water didn't change enough to significantly alter its viscosity, and experiments had shown that what changes did take place were more affected by temperature than they were by pressure.

But in spite of their greater sophistication concerning the compressibility of water, scientists perpetuated a serious misunderstanding concerning its point of greatest density. It was a common error to assume that salt water, like fresh, was densest at 39° F. (4° C.), and this mistaken belief had a much greater effect on early studies of the sea than did the floating remains idea.

Many mariners knew that the temperature of ocean water decreased with depth, and when this observation was combined with the belief that sea water is densest at 4° C., there arose the assumption that the deep basins of the oceans were filled with a static layer of heavy, 4° water. Moreover, it was thought that the water in these deep ponds never moved, for certainly no wind could stir abyssal waters and no density-generated currents of consequence could be caused by the small variations that differences in salinity or pressure (*i.e.*, depth) were known to produce.

The important conclusion that was drawn from all this figuring was that abyssal water never moved. Without movement, there could be no renewal of food particles or dissolved gases, and without these, there could be no life. This conception of a stagnant abyss, added to the more dramatic considerations of the region's tremendous pressures, intense cold, and total absence of sunlight, led to the certainty that life could not exist in the deep sea.

"The popular notion," wrote an oceanographer in 1872, "was, that after arriving at a certain depth the conditions became so peculiar, so entirely different from those . . . of the earth . . . as to preclude any other idea than that of a waste of utter darkness. . . ."♆

And so the abyss remained, uncertain and unproductive, until, in the 1840's, the brilliant Manx naturalist Edward Forbes decided to study the edges of this mysterious realm and to ascertain the limits of its lifeless, or as he called it, azoic, zone.

Botanist, zoologist, and geologist, Forbes gained the reputation during his short lifetime of being an excellent worker in many fields, and among his interests were the distribution of life in the sea and the relationship between modern marine animals and extinct ones, known only as fossils. As early as 1839, when Forbes was only 24 years old, he convinced the British Association for the Advancement of Science to establish a committee "for Researches with the Dredge, with a view to the investigation of the Marine Zoology of Great Britain . . . [its] Geographical Distribution . . . and the more accurate determination of the fossils of the Pliocene period [which ended about one million years ago]."♆

The committee was organized, and its members set out upon many a short excursion using oystermen's dredges. Their work was rarely done in water deeper than 100 fathoms, and obviously was not intended as an examination of deep-sea animals or as an attempt to reach the azoic zone.

Then, in 1841, Forbes got the chance to cruise through the eastern Mediterranean as the naturalist aboard the naval surveying ship *Beacon*. In that part of the sea, which is now known to be a remarkably barren area, he dredged to depths as great as 230 fathoms. He brought up very few animals, and on the basis of this poor collection, and on the richer, but shallower hauls he had made around Great Britain, he

The *Erebus,* shown here surrounded by ice in the Southern Sea, was used with the *Terror* by Sir James Clark Ross for a voyage of discovery to the Antarctic in the years 1839–1843. (*Courtesy National Maritime Museum, Greenwich*)

postulated the existence of eight bands or depth zones in the sea each of which was characterized by a particular assemblage of animals. These zones, he said, extended from the shoreline to an "unknown lower limit," and he thought this limit could be set at about 300 fathoms. Below that depth was a "probable zero of life."

Forbes died in 1854, at the age of thirty-nine, and did not live to participate in the deep-sea expeditions of the 1860's which so completely refuted his idea of an azoic zone but which also owed their existence in some measure to the interest he had generated in the deep sea. The discovery of abyssal animals did not, however, have to wait for the expeditions of the 1860's. Worms and starfish were pulled from the deep sea in earlier years, but these infrequent curiosities were not enough to dislodge the firm and logical belief in an azoic zone.

One investigation of the deep sea which was carried out within a framework of beliefs so unmatched to actual conditions that its results were frequently misinterpreted or entirely ignored, was the Antarctic Expedition of 1839–1843 led by Sir James Clark Ross.*

In 1838 a joint committee from the Royal Society of London and

---

* Sir James, whose expedition was in the Antarctic in about the same years as was the U.S. Exploring Expedition, contested Wilkes' discovery of "Termination Land." Sir James had himself sighted another section of the southern continent's coastline which he had named "Victoria Land," but he had not seen land in the places where Wilkes said that it existed.

the British Association proposed to the Admiralty that a surveying expedition be sent to observe the earth's magnetic properties in the south polar seas and to ascertain the location of the magnetic south pole. A similar project had been supported in 1701, when the astronomer Edmund Halley had sailed in the *Paramour Pink* to collect data for a chart of declinations and, since the earth's magnetic field was known to shift about, it was time to make another survey and construct another chart. The Admiralty had authorized a great number of surveying expeditions since Halley's time and was easily convinced of the importance of the voyage. In 1839, therefore, two ships, the *Erebus* and the *Terror*, were outfitted for their "survey . . . of novel design" and, under the command of Sir James Clark Ross, proceeded southward.

In addition to magnetic observations, which were to be made throughout the three-year voyage, plans for some dredging, sounding, and temperature taking were included in the expedition's schedule at Sir James' suggestion.

Sir James had acquired his taste for scientific experiment from his uncle, Sir John Ross, who had led an expedition to Baffin Bay (west of Greenland) in 1818. His young nephew had accompanied him and had watched excitedly as samples of the deep-sea floor were brought up with a clever contraption which had been constructed by the ship's blacksmith. The "deep-sea Clamm," as the device was called, had been sent down to depths approaching 1,000 fathoms and had, on at least one occasion, closed its hinged jaws around a mass of greenish ooze. The ooze had been examined, and in it were worms. There was great excitement on board, but the simple animals did almost nothing to disturb the azoic theory. They did, however, inspire Sir James to try his own experiments in the Antarctic some 20 years later.

Long surveying expeditions did not put out to sea without their proper complement of doctors or surgeon-naturalists, and Sir James' expedition was no exception. The *Erebus* and the *Terror* carried several of these versatile gentlemen, among them John Robertson and the noted botanist Joseph Hooker. Their duties, in addition to medical care, included a wide and personal choice of scientific studies. Sir James urged these men to dredge when the ship approached land and samples of the bottom with shellfish and with several of the same kinds of worms that had been collected by Sir John Ross were pulled from depths as great as 400 fathoms.

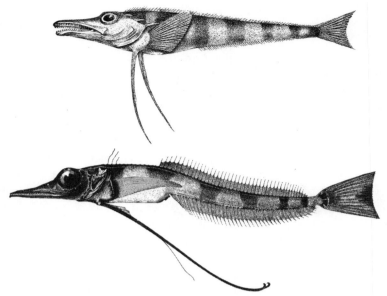

John Robertson's sketch, top, of the frozen fish *Pagetodes* that was eaten by the cat, and *Cryodraco antarcticus*, bottom, the same kind of fish rediscovered sixty years later. (*From the* Résultats du Voyage du S. Y. Belgica, *Louis Dollo*, Poissons)

"Although contrary to the general beliefs of naturalists," wrote Sir James, "I have no doubt that from however great a depth we may be enabled to bring up the mud and stones of the bed of the ocean we shall find them teeming with animal life. . . ."Ψ

Unfortunately, the majority of the biological specimens gathered on the expedition was lost due to bad preservation and worse weather.* A small portion of the collection did survive the voyage, but, like Sir John's worms, did not much affect the azoic theory.

In addition to dredging, Sir James ordered "almost daily experiments

---

* The loss of one specimen—not a deep-sea creature—was particularly well recorded by John Robertson, the surgeon aboard the *Terror*. After a bad storm sailors were breaking ice away from the ship's bow when they came upon a strange fish frozen in the ice. Chipping it free, they gave it to Robertson, who recognized it as a "species new to science" and sketched it briefly before putting it in a pan to thaw. The ship's cat found the specimen before the doctor returned, and the species, so recently introduced to science, was snatched from the category of the officially known. Robertson's sketch, labeled *Pagetodes*, a Greek word meaning "frozen solid," was included in a report on fish. Sixty years later a member of the Belgian Antarctic Expedition found another specimen of the same fish, described it under more favorable circumstances, and included in his report Robertson's original sketch with the caption, "*mangé par le chat de l'équipage de la 'Terror'*."Ψ

on the temperature of the ocean to the depth of six hundred fathoms."[*] The ineffectiveness of these measurements was not due to lost data but to mistaken beliefs and consequent misinterpretation.

One problem in obtaining accurate temperature measurements was a technical one. The expedition had been equipped with unprotected thermometers, even though experiments done earlier in the 1830's had shown that such instruments would be affected by pressure and would give erroneously high readings. (Protected thermometers were available and could have been supplied.)

A second problem was Sir James' belief in a deep layer of 4° C. water. In the south polar seas the temperature of the water actually decreases about 1° C. for every 550 fathoms of depth, and by chance, this decrease is almost exactly canceled by the opposite effect of pressure on an unprotected thermometer, which raises the reading by about 1° C. for every 550 fathoms. As a result, Sir James found exactly what he expected, a ring of 4° water existing from the surface (where it actually was about 4° C.) right down to the bottom (where it was really more like 2° or 3° C.). He surmised that this ring of immobile water circled the bottom of the earth between the latitudes of 40° and 60°, and so sure was he of this, that, on the few occasions when his thermometers gave subsurface readings lower than 4° C., the results were disregarded and the thermometers assumed to be defective.

The expedition's attempts to plumb the depth of the open ocean were more successful than the efforts to collect biological specimens or accurate temperature measurements.

"We have made several fruitless attempts to obtain soundings as we passed through the tropics," wrote Sir James in December of 1839. "These repeated failures were principally occasioned by the want of a proper kind of line, but they served to point out to us that which was most suitable. I accordingly directed one to be made on board, 3,600 fathoms, or rather more than four miles in length, fitted with swivels to prevent it unlaying in its descent, and strong enough to support a weight of seventy-six pounds."[*]

The special sounding line was ready in January, and on January 3 Sir James laconically related that "the weather and all other circumstances being propitious, we succeeded in obtaining soundings with 2,425 fathoms of line. . . ."[*] (These soundings were checked some 125 years later with echo-sounding equipment and found to be remarkably accurate.)

Sir James' account of the expedition, *A Voyage of Discovery and Research in the Southern and Antarctic Regions,* was published within four years of his return, and measured by the standards of its day, the expedition was considered a success. Yet the azoic theory had not been shaken, and the erroneous belief in a vast layer of inert water had been sustained. The second of these misunderstandings was eventually corrected by the increasing use of protected thermometers, but the first offered more resistance.*

By the 1850's animals from several hundred fathoms were being collected by naturalists, and others were being gathered, more or less inadvertently, by cable surveyors who sounded along the routes proposed for submarine telegraph cables. In spite of both groups' collections the suspicion only slowly grew that the deep sea might indeed be inhabited.

At first the boundary of the azoic zone was simply pushed downward from the 300-fathom limit suggested by Forbes to 320 or 330 fathoms, then to 400 fathoms, 450 fathoms, or to wherever an animal could be found. One of the earliest of these rearrangements was made by P. C. Absjørnsen, a Norwegian naturalist and noted collector of fairy tales. In 1853, when Absjørnsen was dredging in 200 to 300 fathoms of water, he brought up several large and luminous starfish. He named the brilliant red animals *Brisinga* after a mythical jewel that was stolen by the Norse god Loki and hidden at the bottom of the sea. His discovery, and others like it, established the existence of life at 300 fathoms. In the following years sea lilies were found a bit deeper, sponges from the Sea of Japan extended the limit to 400, then 500 fathoms, and on it went—but not indefinitely.

While naturalists might chip away at the edges of the azoic zone, they were not willing to allow cable surveyors, working in much deeper water far from shore, to put a hole right in the middle of it. The lifeless region was too well established in the really deep sea to even be questioned, and consequently the tangles of confused starfish that appeared clinging to a sounding line that had been dragged

---

* Two years after the *Erebus* and *Terror* had returned from the Southern Ocean, the ships were used by Sir John Franklin to look for the Northwest Passage. North of Canada's barren Northwest Territories the ships were beset by ice, and the men started south on foot. All perished. Meanwhile, Sir James Clark Ross had been given command of the *Enterprize* and had been sent north with many other rescue parties to find Franklin. In this he did not succeed, but he did make a number of scientific observations while immobilized by winter ice.

through a full mile of inky black water seemed too incredible to be considered seriously. Such animals could not have been living on the bottom.

This reluctance to believe in deep-sea life did not, however, stop the spiny sea urchins and prickly starfish from becoming tangled in the rough hemp lines that were dropped with increasing frequency onto the floor of every gulf, bay, and sea where a cable was to be laid. Early in the 1850's surveys were conducted in the Mediterranean and Red seas, and in 1854 the American Cyrus Field came to Britain to ask that a survey be made of the transatlantic route which Maury's officers had already cursorily examined. The British explored some portions of the route between Newfoundland and Ireland and, with the Americans, successfully laid a cable along it in 1858. When this line failed a short time later, Field returned to Britain to ask that another survey be made. The Admiralty assigned several ships to the task and among them was H.M.S. *Bulldog*, with G. C. Wallich aboard as surgeon-naturalist.

The *Bulldog* made over 100 soundings in the North Atlantic, and while the deepest of these was in 2,032 fathoms of water (still quite a feat in those days), the most interesting was in 1,260 fathoms. From that depth, 13 starfish and a coating of mud were brought up on the line. Wallich reported his findings, but proponents of the azoic theory continued to claim that the animals had attached themselves to the line, not while its slack coils lay on the bottom, but while it was being drawn through surface or intermediate waters.

In spite of the stubbornness of the belief, the azoic theory *was* finally refuted. Certainly one of the most convincing pieces of evidence against it was provided by several clusters of animals that were found growing on a submarine cable pulled up for repairs in 1860. The cable, which had lain along the bed of the Mediterranean near the island of Sardinia in depths exceeding 1,000 fathoms, had been laid in 1857, and when it failed three years later the defective parts were drawn to the surface. Fifteen animals were found, not just attached to the cable, but encrusted upon it. It could not be argued that these creatures, cemented to the cable with their own shell-like materials, had been picked up from surface waters.

The precious piece of cable was given to the Belgian marine naturalist Henri Milne-Edwards, who announced to the French Academy of Sciences the end of the azoic theory. While a few scientists refused

Charles Darwin at the age of thirty-one, a watercolor by George Richmond. Darwin's theory of evolution suggested that marine life might have developed at a particularly slow rate and so might still be able to teach zoologists a great deal about past forms of life. (*By permission of the Royal College of Surgeons of England*)

to respond to this awesome expansion of the naturalists' collecting ground, most finally agreed that the animals on the cable and the starfish on the sounding lines did indicate that life could exist on the floor of the deep sea.

At almost the same time that the cable from the Mediterranean convinced many naturalists of the existence of abyssal life, Charles Darwin's newly published theory of evolution gave them provocative new reasons for studying it. These new reasons, however, did not immediately send scientists out upon the deep sea, for Darwin's ideas took some getting used to: his theory of evolution had badly upset the order of the world.

Before the middle years of the nineteenth century the belief had been that the earth was a young planet whose major features—continents and ocean basins—had been divinely formed in a cataclysmic series of floods and fires. The forms of life which God had created in each of these terrible times had perished in the holocausts, and traces of their former existence remained only as fossils. The earth's present inhabitants, a community which for the first time included man, had also been divinely created and had existed since its creation without a single modification.

Then, in the mid-1800's, disturbing theories were advanced which sought to explain the history of the earth, and the history of the earth's inhabitants, not in terms of divine commands but rather as the continually unfolding results of slow and natural processes.

AN ABBREVIATED GEOLOGIC TIMETABLE*

Era	Period	Epoch	Estimated Time Boundaries in Millions of Years
	Quaternary	Recent Pleistocene	
			1
Cenozoic	Tertiary	Pliocene Miocene Oligocene Eocene Paleocene	
			70
	Cretaceous	Upper Lower	
			135
Mesozoic	Jurassic	Upper Middle Lower	
			180
	Triassic	Upper Middle Lower	
			225
	Carboniferous		270
			350
Paleozoic			600
Precambrian			3,000+

* From the time divisions in use by the U.S. Geological Survey.

In 1830 Charles Lyell, in his *Principles of Geology*, argued that the world had been formed by erosion and sedimentation, tension and compression, by the same forces, in other words, that were known to be remodeling the modern world. These forces, he maintained, were sufficient to explain all the changes which had proceeded (in inanimate nature *only*) throughout the earth's history. Some 30 years later Lyell's friend and admirer Charles Darwin took the idea of natural causes much further and stated in his book *On the Origin of Species by Means of Natural Selection,* that natural forces rather than a series of Divine interventions were responsible for both past and present forms of life.

Darwin insisted that each of the earth's inhabitants, in its struggle to survive and perpetuate its own kind, was imperceptibly changing in order to keep in tune with the demands imposed upon it by a slowly changing environment. If the organism could not adapt to new conditions, it became extinct. This process of adaptation, said Darwin, had been going on for as long as life had existed, and was, like Lyell's geological processes, sufficient to explain all the strange forms of life that had ever existed.

One part of Darwin's argument for evolution was of particular interest to marine naturalists. Darwin said that the rate of evolution depended in large part on the environment's rate of change. The seasons and climates on land were known to change much more rapidly and drastically than those in the sea, and this, said Darwin, apparently explained why "the productions of the land seem to have changed at a quicker rate than those of the sea."꙳

Within the constant sea itself, the most stable region was the sunless and supposedly motionless abyss, and so, according to Darwin's reasoning, its inhabitants should have changed the very least of any animals on earth. This suggested that the deep sea was the most logical place to look for ancient animals, creatures that would clearly link extinct species to modern ones. These missing links or "infinitely numerous connecting links,"꙳ as Darwin called them, were what evolutionists soon felt they needed to find, and marine biologists seemed in a good position to find them.

In the 1860's the enthusiastic search for missing links began to be rewarded by the discovery of a whole series of strangely archaic marine animals which were so similar to fossil forms and so different from modern species that, by almost anyone's reasoning, they ought to have been extinct. Naturalists began thinking of these curious animals as living fossils, and one of their best examples was the sea lily.

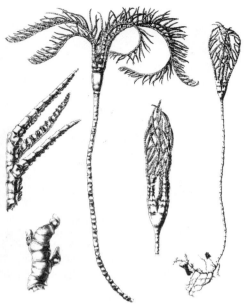

In the 1860's Michael Sars, a Norwegian naturalist and theologian, made a fine collection of bottom-dwelling animals from depths of several hundred fathoms. (*Courtesy Erling Siversten*)

The stalked crinoid *Rhizocrinus lofotensis*, two figures at right, was one of the curious animals discovered by Michael Sars that was apparently more closely related to fossil crinoids than to living species. A similar sea lily, *Rhizocrinus rawsoni*, left, was discovered by Louis François de Pourtalès of the U.S. Coast Survey. (*From the* Challenger *Report*, Zoology, *Vol. 11*)

Sea lilies, or crinoids, were known to have thrived 100 million years ago during the Cretaceous period, and at that time most of the plant-like animals had grown on stalks (hence the term "stalked crinoid"). A few species of crinoids still grew in coastal waters, but almost none of these—as far as nineteenth-century naturalists knew—lived on stalks. Then, in the cold Norwegian Sea and elsewhere, naturalists began finding stalked crinoids which looked a great deal more like fossil crinoids than like the modern, stalkless species.

One of the finest collections of these "old generic types" which had apparently continued to live in the changeless depths of the sea for millions of years was made by the Norwegian naturalist and theologian Michael Sars.

Sars spent his early life as a vicar and diligent naturalist in an isolated

Something "very strange coming from Japan," a Japanese *Hyalonema*, is actually a type of sponge which anchors itself to the muddy sea floor with a wisp of glassy threads. The polyps just below the body of the sponge are parasites. (*From Alexander Agassiz,* Three Cruises of the U.S.C.G.S. Steamer *Blake, Vol. 2*)

seaside village until he received a fellowship which allowed him to travel abroad to further his studies on jellyfish and shallow-water fauna. In 1854, having returned to Norway, he was appointed "professor extraordinary" at the University of Christiania (now Oslo). Some years later he began dredging along the sea floor that dropped abruptly into fjords.

By 1866 he had collected at least 19 different kinds of animals from the region between 300 and 450 fathoms, and among these creatures was a fine example of a stalked crinoid. Previously, this variety had been known only as a fossil.

"The *Crinoidea* has always had a special interest to naturalists," wrote C. Wyville Thomson, a British naturalist who visited Sars in Oslo to see his collection, "and, on the watch as we were for missing links which might connect the present with the past, we eagerly welcomed any indication of their presence."[*]

---

[*] Another one of these missing links which naturalists found so enthralling was a strange type of sponge. Several specimens of incredibly delicate glass sponges with glistening bundles of siliceous bristles were sent from Japan to European naturalists in the 1860's, and one genus in particular, *Hyalonema*, which had supposedly been dredged from about 500 fathoms, seemed so remarkable that its very authenticity was questioned.

"Anything very strange coming from Japan is to be regarded with some distrust," wrote Thomson. "The Japanese are wonderfully ingenious, and one favorite aim of their misdirected industry is the fabrication of impossible monsters

Thomas Henry Huxley, originator of the continuity of chalk theory and discoverer of the nonorganism *Bathybius.* (*Courtesy Trustees of the British Museum* [*Natural History*])

As more of these primitive crinoids, starfish, and sponges were brought up from several hundred fathoms, the belief grew that a community of living fossils did, in fact, inhabit the deep sea. Although erroneous, the idea was an intriguing one and was most effective in stimulating interest in deep-sea explorations. But it wasn't the only idea to do so.

One scientist, who more than most, enhanced the image of the abyss, was Darwin's close friend Thomas Henry Huxley. Huxley had been among the first to discuss the theory of evolution with Darwin, and, impressed by the soundness and apparent inevitability of the idea, he soon became one of the theory's most enthusiastic supporters. One area from which Huxley sought evidence for the theory was the deep sea.

Like Forbes, Darwin, and others, Huxley had become involved in marine zoology while on a naval surveying expedition. As a young

---

by the curious combination of the parts of different animals. It was therefore quite possible that the whole thing might be an imposition; that some beautiful spicules separated from an unknown organism had been twisted into a whisp by the Japanese, and then manipulated so as to have their fibers naturally bound together by the sponges and zoophytes which are doubtless rapidly developed in the Mongolian rock-pools."Ψ

By 1864, however, a similar sponge was found in deep waters off the coast of Portugal, and this verified the Japanese discovery. "Perhaps the most singular circumstance connected with this discussion, was that all this time we had been looking at the sponge upsidedown."Ψ man he had been appointed surgeon-naturalist aboard the *Rattlesnake*, and in 1846 had found himself on the east coast of South America. As was common at that time, little encouragement and no equipment was offered a would-be marine naturalist, but in spite of the others' lack of interest, Huxley and another naturalist on the expedition were determined to dredge. When the *Rattlesnake* rode at anchor in various harbors and bays, the two would try to borrow one of the ship's small boats, and on one occasion, when none could be spared, they "hired one pulled by four negro slaves who, although strong active fellows, had great objections to straining their backs at the oar when the dredge was down. *No sieve having been supplied,* we were obliged to sift the contents of the dredge through our hands,—a tedious and superficial mode of examination. . . . Two days after [we] set to work in Botofogo Bay [near Rio de Janeiro], provided with a *wire-gauze meat-cover,* and a curious machine for cleaning rice; these answered capitally as substitutes for sieves, and enabled us . . . to detect about forty-five species of mollusca and radiata, some of which were new to science."⸸ (Radiata was a former classification which included almost all marine invertebrates that had arms, tentacles, spines, or rays arranged radially around a center.)

By the late 1850's, Huxley was doing most of his work in a laboratory, not a longboat, and among his numerous projects was a study of the deep-sea oozes brought up by the Royal Navy's cable surveyors. From these samples, Huxley ultimately extracted two fascinating but erroneous theories, which, like the belief in living fossils, made men more anxious than ever to explore the deep sea.

The first of these was a theory known as the continuity of chalk, or the theory of modern chalk, which asserted that the chalk deposits found on land had originally been laid down as a deep-sea ooze and, conversely, that the oozes presently forming in the deep sea would someday be part of a continent or island. In 1857 Huxley had been asked to examine and report upon the sediments sampled by H.M.S. *Cyclops* from the Telegraphic Plateau in the North Atlantic, and in examining the samples microscopically, he had been struck by the similarity between the materials comprising the Atlantic muds, as he called them, and the constituents of the chalk cliffs so common in Britain. Both were composed of the calcareous or limelike remains of planktonic organisms. In fact, so close was the resemblance that Huxley found that he could confound his co-workers by pulverizing a bit of chalk, mixing it with water, and inviting his colleagues to try to distinguish between the chalk and the ooze.

Huxley believed that the two materials represented different stages of the same geological process, and he suggested that chalk cliffs and similar formations had been formed in the deep sea as oozes and that these had subsequently been compacted and uplifted to their present positions. As one of his supporters stated a decade or so later, "There can be no doubt whatever that we have forming at the bottom of the present ocean, a vast sheet of rock which very closely resembles chalk; and there can be as little doubt that the old chalk . . . was produced in the same manner."⸸

The second theory that Huxley advanced concerning conditions in the deep sea was a belief in *Bathybius*, an extensive sheet of living slime. Huxley came upon *Bathybius* by accident.

"Quite recently I have had occasion to re-examine specimens of Atlantic mud, which were placed in spirits in 1857 . . .,"⸸ wrote Huxley in 1868, referring to the same samples that had been brought in on the *Cyclops*, 11 years before. In reexamining the contents of the dusty flasks, Huxley noticed, to his surprise, that on top of the oozes lay a thin layer of viscid, mucuslike jelly in which were entangled

The supposed living slime or animal jelly christened *Bathybius haeckelii* was "discovered" by T. H. Huxley in 1868 and was believed to carpet the deep floor of the ocean. (*From C. Wyville Thomson*, The Depths of the Sea)

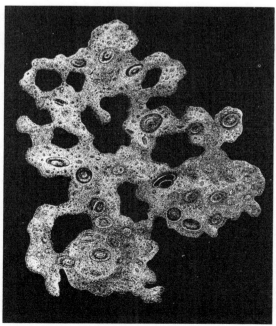

clusters of tiny granules. After watching bits of the slime beneath his microscope and seeing the granules slowly shift about, he decided "that the granule-heaps and the transparent gelatinous matter in which they are imbedded represent masses of protoplasm. . . ."

To this organism, "if," as another scientist mentioned, "a being can be so called which shows no trace of organs," Huxley gave the name *Bathybius haeckelii*, and when the German naturalist Ernst Haeckel examined his namesake, he saw in the jellylike protoplasm even greater possibilities than Huxley had imagined.

First, the primordial ooze provided a way out of the dilemma posed by a foodless abyss. *Bathybius* was thought to grow by engulfing bits of muds and ooze, or whatever happened to be on the bottom, and in turn, the creeping sheet of protoplasm provided an inexhaustible food supply for higher forms of life. Second, and more important, *Bathybius*, or "*Urschleim*" (primordial slime), as it was called by the Germans, was considered the root of the evolutionary tree, the simplest form of life from which all others had evolved. For enthusiastic Darwinists, the essential attribute of the "*Urschleim*" was its "infinite capacity for improvement in every conceivable direction. . . ."

In the decade, then, before Britain launched her first deep-sea expedition in 1868, a very few facts were agitated with a great deal of imagination, with the result that the abyss changed from the most inhospitable of zones to the very cradle of life. Taken together, the belief in *Bathybius*, a community of living fossils, and the theory of modern chalk suggested that the key to the earth's past and to the process of evolution lay in the deep and changeless abyss. From the bottom of the sea to the surface there were thought to be zones or regions within which evolution had proceeded at a faster and faster rate as the environment became more variable. It was even proposed by some to name these zones after geological epochs—the Miocene or Paleocene—according to the types of living fossils that were supposed to inhabit each. Naturalists expected that they would soon be able to trace the branches of the evolutionary tree upward, through watery strata, from the simplest and oldest organism to the most complex, modern ones or, more specifically, from abyssal *Bathybius* to the naturalist himself. Botanists might have argued that plants, not animals, were the first living organisms to appear, and that these had started in the sunny, surface layers of a shallow sea. But that line of reasoning was merely conjecture. *Bathybius* was real.

In the spring of 1868 two naturalists were working on the cold Irish coast trying to piece together the family histories of several spiny-skinned echinoderms. One of the men was Charles Wyville Thomson, a Scot, educated as a botanist at the University of Edinburgh and, in the late 1860's, a professor of natural sciences at Belfast, Ireland. The other was William B. Carpenter, a medical doctor, physiologist, and hypnotist, who, like Thomson, had begun to devote all his spare time and energy to certain departments of zoology. Both men were fascinated by abyssal fauna, and both believed, in a cautious sort of way, in the existence of *Bathybius*, living fossils, and in the theory of modern chalk.

Yet "many heavy questions as to the state of things in the depths of the ocean . . . were still in a state of uncertainty . . . ,"[Ψ] wrote Thomson, and after numerous discussions, in which both he and Carpenter expressed the desire to exchange speculations for observations, the two agreed that their questions could only be answered by the launching of a deep-sea dredging expedition. Because of the expense involved, Thomson "strongly urged Dr. Carpenter [who was a vice-president of the Royal Society] to use his influence at headquarters to induce the Admiralty, probably through the Council of the Royal Society, to give us the use of a vessel properly fitted with dredging gear and all necessary scientific apparatus. . . ."[Ψ]

Carpenter agreed to make the request if Thomson would outline the expedition's objectives, its most "promising line of inquiry." Simply put, this was to investigate the conditions of the deep-sea floor, and what Thomson and Carpenter (and most marine naturalists) meant by that was that they wished to find out what animals lived in the abyss, how they were distributed, and how they were related to ancient fossil animals.

"One or two other questions of the highest scientific interest are to be solved by the proposed investigations,"[Ψ] added Thomson, and he referred to the effects that constant pressure and darkness were assumed to have on the form and structure of abyssal animals. The effects of temperature, either on the animals themselves or on their distribution, were not mentioned as an objective of the expedition, for both Carpenter and Thomson "had adopted the current strange misconception . . . of a universal and constant temperature of 4° C. . . ."[Ψ]

With the purpose of the voyage so outlined by Thomson, Carpenter

The British fleet becalmed in the Baltic with the little paddle-steamer *Lightning* in attendance. The vessel was one of the first steamers built and operated by the British Post Office, and for years she carried the mails across the Irish Sea. In 1868, she was used for the first of a series of deep-sea dredging expeditions. (*Courtesy National Maritime Museum, Greenwich*)

forwarded a plan for a four to six weeks' summer cruise to the Royal Society, and in anticipation of the Society's cooperation and the Admiralty's consent, added that "considerable labour would be spared to the crew if the vessel be provided with a 'donkey-engine' that could be used for pulling up the dredge."[Ψ]

The Admiralty did endorse the plan, and agreed to provide the surveying ship *Lightning* (and the donkey engine). Thomson described the ship as "a cranky little vessel . . . [having] the somewhat doubtful title to respect of being perhaps the very oldest paddle-steamer in Her Majesty's navy."[Ψ] The Royal Society, for its part, supplied the "spirits and jars" that would be needed to preserve the animals that hopefully would be found on the expedition.

On August 11, 1868, the *Lightning* churned away from Stornoway on the Isle of Lewis and headed north then northwest toward the southern end of the Faröe Banks. Bags of extra coal had been stowed all along her deck, and in the space remaining a dredging derrick had been erected over the stern. As the old paddle-steamer rocked toward the open sea, Thomson, Carpenter, and the latter's young son Herbert settled themselves in cabins that had already begun to let in the sea.

On the first afternoon, several hauls were made in shallow waters to test the equipment. "All the appliances worked well," reported Thomson, "but the dredge brought up few animal forms, and all of them well-known inhabitants of the seas of the Hebrides."[*]

For the remainder of the month the unseaworthy *Lightning* spent her time hove to in gales or steaming over shallow waters where hauls of relatively familiar animals were made. Then, at the very end of the month, when the *Lightning* rode over 510 fathoms of water, "the weather was sufficiently moderate to allow us to work our dredging gear, and the first trials were of great interest, as it was our first opportunity of making the attempt in so great a depth of water."[*] The dredge was let down on a sandy, gravelly bottom, and after being pulled along by the slowly drifting ship, it returned to the surface with some shrimplike creatures, a scarlet starfish, and a few rhizopods (small and simple creeping organisms). Thomson had hoped for a much richer haul, but before the dredge could be tried again, bad weather closed in, which made even sounding impossible.

The weather cleared after several days, and since the ship was still over deep, cold water—the thermometer registered an almost unbelievable 0° C. at the bottom—additional hauls were made. Here Thomson and Carpenter began to find the "very wonderful creatures" that they had thought so much about the previous spring. The most striking animal to emerge from the dredge's rough bag was "a magnificent specimen of a new starfish. . . ."[*] It was a relative of the one Absjørnsen had found off Norway, and like his, the large, multiarmed creature was of deep crimson and fiery orange colors. In the same haul were sponges, sea urchins, and shellfish, and the best of each were pickled in spirits and carefully crated for the rough journey home.[*]

Only three days after the *Lightning* had dredged in this area of cold water, temperature measurements indicated that at a similar depth

---

[*] Thomson had expected that many of the animals found living in the perpetual darkness of the deep sea would be blind, as were certain cave-dwelling creatures. Consequently, the discovery of animals such as deep-sea scallops, found living in total darkness but equipped with a hundred or more eyes, was very disturbing to naturalists who were convinced that nature admitted of no superfluity. One such newly discovered mollusc, equipped with a "pair of well-developed eyes on short foot-stalks," prompted a London newspaper reporter to conjecture that "where there are eyes, there must one would suppose—equally on Darwinian or on teleological principles—be light. But it is scarcely possible that sunlight can visit these abysses, and as Mr. W. Thomson is accordingly inclined to believe that whatever light affords an exercise to these deeply buried organs of sight may be due only to phosphorescence."[*]

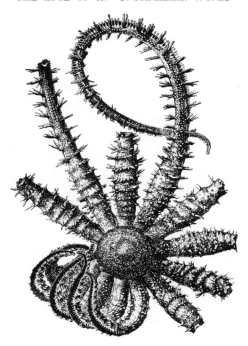

The magnificent specimen of a new starfish, *Brisinga coronata*, that was collected on the dredging cruise of the *Lightning* in 1868. This type of starfish, although not this particular species, had been discovered by P. C. Absjørnsen in 1853. (*From C. Wyville Thomson*, The Depths of the Sea)

(about 500 fathoms) the ship now rode over relatively warm water. When a reading above 9° C. was obtained, the thermometer's accuracy was suspected, but subsequent readings showed consistently high temperatures. The dredge was of course let down, and "the dredging here was most interesting. The bottom was for the first time 'Atlantic ooze' . . . . Imbedded in this mud there came up an extraordinary number of silicious [glassy] sponges of most remarkable and novel forms. . . . The most characteristic forms which we met with on this occasion were the beautiful sea-nests . . . and the even more strange [ones which are] . . . closely related to the glass-rope sponges of Japan which have so long perplexed naturalists. . . ."⸸

Also in the muddy ooze Thomson and Carpenter found sea pens and fine branching corals, a curious variety of sea urchins, brittle stars, and "some remarkable crustaceans." One of these was bright red with "large brilliant eyes, of the colour and lustre of burnished copper."⸸

After another run into the unseasonably stormy North Atlantic, the *Lightning* returned to Scotland. She had been out for approximately six weeks and in that time "only ten days were available for dredging in the open sea, and on four of these only we were in water over 500 fathoms deep."⸸ The deepest haul of the *Lightning* had been made in 650 fathoms, or less than a mile of water. (The depth of abyssal plains, which except for trenches are the deepest portions of the oceans, averages slightly over three miles.)

In spite of the few and relatively shallow dredge hauls, the cruise of the *Lightning* had shown that animal life was varied and abundant, at least to a depth of 650 fathoms, and that the kinds of animals found on the bottom varied with the temperature of the water and the nature of the sea floor. The idea of a uniform water temperature of 4° C. had been disproven "and it had been shown that great masses of water at different temperatures are moving about, each in its particular course, maintaining a remarkable system of oceanic circulation, and yet keeping so distinct from one another that an hour's sail may be sufficient to pass from the extreme of heat to the extreme of cold."⸸

Dr. Carpenter submitted a preliminary report of the *Lightning's* work to the Royal Society, and the expedition's results were regarded as "sufficiently new and valuable to justify a strong representation to the Admiralty urging the importance of continuing an investigation which had already . . . achieved a fair measure of success."⸸

The Admiralty was suitably impressed and two, more able vessels, the *Porcupine* and the *Shearwater*, were lent for extensive cruises made in the summers of 1869, 1870, and 1871. On these cruises, dredging was extended to a depth of 2,435 fathoms, and still an abundant variety of animals was found.

"A grand new field of inquiry has been opened up, but its culture is terribly laborious,"⸸ wrote Thomson in *The Depths of the Sea*, a book summarizing the accomplishments of four of the cruises. "Every haul of the dredge brings to light new and unfamiliar forms—forms which link themselves strangely with the inhabitants of past periods in the earth's history; but as yet we have not the data for . . . speculating on its geological and biological relations; for notwithstanding all our strength and will, the area of the bottom of the deep sea which has been fairly dredged may still be reckoned by the square yard."⸸

While Thomson, Carpenter and their compatriots were thus busily engaged in exploring the deep sea and popularizing their discoveries, American scientists could have argued, although few of them did, that the deep-sea dredging conducted by the Coast Survey had anticipated the investigations of the British by at least a year. Under the direction of Survey Superintendent Benjamin Peirce and the influence of Louis Agassiz, Pourtalès had led three short dredging cruises off the east coast of Florida in 1867, 1868, and 1869.

On the first of these, the Coast Survey steamer *Corwin* had been sent to survey the floor of the Straits of Florida, not solely in the interests of science, but to find an appropriate route for a submarine cable which was to be laid between Key West and Havana. In addition to sounding between the two points, Pourtalès had been directed to experiment with a half-sized dredge, and this he did in water between 90 and 350 fathoms deep. Only four successful casts of the dredge had been made, and from the deepest were collected the broken remains of corals and shells and a few live worms. In spite of these meager results Pourtalès had been greatly impressed, for he had, with many others, believed life incapable of existing below 300 fathoms.

In 1868 and 1869 similar dredging operations had been carried out from the *Bibb*. As on the other side of the Atlantic, sea urchins, sponges, and deep-sea corals had been discovered and traces of life had been obtained from depths as great as 850 fathoms. Louis Agassiz, who had accompanied Pourtalès on the last of these expeditions, predicted that "it will be seen that we owe to the Coast Survey the first broad and comprehensive basis for an exploration of the sea-bottom on a large scale, opening a new era in zoological and geological research."⸙ But the Survey was not as interested in deep-sea animals and oozes as Agassiz had hoped they would be, and as if to disprove his prediction, the Survey suspended deep-sea dredging operations (with one exception of only marginal success) for the next eight years.

Accounts of the work done aboard the *Corwin* and *Bibb* were published as appendices to the Survey's annual reports and were presented in greater detail in the *Bulletin of the Museum of Comparative Zoology*. In spite of this, they received less notice than did the accounts of similar operations undertaken on British gunboats, nor could a few specimens from 850 fathoms compare with the rich northern hauls pulled aboard the *Porcupine* from well over two miles down. But it was not the depth, the richness of the hauls, or the quality of the reports which enabled British naturalists to follow their summer cruises with the larger, more ambitious *Challenger* expedition, while in America dredging became the hobby of a few wealthy scientists. The difference lay with the governments' dissimilar attitudes toward science and with the ability of each country's scientific organizations to shape and exploit these attitudes.

In the United States, requests for the support of a subject as esoteric as deep-sea dredging would not be likely to win favor, for no matter how eloquent the appeal, there was little practical advantage in knowing the deep-sea distribution of sea urchins or starfish. The country's national scientific organizations, the American Association for the Advancement of Science, founded in the middle of the century, and the National Academy of Sciences, begun in 1863, did not much influence scientific development at this time, and would not until World War I.

In Britain, however, scientific societies had a long tradition of sponsoring expeditions—at government expense—and it was not difficult for them to continue persuading Her Majesty's Government to provide support for the glory of science and country. The Royal Society, for example, had, over the past 100 or more years, successfully urged upon the attention of the government the equipment of the Royal Observatory, the Geodetic Survey of 1784, Captain James Cook's expeditions to observe the transits of Venus, the Antarctic expeditions of Cook and later of Sir James Clark Ross, a number of Arctic expeditions, eclipse expeditions, tidal observations, and of course the summer cruises of the *Lightning, Porcupine,* and *Shearwater.*

A second difference between the countries on either side of the Atlantic was that in Europe interest in the theory of evolution and all its ramifications was intense, while the American attitude toward the same issue was somewhat diffident (with some exceptions, of course).

The paths that Americans and Europeans were following in studying the sea were certainly different, but in 1871 they did not appear to contrast so sharply. In that year the U.S. Coast Survey announced plans for sending Louis Agassiz and the Survey vessel *Hassler* around South America on a grand deep-sea expedition. For a number of reasons this was not a successful voyage, but no one could know that before it had started and the very thought of the expedition so upset Carpenter that he pronounced, with calculated overemphasis, that "having shown other nations the way to the treasures of knowledge which lie hid in the recesses of the ocean, we are falling from the van into the rear, and leaving our rivals to gather everything up. Is this creditable to the Power which claims to be Mistress of the seas?"⊬

Certainly not. Plans for the *Challenger* expedition were approved within a year. That these plans were made and carried out speaks well not only for the effectiveness of the Royal Society and the British Association for the Advancement of Science, but for the enthusiasm generated by the *Lightning, Porcupine,* and *Shearwater* cruises.

Yet in spite of the great interest that these summer expeditions had inspired, the understanding of the deep sea was still based more on the speculations of the early 1860's than on the subsequent but modest accumulation of data. When the *Challenger* set out in 1872, *Bathybius* was still thought to quiver on the sea floor, and living fossils of grotesque and curious form were believed to creep and glide above it. For scientists like Thomson and Huxley, the white chalk cliffs of some future continent were forming in the deep sea, and conversely, it seemed possible that the rubble from an ancient land—Mu, Atlantis, or Lemuria—might be lying in the depths.

The area of the deep sea sampled before 1872 was infinitesimal, and naturalists knew that to prove or disprove any of their hypotheses they would need to collect much more data from many more parts of the deep sea.

What animals, they wanted to know, might live three miles or more below the surface in the cold abyss? What oozes carpeted, or rocks paved, these great dark basins? And in what grand and timeless pattern did the waters of the oceans sink and rise, and sink again?

Confident that these questions, however awesome, were no more than a fair match for Britain's acknowledged prowess on the sea and her scientists' ability in the laboratory, a Circumnavigating Committee was established by the Royal Society in conjunction with other associations, and through this committee, support for an extensive deep-sea expedition was sought and obtained from the government. The Admiralty was requested to provide a suitable ship and crew while the Committee selected the expedition's scholarly civilian staff—four naturalists, one chemist, and an artist-secretary.

CHAPTER III

Naturalists Explore the Deep Sea:
The *Challenger* Expedition

O what an endlesse worke have I in hand,
To count the sea's abundant progeny,
Whose fruitful seede farre passeth those in land.
        —Edmund Spenser, *The Faerie Queene*

Four days before Christmas, 1872, H.M.S. *Challenger* stood out from Portsmouth and beat her way against a strong southwest wind toward the mouth of the English Channel and the deep sea beyond. For a week she fought the gale, rolling over a lumpy sea, and for a week her crew and the scientists on board worked to secure her enviable cargo of delicate laboratory equipment and heavy deep-sea gear. The ship and the 240 men aboard her were embarked upon a three-and-one-half-year expedition with the modest charge of investigating all aspects of the deep sea. But what "all aspects" would include, and how much of the sea was deep, was not nearly as clear to anyone that windy December as it would be upon the *Challenger*'s return.

The ship chosen by the Admiralty for this important expedition was the 2,300-ton naval corvette *Challenger*. She was 226 feet long with a 30-foot beam, and had an auxiliary engine of 1,200 horsepower. The only other sign of mechanization was an 18 h.p. steam winch fitted on her deck, which would, when aided by the strenuous heaving of many mariners, pull in the heavy trawls and dredges from the deep sea.

The *Challenger* at Bermuda in 1873. (*Courtesy Trustees of the British Museum* [*Natural History*])

Space aboard the warship had been modified considerably to accommodate the scientists and their equipment, although neither scholars nor spirit jars presented storage problems comparable to those involved in the stowage of hundreds of miles of hempen line. Bridges or platforms had been erected on the main deck for handling dredges and trawls, and extra cabin and laboratory space had been constructed aft. All but two of the *Challenger*'s guns had been landed to make room for her new and more pacific operations.

As director of the scientific staff, the Circumnavigating Committee had chosen C. Wyville Thomson, who, with Dr. Carpenter, had organized the summer cruises aboard the *Lightning* and the *Porcupine*. At first, Thomson had refused the position in consideration of his advanced age—forty-two—and his family responsibilities, but after a time he had persuaded himself that the proper conduct of the expedition was his own responsibility, and, so convinced, had launched himself tirelessly into its preparation.

C. Wyville Thomson, right, leader of the scientific staff on the *Challenger*, Rudolf von Willemoës-Suhm, center, a young German naturalist with the expedition, and a member of the ship's crew all rest for a while at St. Thomas in the Virgin Islands after the expedition's first ocean crossing. (*Courtesy Trustees of the British Museum* [*Natural History*])

Two other scientists aboard the *Challenger* came from the University of Edinburgh: John Murray, a driving young naturalist of broad interests (who later became director of the Challenger Office), and the expedition's chemist, John Young Buchanan. Murray, who was thirty-one when the *Challenger* departed, had been born in Canada of Scottish parents and, as a young man, had put to sea as a surgeon on the whaler *Jan Mayen.* He had made what oceanographic measurements he could on that voyage, and because of that experience—and because he was a student at Edinburgh—he had been asked to help prepare the scientific apparatus for the *Challenger* expedition. He had not, however, been asked to accompany the voyage. Then, at the last minute, a vacancy opened, and Murray was appointed.

The chemist, Buchanan, was the only physical scientist on the civilian staff, and the coolness of his manner, and perhaps the exactness of his discipline, combined to set him apart from the naturalists. He was "an eager investigator," wrote an acquaintance, but a man "with no tendency towards friendship."✢

The youngest scientist on the staff was a twenty-five-year-old German naturalist, Rudolf von Willemoës-Suhm. He had left a position as lecturer at the University of Munich to join the expedition. Henry Nottidge Moseley, a British naturalist who like so many others had studied both medicine and science, joined the *Challenger* after returning from a Government Eclipse Expedition to Ceylon. And, completing the staff was the expedition's artist and secretary, James John Wild.

The Admiralty had been as careful in the selection of officers and crew as the Society had been in its choice of scientists, and many of the seamen assigned to the *Challenger* were veterans of surveying expeditions. The ship's commanding officer was Captain George S. Nares, and he and the navigating officer, Lieutenant Thomas Tizard, had both served aboard the *Shearwater* and were familiar with the work and responsibilities of a scientific cruise. The naval personnel were

John Murray, a member of the *Challenger* expedition, later became director of the *Challenger* Office. He is shown here at the age of forty. (*Courtesy Museum of Comparative Zoology, Harvard University*)

to operate both the ship and the deep-sea machinery, and were also in charge of making all magnetic and meteorological observations.

The real work of the *Challenger* expedition began with a month-long voyage westward across the Atlantic from the Canary Islands to the West Indies, for the stormy passage from Portsmouth to the Canaries had been meant as a trial or shakedown cruise.

On February 14 the *Challenger* left the harbor of Santa Cruz under clear skies and a light breeze and twelve days later was riding over the deep Canaries basin. On that morning, as on hundreds of other mornings during the next years, the sailors rose with the sun to shorten and furl the *Challenger*'s sails. Steam was got up, and the ship, headed into the constant Trades, was kept as still as possible as a Baillie sounding machine was let over the side.*

The depth at this station was 3,150 fathoms, over half a mile deeper than any water in which dredging had been tried before. There was an air of anticipation as one of the *Challenger*'s forty dredges dropped out of sight alongside the ship. It was calculated, however, that the extra hemp line required to reach this great depth would weigh nearly 1,000 pounds, and if this addition proved too much for the ship's lifting apparatus, the expedition's oceanwide ambitions would be restricted to more moderate depths.

There were many hours to wait before the dredge would be hauled in, and the ship, as Moseley described it, lay "rolling about all day, drifting along with the wind, and dragging the dredge over the bottom. From daybreak to night the winding-in engine was heard grinding away with a painful noise, as the sounding-line and thermometers were being reeled in.

"At last . . . , the dredge-rope was placed on the drum, and wound in for three or four hours. . . ."* The inescapable complaint of the winch ceased, and, by the light of the deck lamps, the dredge, streaming muddy, brown water from its sides, was pulled onto the dredging platform.

The floor of the ocean, some three and one half miles below, had

---

* This machine, similar to the one designed by American Passed Midshipman Brooke in 1853, was used throughout the voyage of the *Challenger*. A model of Sir William Thomson's (Lord Kelvin's) sounding machine, which used piano wire rather than hemp, was used once aboard the ship, but on its first trial the reel collapsed and the wire tangled so badly that it was never tried again. (At about the same time, sailors aboard the American ship *Tuscarora* began using a wire sounding machine with good results.)

parted with 100 pounds of pasty, chocolate-colored clay, a deposit which had never been sampled before. Murray, describing it as dark brown, called it "red clay," and when thoroughly examined, the new deposit yielded not a sign of an animal larger than a microscopic plankton.

This was a disappointment, and an unexpected one, for there was no reason to believe that animals could not adapt to the conditions that existed three and a half miles below the surface if they could live in the frigid darkness of two or two and a half miles' depth. The dredge was not successfully cast in such deep water again until the *Challenger* had crossed the Dolphin Rise (now called the Mid-Atlantic Ridge) and was approaching the West Indies. Here, in mid-March, the dredge was again put over the side, and toward evening it returned from its long journey carrying another watery load of precious mud. The goop was shoveled into nesting sieves of graduated sizes, and as the scientists looked on, the sailors washed the ooze through the sieves. Sometime that evening a shovelful was dropped into the sieve, and, refusing to pass through even the coarsest meshes, were worm tubes. Two of them had limp and ugly annelid worms inside—treasures.

"The most . . . remarkable biological result of the recent investigations is the final establishment of the fact that the distribution of living beings has no depth-limit," proclaimed Thomson. ". . . Animals . . . exist over the whole floor of the ocean."⸸

But the number of animals dredged from the very depths of the abyss were poor indeed compared to the astounding assortments brought up from depths of less than a mile. In the South Atlantic, for example, off the coast of Argentina, a wide-mouthed trawl was shot in 600 fathoms of water and hours later returned with one of the richest hauls ever taken aboard the *Challenger*.

Caught in the meshes of the net were dozens of sea anemones with crowns of colorful tentacles, long-armed brittle stars, tubular sea cucumbers, a prickly mass of familiar and unfamiliar sea urchins, a strange feather star, a glistening, translucent jellyfish, and, of course, the silvery fish that had swum through the waters above. Deeper in the trawl, partly hidden in the ooze that had been scraped from the sea floor, were smaller creatures—snails and sea slugs, a single squid, seven kinds of worms, two kinds of barnacles, shrimplike amphipods, beetle-like isopods, lumpy tunicates, and the broken fragments of 10 different kinds of deep-sea corals.

"Excluding Protozoa [microscopic, single-celled animals], nearly

500 specimens of invertebrates and fishes were obtained at this station, belonging to about 127 species, of which 103 are new to science. . . ."✠

The pronouncement that life—varied or monotonous—had no depth limit in the oceans was one of the few statements that Thomson could make with certainty. The ocean did not expose her secrets easily, and her laws were often obscured rather than explained by the mass of prior speculations which had put to sea with the expedition. Several erroneous hypotheses were exposed as a result of the *Challenger's* journey, while others returned, often in an altered form. One such questionable proposition, which fascinated the scientists aboard the *Challenger* and was quickly taken up by other oceanographers, was a variation of the azoic theory. In spite of the *Challenger's* success in deep-water dredging, scientists questioned the existence of life, not on the bottom, but in the middle depths between the rich surface layer and the sea floor.

"There is every reason to believe," wrote Thomson, "that the fauna of deep water is confined principally to two belts, one at and near the surface, the other on and near the bottom. . . ." This left, in his mind at least, "an intermediate zone in which the larger animal forms . . . are nearly or entirely absent."✠

In this Thomson was mistaken, but understandably so, for at the time of the *Challenger* expedition, naturalists had not yet thought of a way to collect animals from the intermediate zone without inadvertently sampling the surface waters as well.

On board the *Challenger*, free-swimming animals from the questionable middle depths were strained from the water through conical nets of silk. These could be towed along the surface or attached at intervals to a weighted line and drawn through deeper waters. The nets filtered the sea water, trapping everything larger than the mesh size, and washed these creatures into a small tube or jar at the apex of the net. The other end of the net, the wide mouth, was held open by a hoop. Consequently, as the nets were sent down or hauled in, they passed, mouths rigidly open, through the richly populated surface zone. In spite of this obvious disadvantage, it was found that nets drawn through deep water yielded animals that were never taken in the surface nets. This suggested that a middle water fauna did exist, but, as Moseley noted, "a [closing] net of some kind is required to settle this question."✠

Scientists from many countries applied themselves to this problem, and closing nets and traps began to appear early in the 1880's. These devices were sent down to the desired depth in a closed position, opened for sampling, then closed for ascent to the surface. Paradoxically, the results of these early trials were used to support both sides of the argument.

Alexander Agassiz, on a cruise aboard the *Blake* in 1880, used a "gravitating trap" designed by Captain Charles Sigsbee. These experiments, he felt sure, "serve to prove that . . . there is at sea an immense intermediate belt in which no living animals are found."᛭

Three or four years later, however, on the other side of the Atlantic, Captain Giuseppe Palumbo on the Italian corvette *Vettor Pisani* designed two closing nets which, unlike the gravitating trap, retrieved animals time and time again from the same intermediate zone.

In the following decades a great variety of nets and traps was designed (three or four new models appearing each year), and the growing use of such devices convinced most scientists that life existed from the very top to the very bottom of the ocean. Later, this evidence led Moseley and Murray, of those scientists aboard the *Challenger*, to reinterpret the results of the expedition's tow-net collections as favoring the belief in life at all depths.

A month after her departure from Santa Cruz, the *Challenger* had completed her first crossing. She had occupied 22 stations and had made 13 dredge hauls, nine of them successful. Life had been found to exist at the greatest depths sampled, and already several new species of animals had been discovered and new kinds of deposits found. This was just the beginning.

From the West Indies the *Challenger* proceeded northwest to St. Thomas in the Virgin Islands, and from there headed north toward Bermuda. She was sailing now across the Sargasso Sea, and from the floating rafts of sargasso weed (mere wisps of the legendary masses from which ships seldom reappeared) crabs, fish, and other animals were collected. Each member of this strange, drifting community was colored and patterned like the weed itself, and some even copied the weed's characteristic blemishes and imperfections.

This was the dredging ground of Louis François de Pourtalès of the Coast Survey and of Louis and Alexander Agassiz, whose joint collections, installed in the Museum of Comparative Zoology at Cambridge, were already on a par with European collections in several categories.

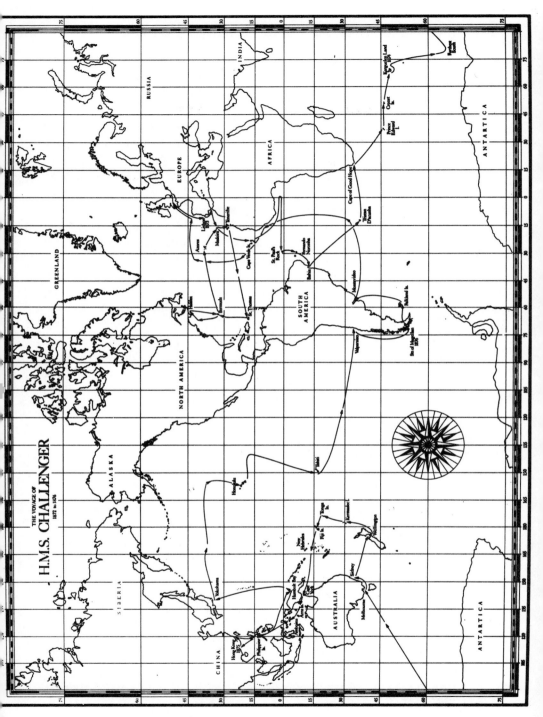

The general track of the *Challenger* on her trip around the world in the years 1872–1876. (*From Daniel Merriman,* Proceedings of the Royal Society of Edinburgh, *in press*)

Moseley was particularly anxious to collect deep-sea corals here, many of which had already been described by Pourtalès, but almost none of which had yet been compared to European varieties. Pulling the stony coral branches from the rough and rocky floor was a difficult task, as the dredge had a tendency to catch on every ledge or clump of coral and a trawl would have been lost or torn to pieces in a matter of minutes.

With the irregularities of the terrain in mind, a heavy iron dredge was lowered over the side, but hardly had it hit the bottom when, in spite of its weight and strength, it was caught on an outcrop and held fast. Before the sudden strain on the line could be eased, a block and tackle tore loose, flew across the deck, and dashed a young sailor against the hammock fittings on the far side of the deck.

William Stokes never recovered consciousness, and after a solemn service on the following day, his body was committed to the deep. The scientific nature of the cruise did not prevent Bill's shipmates from sending a deputation to Thomson to ask if Bill's body, with weights attached, would sink all the way to the bottom, or, as tradition had long had it, would float at some intermediate depth. Thomson assured them that Bill would rest on the bottom, and a long trip it would be, for, just before the funeral, a sounding had indicated that the sailor's grave would be nearly four miles down, the deepest spot that had ever been fathomed in the ocean.

> Then a splash and a plunge and our task was o'er,
> And the Billows rolled as they rolled before,
> And many a wild prayer followed the wave,
> as he sunk into a sailor's grave.*

The *Challenger* proceeded under all plain sail for the Bermudas, where a cross for Bill was placed in a crowded cemetery on Ireland Island. His crewmates presumably recovered their spirits by augmenting their daily ration of rum, while the scientists set out to study the islands' coral reef and collect the local fauna.

While Thomson waded along the shallow reef, peering at its plant-like animals through a glass-bottomed bucket, and Von Willemoës-

---

* Few such graves marked the track of the *Challenger*. During her three-and-one-half-year voyage there were two drownings, the same number of insanities (one of them fatal), one suicide, an accidental poisoning, and a case of erysipelas which killed the young German naturalist Von Willemoës-Suhm. In addition, there were 28 cases of nonfatal malaria, 10 of syphilis, and innumerable complaints of coughs and sore muscles.

Suhm hunted beetles in the swamps, the *Challenger* was tied alongside the wharf in the dockyard awaiting repairs and supplies. As was the custom, boards were placed on all the mooring chains to discourage the adventurous waterfront rats.

"Rats, nevertheless appeared in the ship," wrote Moseley in his journal, "and were all curiously enough of the old species, the Black Rat. One night as we were sitting at whist, Mr. J. Hynes, the Assistant Paymaster, suddenly started up with a yell, and danced about as if gone mad, clutching one of his legs with both hands. A rat had mistaken his trousers for a pipe or wind-sail, and had gone up."♀

The *Challenger* sailed north from Bermuda, sampled the rocks on George's Bank, and returned to Bermuda. There she waited for a mail boat from England before heading back across the Atlantic. The ship crossed this ocean four times in the first ten months of her voyage, twice north and twice south of the equator. After the fourth crossing she anchored in Cape Town harbor at the southern tip of Africa.

The expedition remained in the lee of Table Mountain for two months. The animals and plants already collected were catalogued and packed. Sixty-four crates, addressed to the Hydrographer of the Admiralty, were deposited on the dock, bound for England.

On December 17, 1873, the *Challenger* sailed from Cape Town, and leaving behind the familiar Atlantic, steered for the little-known regions of the Antarctic. The prospect of cold and stormy weather, of sailing among menacing but beautifully carved icebergs, was antici-

The zoological laboratory on the main deck of the *Challenger*. (*From the* Challenger *Report*, Narrative, *Vol. 1, First Part*)

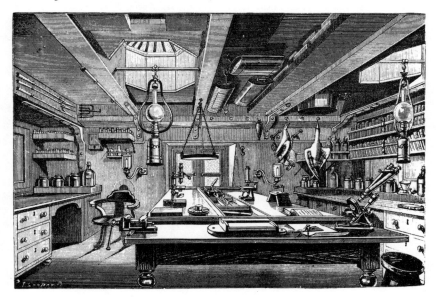

pated by many, for the routine of sounding and dredging and sampling the surface waters had, in less than a year, become monotonous for all but the most persistently curious.

"At first, when the dredge came up, every man and boy in the ship who could possibly slip away, crowded round it, to see what had been fished up," recalled Moseley. "Gradually, as the novelty of the thing wore off, the crowd became smaller and smaller, until at last only the scientific staff, and perhaps one or two other officers besides the one on duty, awaited the arrival of the net on the dredging bridge, and as the same tedious animals kept appearing from the depths in all parts of the world, the ardour of the scientific staff even, abated somewhat, and on some occasions the members were not all present at the critical moment, especially when this occurred in the middle of dinner-time, as it had an unfortunate propensity of doing. It is possible even for a naturalist to get weary even of deep-sea dredging."*

As the ship made her way south, the weather did indeed become colder, the sea rougher, and the routine upset. Lookouts were posted to watch for icebergs, and in the pervasive light of the southern summer's midnight sun, the first was sighted at three o'clock one morning. Sudden storms were more dangerous than an occasional iceberg, however, and one short but furious gale whipped up a wave of unusual size which struck the *Challenger*, stove in two ports on the starboard side, and floated everything out of the sick bay. (The hammocks carried off by the receding wave bore no passengers, for among the seasick mariners none had felt well enough to stagger to the sick bay that morning.)

By the middle of February, the unprotected ship—the first steam vessel to cross the Antarctic Circle—was in brash ice. More than 50 bergs were in sight, and the treacherous pack ice lay close to the south. Ten days later the *Challenger* lay within an estimated 25 miles of "Termination Land," a feature placed on the still incomplete charts more than 20 years before by Lieutenant Charles Wilkes of the United States Exploring Expedition. But no land was visible to the men aboard the *Challenger*, and as the ship skirted the edge of the pack, their view of even the nearest icebergs was obscured by a heavy fog that rolled in from the north. A storm came on, with gusty winds of 40 miles per

---

* It was not only possible but probable that most of the naval staff would become bored by the monotonous routines. Sixty-one sailors deserted during the voyage. The rest became chronic complainers.

The *Challenger* after colliding with an iceberg in the Southern Ocean in February, 1874. (*From the* Challenger *Report*, Narrative, *Vol. 1, First Part*)

hour, which blew the ship still closer to the grinding blocks of pack ice. Captain Nares ordered steam up and positioned the *Challenger* in the lee of an enormous berg. Snow, now falling hard, closed in about the heaving ship as she steamed toward the sheltering iceberg with just enough speed to balance the forces of wind and current which pushed her toward the pack. Then, for a moment, the wind dropped, and with a shivering thud the *Challenger* plowed into the iceberg, gouging it with her jib boom and losing the same to the ice.

No serious damage was sustained, but when, after an uneasy day, the winds subsided, the *Challenger* did not linger near the pack to catch a glimpse of "Termination Land."

It was the rule, rather than the exception, in these waters for the weather to prevent all deep-sea work except a cast of the lead. This did not seriously inconvenience Buchanan, for he could attach his thermometers to the sounding line at 100-fathom intervals and thereby obtain a temperature profile of the water from top to bottom. The instruments he used were Miller-Casella self-registering thermometers, a variation of James Six's maximum-minimum thermometers, which were encased in copper tubes to protect them from the great pressure of deep waters. Buchanan was aware, of course, that this type of instrument could only register the extremes of temperature which were encountered, and he had to assume that the lowest readings were made near the bottom and the highest near the top. A temperature inversion—a layer of warmer water lying beneath a colder one—could not be perceived with a maximum-minimum thermometer unless the inversion occurred right at the surface. In such a case, the Miller-Casella thermometer sent down to a depth of several hundred fathoms would return with a maximum reading some two or three degrees (Fahrenheit) higher than the reading made of the surface water that gurgled and splashed along the side of the ship. Buchanan discovered this type of inversion at several stations in the Southern Ocean, and he correctly surmised that the abnormally cold surface water was due in part to the melting of icebergs. This melt water, being less saline than the underlying water, would therefore be lighter in spite of its lower temperature, and would float at the surface.

Buchanan also measured the salinity of water samples drawn from different depths and used the determinations to distinguish among the masses of water that moved through the sea. He found that these masses each differed from the others in both temperature and salinity, and this implied a general circulation of water, for the natural mixing processes, however slow, would eventually erase the unseen boundaries unless the different types of water were constantly renewed.

"To take an example," wrote Thomson, "the bottom temperature near Fernando Noronha, almost under the equator, is 0°.2 C., close upon the freezing point; it is obvious that this temperature was not acquired at the equator, where the mean annual temperature of the surface-layer of the water is 21° C. [almost 70° Fahrenheit], and we may take the mean normal temperature of the crust of the earth as not lower at all events than 8° C. [46.4° F.]. The water must therefore have come from a place where the conditions were such as to impart to it a freezing temperature. . . . Across the whole of the North Atlantic the bottom temperature is considerably higher, so that the cold water cannot be coming from that direction; on the other hand, we can trace a band of water at a like temperature at nearly the same depth continuously to the Antarctic Sea, where the conditions are normally such as to impart to it its low temperature. There seems therefore to be no room for doubt that the cold water is welling up into the Atlantic from the Southern Sea. . . ."[*] (And indeed this seems to be the case.)

In addition to measuring temperatures as an indirect means of examining the ocean's circulation, the men on board the *Challenger* also had a variety of unsophisticated devices with which to directly study both surface and subsurface currents. Drift bottles, spars, and the ship herself could all be used to measure the drift of surface waters, and current drogues, like underwater weathervanes, were weighted and let go to try to estimate the deeper flows. Thomson eventually admitted that these drogues were useless, but consoled himself with the popular belief "that the movement of masses of underlying water are so slow, that, even if we had some feasible method of observation, the . . . movements . . . would be too slight to be measured with any degree of accuracy."⚓

These, and other physical studies of the ocean, which were recognized at the end of the nineteenth century to be of great importance, were passed over lightly on the *Challenger* expedition, and only two volumes of the fifty-volume scientific report were devoted to physics and chemistry. The first of these volumes, published in 1884, contains Buchanan's "Report on the Specific Gravity of Samples of Ocean Water" and Professor William Dittmar's well-known "Report on Researches into the Composition of Ocean-Water."* The second volume, issued five years later, is a catch-all for reports on magnetic studies, rocks, and atmospheric circulation. An uninspired report on oceanic circulation, written by a meteorologist, finally appeared in 1895, and was included as an appendix to the last *Challenger* volume. The expedition's contributions to the nascent disciplines of physical and chemical oceanography were not great, especially in view of the opportunities afforded by the voyage. These aspects of oceanography were not emphasized until the end of the century, and were then developed by a group of Scandinavians.

From the place where Wilkes' "Termination Land" ought to have been, the *Challenger* sailed northeast to Australia. From Sydney she looped eastward into the Pacific, calling at New Zealand, the Friendly Islands, and the Fijis. (Her arrival at Rewa in the Fijis caused the

---

* This analysis of 77 of the *Challenger*'s water samples showed that while the concentration of salts in sea water varies, the ratio among the dozen or so major constituents remain constant. This meant that a single ingredient (most often chloride) could be used to determine the amounts of the other major constituents present. Dittmar's analyses, accepted by most scientists until the turn of the century and then modified only slightly, are considered remarkably accurate even by present-day standards. The theory of constant proportions that he advanced had already been suggested by Alexander Marcet in 1819 and by the Danish geologist J. G. Forchhammer in 1859, but neither man had gone into the problem so thoroughly as Dittmar.

natives to flee, for the ship, some years before becoming a research vessel, had shelled Rewa in retaliation for the murder of a missionary.)

The *Challenger* returned to the coast of Australia, then bore north into the Malay Archipelago. She anchored and weighed many times among the islands and atolls of Arrou, Banda, Amboina, and Gillolo.

It was in the Banda Islands, during the hot, tropical September of 1874, that the expedition brought up from 360 fathoms a single, dead specimen of spirula. The small squid, with its delicate coiled shell partly inside and partly outside its body, was one of the few animals examined aboard the *Challenger* that could be considered a missing link.

Before the *Challenger* set sail, the belief had been that animals would be found in the deep and changeless abyss that had scarcely evolved for hundreds of thousands of years. This idea had formed in the minds of imaginative evolutionists and had seemed to receive support from the infrequent but tantalizing discoveries of the few deep-sea animals that more closely resembled fossils than living species. In the 1860's, such discoveries as the stalked crinoids dredged up by Norwegian theologian-scientist Michael Sars and the curiously flexible sea urchin found by Thomson during one of the summer cruises seemed links that might help bridge the gap between ancient and modern fauna.

This hope was not restricted to scientists east of the Atlantic, nor to proponents of Darwinian evolution. Louis Agassiz, an adamant anti-evolutionist, was sure he would find a community of living fossils existing serenely on the deep-sea floor. Before setting off around South America on the Coast Survey steamer *Hassler* in 1871, he had written a letter to the Survey's superintendent anticipating the results of the expedition's dredging. He believed that older forms of sea urchins and starfish, "near-Trilobites," and variations of the Chambered Nautilus would all number among the *Hassler*'s trophies.* None of these was found, although the test of Agassiz' hypothesis was hardly conclusive, as little deep dredging was accomplished on the *Hassler,* primarily due to faulty equipment.

Agassiz' disappointment anticipated the *Challenger*'s, for after almost two years of vainly searching for living fossils and missing links, the discovery of spirula seemed the exception that proved the rule.*

---

* An account of spirula was included in the last volume of the *Challenger* Report and was based on the study of two spirula, the other collected by Alexander

Aboard the *Challenger* it had been Thomson who had been the most interested in the missing links, and during the summer cruises that had preceded the expedition, it had seemed to him that "every haul of the dredge brings to light new and unfamiliar forms—forms which link themselves strangely with the inhabitants of past periods in the earth's history."[Y]

"Such hopes were doomed to disappointment," wrote Moseley, "but even to the very last, every Cuttlefish that came up in our deep-sea net was squeezed to see if it had a Belemnite's bone in its back [Belemnites, ancestral relatives of the squid, had large internal bones that often became fossilized] and Trilobites were eagerly looked out for. . . .

"Large numbers of interesting new animals were obtained . . . but . . . we picked up no missing links to fill up the gaps in the great zoological family tree."[Y]

Thomson modified his ideas on living fossils but never abandoned the idea completely. Shortly before declining health curtailed his work, he stated his intention of writing an article "On Abyssal Fauna and the Origin of Species," but the article was never written.

Murray, less affected by the idea's demise, theorized that "the colonizing of the deep sea seems to have been effected by successive migrations from the shallower reaches of the ocean. . . . The idea that a universal and peculiar fauna of great antiquity overspreads the deep ocean-floor has not been supported by systematic investigations in deep water."[Y]

From the Banda Islands the *Challenger* proceeded north, sailed through the South China Sea, and arrived at Hong Kong in January, 1875. The expedition had been under way for over two years, and during this time many of the muds and oozes collected from the sea floor had been examined under a microscope with the hope and expectation of discovering *Bathybius* "in a fresh state." But the "living slime" could not be found. Then, as the ship proceeded toward Japan, Buchanan noticed that there was indeed a gelatinous coating on the sediments that had been collected early in the voyage and preserved in

---

Agassiz. Specimens were exceedingly rare, even though piles of the squid's buoyant white shell were found on many tropical beaches. These are known on some South Pacific islands as "the shell that has no animal."

alcohol, but there was none on the muds and oozes bottled in sea water. Having grown increasingly suspicious, Buchanan finally tested the "animal jelly" and found it to be nothing more than a precipitate of calcium sulphate—the result of mixing alcohol with sea water.

"Prof. Wyville Thomson . . . informs me that the best efforts of the *Challenger*'s staff have failed to discover *Bathybius* in a fresh state," wrote Huxley in an open letter to the scientific journal *Nature*, when he had learned of Buchanan's experiments, "and that it is seriously suspected that the thing to which I gave that name is little more than sulphate of lime. . . . Prof. Thomson . . . does not consider the fate of *Bathybius* to be as yet absolutely decided. But since I am mainly responsible for the mistake, if it be one . . . I think I shall err on the right side in attaching even greater weight than he does to the view which he suggests."⚹

And with that, *Bathybius* was gone.

During its third year, the *Challenger* expedition recrossed the eastern Pacific several times, zigzagged around the islands of Japan studying the Kuroshio (or Kuro Shiwo—the black stream—as it was called then), then headed east toward Hawaii. Here quantities of fresh pineapples were taken aboard, and a gift of two huge tortoises from the Galapagos was accepted in spite of the threat that they posed to any pineapples left within their reach.

The *Challenger* proceeded south to Tahiti, crossed to the west coast of South America, and bore south for Cape Horn.

"On the morning of the 20th of January, 1876, the 'Challenger' passed through the 'first narrows' of the Strait of Magellan, wind and tide in her favor, at the rate of about 17 knots an hour; shortly after mid-day she rounded Cape Virgins, and a long, uneasy swell gave somewhat unpleasant evidence . . . that we were once more yielding to the pulses of the broad Atlantic."⚹

The *Challenger* was homeward bound, but not directly. There were stations to be made near the Falkland Islands, and still more to occupy as the ship crossed the Atlantic for the fifth time. From the Cape Verde Islands off Africa she sailed for the Azores, and just south of these islands occupied her 345th and final deep-sea station.

"From this point we made our way home as speedily as we could," wrote Thomson, "[but] north-east winds . . . were dead in our teeth, and as our coal and fresh provisions began to get low, we in our weariness and impatience were driven to the verge of despair. At length, hopeless of any relenting, we resolved to go in to Vigo [Portugal] and get some coal and some fresh provisions. . . . As we steamed up Vigo Bay on the 20th of May the Channel Fleet under the command of Captain Beauchamp Seymour, one of the finest squadrons of ironclads ever afloat, gradually resolved itself, ship after ship, out of the mist. . . . The fleet consisted of Her Majesty's ships 'Minotaur,' 'Iron Duke,' 'Monarch,' 'Resistance,' 'Defence,' 'Black Prince,' 'Hector,' and the despatch boat 'Lively,' in attendance. As we rounded the stern of the 'Defence' to our anchorage, her band struck up the air 'Home, Sweet Home,' and tried the nerves of some of us far more than they had ever been tried among the savages or the icebergs. . . ."⍦

On the following afternoon the winds, shifting, sent low clouds scudding across the sky from the southwest. The ship stood out from Vigo Bay and in "half a gale . . . the old 'Challenger' sped across the Bay of Biscay and up Channel at a pace very unusual to her. On the evening of the 23rd we passed Ushant Light, and at 9:15 p.m. on the 24th of May, 1876, after an absence of three years and a half, we stopped and came to an anchor in seven fathoms' water at Spithead."⍦

The *Challenger* had traveled 68,930 miles, at an average speed just a little better than a walking pace, to bring home to England more than 13,000 kinds of animals and plants, 1,441 water samples, and hundreds of sea-floor deposits. Her crew and staff had been working at sea for three years and five months, yet for some of them, Thomson and Murray in particular, the *Challenger* expedition had just begun.

Thomson, weary from the long sea voyage, estimated that it would take another five years to have the expedition's vast collection sorted, distributed, examined, and described. It required 19, but Thomson lived through the tensions and bickering of only six. Upon his death in 1882, Murray took on the job of publishing the now famous *Challenger* Report.

Before the *Challenger* sailed, the fate of her collection had not been definitely decided. The Treasury had paid most of the bills and had an interest in the matter, as did the Admiralty, the Royal Society, and the British Museum. For a time a four-way tug-of-war threatened to divide the collection before the *Challenger* staff had been given a chance to examine their booty.

To Thomson's profound relief, a temporary government department, the *Challenger* Expedition Commission, was established and given quarters in Edinburgh. The new office was allotted a small amount of time, moderate funds, and the immense task of publishing the expedition's results.

With crates and boxes, bottles and jars piled to the ceilings in the new office, the first task was to distribute the collection among specialists, and in this Thomson asked Alexander Agassiz to help. The younger Agassiz was familiar with British deep-sea research, for he had visited Thomson in Ireland in 1869 to see the specimens brought in by the *Lightning* and *Porcupine,* and had also visited the *Challenger* when the ship was in Halifax. Equally important, Agassiz, director of the Museum of Comparative Zoology in Cambridge, was experienced in the care and handling of large collections.*

Seventy-six scientists received animals, oozes, or water samples from the *Challenger.* Ernst Haeckel in Germany produced a three-volume report on Radiolarians (microscopic animals whose remains comprise certain deep-sea oozes); Norwegian naturalist Georg Sars, son of Michael Sars, studied several crustaceans; the Marquis de Folin (organizer of the French *Travailleur* cruises) examined a collection of tiny snails; and in the United States Theodore Lyman and Agassiz took on the *Challenger*'s brittle stars and sea urchins. Agassiz, after four years' work with hundreds of large, small, light, dark, long-spined, and short-spined sea urchins, expressed the hope that he would *never* see another sea urchin again and that the entire race would gradually become extinct.

The *Challenger* Report was heir to far more than its own collection. As the lengthy examinations continued, other expeditions brought in additional specimens of shrimp and jellyfish or rocks and oozes. It was natural that many of these should be sent to the specialists already involved with the *Challenger* collection, for it was only these men who could tell whether the most recent finds were truly new to science or had already been discovered aboard the *Challenger.* In some cases, results of the later cruises were incorporated within the *Challenger* volumes, and many of the monographs in the expedition's report include all that was known on their subjects at the time.

---

* The plants, birds, bones, and other land collections were sent to the British Museum of Natural History, but the marine collection remained in Edinburgh until sorted and distributed. Later, many of the *Challenger*'s sediment samples and a large part of her fish collection were also deposited in the museum, and have remained there.

The first volume (but not Volume I) was published in 1880, the fiftieth, and last, in 1895. Seven hundred and fifty sets were assembled throughout the world. The report was the most complete expression of man's knowledge of the deep sea, for although heavily weighted in favor of zoological studies, the fifty volumes contained some account of almost everything (with the exception of waves and tides) that had been studied in the ocean.

The *Challenger*'s effect on oceanography was, like the expedition's Report, delivered in installments. Even before the ship had returned to England she had become a symbol of deep-sea exploration and a model of naval and civilian cooperation. Although in the 1870's the venture appeared less dramatic and its role less unique than is indicated in most current accounts, the expedition nevertheless set an enviable record and was an example to be followed. And followed she was. Before she had completed her journey the U.S.S. *Tuscarora* (using wire instead of hemp line to sound) was sent to criss-cross portions of the Pacific; the German warship *Gazelle* started upon her circumnavigation of the globe; and the Norwegian North Atlantic Expedition prepared for a three-summer study of the Norwegian Sea aboard the *Voringen*.

In the decade after the *Challenger*'s return, still more cruises were launched. There were the globe-circling voyages of the Russian ship *Vitiaz* and the Italian corvette *Vettor Pisani*, and the shorter expeditions of the *Blake* and the *Albatross* from the United States; the *Drache* and the *National* from Germany; the *Knight Errant, Triton, Rambler,* and *Penguin* from Great Britain; the *Washington* from Italy; and the *Travailleur* and *Talisman* from France. Scandinavians were active at this time, too, and expeditions sailed into the cold Arctic Sea as well as back and forth across the Baltic, North, and Norwegian seas. Also in the 1880's the Prince of Monaco began his career as a promoter of oceanographic research, and the first of his four yachts, the *Hirondelle*, began her scientific cruises.

Important as the *Challenger*'s voyage may have been in stimulating further deep-sea research (and the effect is a difficult one to weigh, for although her name was reverently invoked by scientists in need of funds, the desire to explore the deep sea was probably already present), it was the Report, rather than the cruise, that provided the new science of oceanography with a framework. The bare bones that the *Challenger* staff had to some extent discovered and to a greater degree arranged had been fashioned into a skeleton that would support the further discoveries of oceanographers for many years with remarkably few alterations. Especially in the fields of zoology and geology, subsequent cruises filled in the blanks—discovered deeper trenches, new animals—but left the picture of the oceans much as it had been drawn by the *Challenger* staff and expressed in that expedition's Report.

"While any new expedition will undoubtedly clear up many of the points left doubtful by the *Challenger*," wrote Alexander Agassiz, expressing the consensus of scientific opinion, "we can hardly expect them to do more than fill out the grand outlines laid down by the great English expedition."✟

An outline encompassing some 138 million square miles of ocean, however, requires a lot of filling in, and in terms of miles logged over this vast area, no one did more to clear up the doubtful points than Agassiz himself. From 1877, at the age of forty-two, until his death in 1910, Agassiz worked in many parts of the world's oceans. Only the first of these years were spent in the Atlantic, for his interests soon turned to coral reefs, and to visit these he sailed throughout the tropical portions of the Pacific and Indian oceans.

His first series of expeditions began when he accepted an invitation from the Coast Survey to take charge of two dredging cruises in the Caribbean and a third off the Atlantic east coast aboard the steamer *Blake*. These were no carbon copies of the *Challenger*'s voyage, for the engineer-naturalist (he had degrees in both from Harvard) had new ideas on how to rig a ship for deep-sea exploration.

Agassiz had gotten these ideas, and his fortune, late in the 1860's, from the Calumet and Hecla copper mines in northern Michigan. There, at the expense of his health and his family, he acquired much experience in hoisting pieces of heavy machinery and loads of ore up and down mine shafts, and this knowledge he applied to the similar problems of raising loads of deep-sea animals aboard the *Blake*. He worked closely with Charles Sigsbee, the steamer's able and inventive captain, and the two of them rearranged the drums and winches aboard the 139-foot ship, replaced her bulky hemp lines with thinner but stronger wire rope, and equipped the vessel with a double-edged dredge (that could land on either edge and still work), a better trawl, and Sigsbee's own improved version of Lord Kelvin's sounding machine.

With Agassiz in charge, the regular routine aboard the *Blake* was disrupted, and the customary soundings and samplings done for the

Coast Survey's Gulf Stream studies were replaced by a strenuous schedule of dredging. Work began each morning before the low outline of the steamer could be distinguished from the dark waters on which she rode. Her navigators came up from below decks, through a maze of shadowy lines and machinery, and in the dimly lighted chart room reckoned the vessel's position. If necessary, her course was corrected, and soon afterward a clanging and banging and gnashing of gears signaled the start of the day's dredging.

"A new crew requires a little practice to become familiar with the working of the machinery," wrote Agassiz, "and in our first attempt . . . we came to grief by paying out the dredge-rope too fast. This produced a tangle of about two hundred fathoms of steel wire. . . ."⅄

After cutting away the kinked and knotted line, a successful routine was established whereby the dredge was lowered as the ship backed slowly away. This kept the line taut and free of tangles. Once the dredge was down, Captain Sigsbee would periodically grip the wire with one hand, for he had found that "by keeping hold of the wire rope on deck, the least movement of the dredge or trawl on the bottom is transmitted with absolute certainty and it soon becomes an

The U.S. Coast and Geodetic Survey steamer *Blake*, which was used by Alexander Agassiz for a series of three cruises from 1877 to 1880. (*Courtesy National Ocean Survey*)

The Sigsbee Sounding Machine, a modification of Lord Kelvin's sounder, was designed by Captain Charles D. Sigsbee. With this machine, wire soundings could be made even from a ship under way. It is shown here aboard the fisheries steamer *Albatross*. (*From the* Fishery Bulletin of the U.S. Fish Commission, *Vol. 16*)

easy matter . . . not only to tell whether the trawl is dragging well, but also the kind of bottom over which it is passing."⟊

Improved methods and equipment notwithstanding, Agassiz brought back the same kinds of animals and oozes that had been taken aboard the *Challenger*. These he distributed in much the same manner as Thomson had divided the *Challenger*'s booty and, in a few years' time, when the reports were completed, Agassiz gathered the charts, illustrations, and reports together and in 1888 published a two-volume work, *Three Cruises of the U.S.C.G.S. Steamer "Blake."*

The work was the only one of its kind in America, and the collections themselves were, in Agassiz' opinion, "but little inferior to those of the 'Challenger.' "⟊

In other aspects, however, the cruises of the *Blake* were less successful. In 1880 she steamed along the northeast coast of America from

George's Bank south to Charleston, and on this voyage Captain Sigs-bee's gravitating trap was first used. The lifeless samples of water it returned to the deck of the *Blake* seemed to justify Agassiz' belief in an azoic zone.

Agassiz maintained this mistaken belief throughout his life and could not be swayed by evidence to the contrary collected in the closing nets of the Italians, British, French, or Scandinavians.

In Europe, the growing number of exploratory cruises launched each year were more often sponsored by a government office rather than by a wealthy individual, but like the cruises of the *Blake,* their results complemented the work already done by the *Challenger.*

The British continued to undertake the lion's share of deep-sea exploration, and even as the Treasury underwrote the *Challenger* Report, it was persuaded to partially finance the summer cruises of the *Knight Errant* in 1880 and the *Triton* in 1882. Both were sent out to investigate one particular problem—an approach which indicated a degree of sophistication which had not existed even ten years before.

During the cruises of the *Lightning* and the *Porcupine* two masses of water had been sampled, deep in the Faroe Channel, one with a mini-mum temperature of 32° F., the other with a minimum temperature no lower than 46° F. What kept the two masses from mixing? The cause of this curious pattern had not yet been discovered when the *Chal-lenger* left England, but by the time she returned, Thomson and the others on board had a good idea of what it must be, for they had encountered and explained similar circumstances in other parts of the ocean.

"If our generalization . . . be correct," wrote Thomson, before the *Knight Errant* set forth, "a submarine ridge rising to within about 200 fathoms of the surface must extend across the mouth of the channel between the coast of Scotland and the Faroe banks."[*]

When the *Knight Errant* returned (Thomson's failing health pre-vented him from joining all but the first leg of the cruise), her sound-ings showed just such a ridge. Aptly it was named after Thomson and was further explored by the *Triton* two years later.

The French government also dispatched a series of summer cruises, although only after considerable persuasion. Alexandre Guillaume Léopold, the Marquis de Folin, independently wealthy and avidly interested in the floor of the sea, had been exploring the sea bed off the coast of Brittany in a chartered fishing boat since the late 1860's. The results of these efforts he had published as *Les Fonds de la Mer*. In 1879, encouraged by all the money that the British government was putting into deep-sea research, he requested that the French government start doing the same. After much correspondence and cajolery, a commission was established for marine exploration. The French Navy was asked to provide a paddle-wheel steamer, the *Travailleur*, and this vessel was suitably modified for her new but temporary tasks. With the British naturalist Gwyn Jeffreys aboard as adviser (he had been chief scientist aboard the *Porcupine* in 1869 and 1870 and on the *Valorous* in 1875), the steamer set out for the Bay of Biscay. Her first cruise lasted only a few weeks, but there was time enough to sound, dredge, and trawl. The depths recorded and the animals collected were neither new nor startling but agreed well with the earlier findings made by the British. Corroboration, not to mention the acquisition of a new collection for the Museum of Natural History in Paris, seemed an adequate reward, and the *Travailleur* was sent out again. In subsequent summers she was replaced by the larger vessel *Talisman*.

Regardless of the size of the ships used, a series of summer cruises seemed, in the opinion of Prince Albert the First of Monaco, an insufficient commitment for the French (including the people of Monaco) to make to the exciting field of oceanography. Although he did not seek personally to remedy this situation until the turn of the century, he had by the 1880's already embarked upon a remarkably varied career in the marine sciences.

In September, 1873, at the age of twenty-five, Prince Albert bought a schooner against the wishes of his father, and for reasons not even remotely connected to oceanography. The young prince, trained as a navigator and mechanical engineer, was badly in need of something to do. He was, at the time, a prince without a function, a husband without a wife (his first marriage was being annulled), and a father without a son. At least he was no longer a navigator without a ship.

For 12 years Prince Albert sailed the *Hirondelle* through the seas and coastal waters of Europe, earning a reputation as a skillful navigator. In 1885 the *Hirondelle* again slipped from the lee of the Rock, but this time she carried in her hold 20 beer barrels, 10 copper spheres, and 150 glass bottles. Prince Albert had combined forces with Georges Pouchet, a professor at the Museum of Natural History in Paris, and, with the latter's plan and some museum money, the *Hirondelle* was off to study the Gulf Stream.

The cause of ocean currents was still being debated, and the proponents of various theories centered their attentions on the Gulf Stream. Their studies produced as many questions as answers, and one of the former was the extent of the Stream. Some maintained that the current never crossed the Atlantic, while others insisted that it did, dividing somewhere in the North Atlantic, to send one arm north along the British Isles and another south toward the Iberian Peninsula. From time to time studies aimed at settling this question and others related to it were initiated but in spite of these, in 1885 no one really knew where the Gulf Stream water went once it left the more familiar coastal zone off New England and Canada. There were, of course, conflicting theories which explained where the water ought to go, but the prince decided that he would launch drift bottles to try to see where, in fact, it did go.

The *Hirondelle* with her strange cargo sailed through the Strait of Gibraltar, into the Atlantic, and, some miles northwest of the Azores, released 169 floats, each of which carried a polite message in 10 languages. The following year, 510 bottles were set adrift, this time between the southern coast of England and Cap Finisterre, and on the last of his drift-bottle cruises 931 floats were released as the *Hirondelle* sailed from the Azores to the Grand Banks off Newfoundland.

Of the 1,675 floats set adrift, only 227 were recovered, but the positions of these royal bottles and beer barrels seemed to indicate the existence both of a clockwise North Atlantic gyre and of a current (the North Atlantic Drift) which branched northward from it.* Those arguing for a transatlantic Gulf Stream considered the prince's evidence decisive.

The prince of Monaco confined neither his energies nor his fortune to the study of ocean currents, and within the next 20 years his attention had turned to questions of biological oceanography, bathymetry, marine meteorology, and education. He was, in fact, the epitome of oceanography's early benefactors, for his projects—inventive, unorthodox, and often dramatic—stirred interest in all aspects of the new

---

* In the years following World War I the flow of the North Atlantic gyre was further confirmed by the drift of German mines. On the basis of his drift-bottle experiments the prince predicted that mines set in European harbors and estuaries would, if torn loose, drift southward, and then westward to the American coast, then return via the Gulf Stream. Some 60 mines were found which had apparently completed this circuit.

science and were often designed to further and encourage the work of others.

Among his early concerns was the disputed existence of life in the middle depths, and to this problem he brought new equipment and techniques without precedent. In 1892 the prince took command of his new 175-foot schooner the *Princesse Alice* (the first of his ships to be built specially as a research vessel) and set out to capture the elusive creatures of the intermediate zone. The prince, and the scientific advisers who accompanied him, believed that fish, squid, and other fast-swimming animals inhabited all layers of ocean water but were not caught because the trawls then in use were too slow. To prove this hypothesis, the prince designed a high-speed trawl and, experimenting with a less aggressive approach, also had a number of wicker fish traps made. To dangle baited fish traps a mile or so beneath a ship and let them hang there for hours at a time required a leisurely cruise schedule, which few oceanographers besides the prince enjoyed. If time was available, however, the traps produced results, and from a single trap the prince once took as many as 328 fish.

Still another way to capture rapidly swimming animals was to collect them second hand after a whale or a dolphin had already hunted and swallowed them. The prince first stumbled upon this bizarre technique in 1895 among the islands of the Azores and, like the traps, second-hand capture was rarely used except on royal expeditions.

Prince Albert had been collecting specimens in a more conventional manner near the island of Terceira when he spied a pair of fluttering sails moving from the lee of the island and guessed correctly that the small boats belonged to local whale fishermen. From a tall tower on the island a lookout had sighted the blows of whales, and the small boats had been launched in pursuit. The islanders' method of whaling was to sail as close to the herd as possible, drop sails, and row hard after a single whale. The animal was harpooned and allowed to drag the boat until it tired and finally died. If the harpoon did not hold, the fishermen rowed after the wounded animal and tried to kill it with a lance.

Prince Albert was as ardent a huntsman as he was a promoter of oceanography, and upon sighting the sails he ordered work halted aboard the *Princesse Alice* and the ship brought into position near the whales. The royal yacht stood clear until a 40-foot sperm whale was harpooned, then moved in for a closer look. The passengers aboard the yacht watched in silence as the bleeding whale beat the water with its powerful tail, sending geysers of rose-colored water into the air.

"Suddenly," related the prince, "the whale stopped thrashing about and, as if revived by our very proximity, threw himself straight toward us at a great speed.

"In a flash, I tried to imagine what damage would be done if his body smashed into the side of the ship . . . when the animal, still 60 or 70 feet away from us, disappeared. . . . Ten long seconds later, his writhing body emerged on the opposite side of the ship. It stopped and did not move again."✠

As the dead whale lay floating next to the *Princesse Alice,* its heavily toothed lower jaw swinging open and shut with the gentle swell, the prince saw flowing from its mouth the remnants of its last abyssal meal. Pieces of cuttlefish and squid of an enormous size washed from the whale's maw, and the prince, recognizing the scientific value of these "precious regurgitations," ordered a dinghy lowered over the side to collect them. As the small boat was prepared, the waving tentacles began sinking slowly through the bloody water.

"Full astern!" shouted the prince, and, as he had hoped, the turbulence of the ship's propellers churned the squid to the surface, where they were quickly scooped into the dinghy. Crew and passengers crowded along the rail to see the armless body of a squid, seven feet long, and the thick tentacles from another armed with horny suckers as they were pulled into the boat. Once on deck and carefully examined, it was discovered that the whale had provided the expedition with five different kinds of squid, all new to science.

The following year the prince had two small whaling boats with power harpoons put aboard the *Princesse Alice.* He hired a Scottish harpooner to command one, directed the other himself, and, between the two of them, many whales and dolphins were captured and many strange remains taken from their stomachs. These, and other royal bounty, were often studied by the same community of scholars who had turned their efforts so effectively to the examination of the *Challenger*'s collection.*

Oceanography was growing, and by the turn of the century many nations had found it necessary to establish commissions or committees to direct their expanding programs of marine research. More and more ships were being converted for oceanographic use, and cooperative

---

* The chemist aboard the *Challenger*, J. Y. Buchanan, accompanied Prince Albert on many of his cruises.

The prince of Monaco's fourth yacht, *Hirondelle II*, was, like her predecessors, fully equipped for oceanographic research. Prince Albert took command of this vessel in 1911 and after his death eleven years later there was an unsuccessful plan to purchase the ship for international investigations. (*Courtesy Musée Océanographique de Monaco*)

programs of marine investigation had already begun among several countries. As interest grew and the field expanded, a need arose for special schools, laboratories, libraries, and museums, and in anticipating and meeting these needs, as well as in exploring the sea, the prince of Monaco was a leader.

By 1898 Prince Albert had decided to build an oceanographic museum in Monaco that could be used in conjunction with an institute of oceanography which he intended to establish in Paris. With plans of magnificent proportions in his mind's eye, and a state budget fattened by the profitable casino at Monte Carlo, the prince proceeded to have his double institution built. In the museum there would be aquariums stocked with fish from many seas, collections of preserved animals, and exhibits of oceanographic equipment. For scientists, and especially for students of the Paris institute, there would be laboratories, a library, conference rooms, and a small collecting vessel.

Work on the museum was begun in 1899, and the structure, as commanding as his vision, was formally and lavishly dedicated in 1910. Complementing its progress, courses in oceanography had been started in Paris in 1903. The Institut Océanographique de Paris was officially opened in 1911, and, reflecting the current interests of French and British marine scientists, included three professorships, one each in marine biology, the physiology of marine life, and physical oceanography. Students were to receive formal training at the institute and gain field experience by using the facilities at the museum. Both institutions had a journal in which to publish the results of their work, and the two publications were soon merged into one, the *Bulletin de l'Institut Océanographique*, which is still being published today. This was the first periodical established exclusively for oceanographic studies.

As was mentioned, Prince Albert felt that the French were seriously neglecting oceanography. The establishment of the museum and institute in Paris would, he hoped, be a step toward correcting this situation, but in this he was disappointed. The French government did not build upon the foundation he had laid. The institute was not expanded, nor were others like it begun.*

---

* This is not to say that France neglected all aspects of marine science. Marine biology was a popular subject after the 1870's, as it was in most European countries, and numerous seaside laboratories, most of them founded by universities

Prince Albert the First of Monaco at the age of sixty-two. In 1910, the year this photo was taken, he dedicated the Oceanographic Museum of Monaco which he had planned and built. (*Courtesy Musée Océanographique de Monaco*)

It was in Scandinavia, not France, that the idea of teaching oceanography took hold more strongly. Here interest was directed more toward physical and chemical oceanography. By 1907 the University of Oslo in Norway offered courses in physical oceanography, the Göteborg Högskola in Sweden began the same in 1914, and the Bergen Museum's Geophysical Institute, established in 1917, had chairs in physical oceanography and dynamical meteorology.

Special interests were beginning to emerge within the broad field of oceanography, and although the men who sailed aboard the *Challenger*, *Blake*, and other research vessels brought back diverse collections of plankton, fish, sediments, water samples, and measurements, such specimens and data were increasingly divided among scientists of special interest and experience for final study and report.

One such special interest or aspect of oceanography to emerge in the last quarter of the nineteenth century was what is now spoken of as geological oceanography, or marine geology. It was begun as the study of bathymetry—the shape of the ocean basins—and of marine sediments. As with marine zoology, the investigation of the sea floor was stimulated by a variety of fascinating speculations, and was most thoroughly undertaken by the *Challenger* expedition.

---

and open only during the summer months, were established. These were both the workshops of naturalists concerned with the description and classification of marine organisms, and the laboratories of physiologists and embryologists who found the simple forms of marine life easier to study than more complicated land forms.

Among the hundred or so laboratories founded in Europe by the turn of the century the Zoological Station at Naples, started in 1872, and serving scientists from Europe and America, was the most famous. In France, Henri Lacaze-Duthiers founded the zoological station at Roscoff in 1871 and a laboratory ten years later. These same dates marked the establishment of two Russian facilities, one each on the Black and White seas. German naturalists and physicians established botanical-zoological stations at Kiel and Heligoland, and the Dutch constructed a portable laboratory which was moved to a new spot each summer. Several stations were begun in Britain and the United States, one of the first here being Louis Agassiz' Anderson School for Natural History, which as described in Chapter I, was run on Penikese Island during the summers of 1873 and 1874. Few of these laboratories were very closely related to the oceanographic efforts of the time, and it was only after World War I that an occasional marine biological laboratory grew into an oceanographic institution.

CHAPTER IV

Reefs, Rocks, and Oozes:
Geological Oceanography

There rolls the deep where grew the tree.
O earth what changes has thou seen!
There where the long street roars, hath been
The stillness of the Central Sea.

The hills are shadows and they flow
From form to form and nothing stands:
They melt like mists, the solid lands,
Like clouds they shape themselves and go.
                    —Alfred Lord Tennyson,
                    "In Memoriam A.H.H."

When the *Challenger* put out to sea in 1872, the sea floor that lay off
the shores of busy maritime countries had already been mapped, and
the sediments that lay upon it had been collected, studied, and given
four or five names apiece. The interests prompting such a collection of
facts were neither new nor, in most cases, scientific. The mapping of
navigable waterways had a long history and at least as early as the
seventeenth century the depth of the water was indicated on charts by
a scattering of individual measurements. Then, in the eighteenth
century, the practice was begun of connecting the sounding notations
of like depth to make contour lines which expressed in a much clearer
way the general configuration of the sea floor. To these maps were
added brief notices of sediments, for experienced seamen could ascertain their approximate whereabouts by the oozes, muds, or "Yorkshire
beans" that came up with the sounding lead.

Later still, the job of charting was passed from private hands to
public, and the endless work was industriously undertaken by national
hydrographic offices and surveying bureaus. Many of these agencies
began collecting samples of the sea bottom as a routine part of each

Louis François de Pourtalès examined marine sediments for the Coast Survey, and in the late 1860's mapped their distribution off the East Coast of the United States. (*Courtesy Museum of Comparative Zoology, Harvard University*)

survey. In the United States, for example, more than 9,000 sediment samples were gathered in the years between 1844, when Coast Survey Superintendent Bache ordered such samples preserved, and 1870, when Louis François de Pourtalès put them in order. Pourtalès, who had gone to work for the Coast Survey shortly after emigrating from Switzerland, had been placed in charge of the rapidly growing sediment collection. By the late 1860's it seemed to him—and certainly to his eminent friend and adviser, Louis Agassiz—that a great deal might be learned of the geology of North America's eastern coastal regions if a map were made showing the submarine distribution of sands, clays, gravels, muds, and rocks. Pourtalès drew up such a map which covered an area extending from Cape Cod south to Florida and from the coastline east across the continental shelf and slope. The kinds of sediments were indicated by different colors, and the shape of the sea floor, including such details as the deep Hudson Canyon, was represented by contour lines. From the completed map it was easy to see that portions of the relatively shallow border regions or continental margins had been subjected to very different geological conditions in the past. As Agassiz had predicted from his study of land geology, the sea floor off southern New England was scattered with sand and gravel and peppered with erratic boulders. This was the characteristic jumble of glacial debris, and Pourtalès correctly surmised that glaciers had once extended over this area as far south as New Jersey and had spread rocks and gravels eastward as well across large portions of the continental shelf.

Pourtalès' map of marine sediments off the East Coast of the United States. The map was drawn for the Coast Survey but published in a German scientific journal. (*From* Petermann's Geographische Mitteilungen, *1870*)

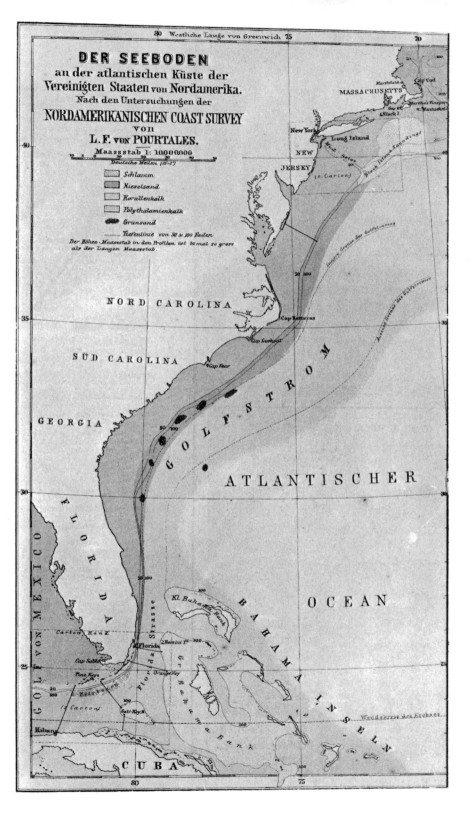

DER SEEBODEN
an der atlantischen Küste der
Vereinigten Staaten von Nordamerika.
Nach den Untersuchungen der
NORDAMERIKANISCHEN COAST SURVEY
von
L. F. von POURTALES.

Maassstab 1: 10000000

Off the Middle Atlantic states Pourtalès' map showed a wide, featureless shelf, while close off southern Florida lay a narrow belt of living coral bordered by a rocky plateau now known as the Pourtalès Terrace.

Unfortunately, the Coast Survey's chronic shortage of funds prevented Pourtalès' map from being published in the country whose borders it was designed to portray. It was printed instead in *Pettermann's Mitteilungen* (1870), a German geographic journal, and two years later, a report, *sans* map, appeared in the Survey's annual report.*

In Europe, scientists with less official ties to surveying bureaus were concerned with similar projects. In the 1840's and 1850's the "zoo geologist" Edward Forbes described both the animals and the sediments found around the British Isles, and in 1871 the French geologist and engineer Achille Delesse published a two-volume work, with atlas, on the *Lithologie du Fond des Mers*, which included maps of marine sediments found along the coast of France, in the English Channel, and in parts of the Mediterranean.

Information from the deep sea was slower to accumulate due to the greater difficulty of sampling and to the lack of practical motivation. By the mid-1800's, however, the possibility of laying trans-oceanic telegraph cables had begun to provide the necessary stimulus for deep-sea work and enough soundings were made and samples taken to permit the construction of a topographical map of an entire ocean basin—Maury's "Map of the Basin of the North Atlantic"—and to start rumors running as to the nature and significance of the deep-sea floor.

There were two especially fascinating rumors or speculations which came up with the early deep-sea casts. One concerned the possibility of drowned continents and was in part the result of soundings made over unsuspected submarine mountains and ridges. The other, as was mentioned in Chapter II, was the continuity of chalk theory, which posited a father-to-son relationship between deep-sea oozes and land deposits. The theory had been suggested by the comparisons made between two very similar substances, globigerina ooze and continental chalk, and it implied that at some time in the past, portions of the deep sea had been uplifted many miles and were now parts of continents.

---

* Pourtalès' data were later used by Alexander Agassiz for a map which appeared in *Three Cruises of the U.S.C.G.S. Steamer "Blake"* (1888).

Both the drowned continent and continuity of chalk theories argued for a world in which deep ocean basins and high continents could slowly trade positions, and both ideas were vehemently opposed by scientists who believed that this was not the case and that ocean basins and continents were permanent features of the earth's crust.

The permanence *versus* impermanence of continents and ocean basins, as this debate was called, was not concerned with continental drift or other forms of horizontal movement, for in the nineteenth century the only motions considered possible within the earth's crust were vertical ones—the upwarping of a mountain range or the submergence of a valley. The question that was boldly argued was *how much* vertical movement had taken place since the earth had cooled. This problem was approached by scientists of many disciplines, most of whom gathered their evidence on land and a few of whom went to sea for their answers. The debate on permanence was already well defined when oceanographers began collecting pertinent data and the conditions that they expected to encounter in the deep sea depended in part on the position they took in this debate.

Some scientists, like Sir Charles Lyell, C. Wyville Thomson, and Thomas Henry Huxley, believed that whole continents had sunk some three or more miles and now lay at the bottom of the sea and, conversely, that the deep-sea floor with its carpet of oozes had risen a similar distance to become a part of modern continents. Consequently, these men thought that continental rocks might be dredged from a sunken land and that deep-ocean oozes could be collected that would be similar—if not identical—to land deposits.

In the opposite camp, and, at the start of the argument, equally uninformed as to the actual conditions in the deep sea, were Louis and Alexander Agassiz, John Murray, and many physicists—notably Sir William Thomson, Lord Kelvin. For a variety of reasons these men believed that continents and ocean basins had always existed as first formed and had undergone only minor swampings and upliftings. They were sure that the rocks or oozes in the deep sea would be quite different from those found on land. Such were the conflicting views that went to sea with the *Challenger*.

The scientists aboard the *Challenger* used dredges and trawls of several designs as well as a conventional sounding machine to gather pounds instead of ounces of sediment from the deep-sea bottom. Prior

The dredging and trawling apparatus used aboard the *Challenger*. (*From the* Challenger *Report*, Deep-Sea Deposits)

to the expedition, marine sediments had been sampled inadequately with narrow sounding tubes which at best brought up a half-cupful of ooze and at worst clanged ineffectually on the rocky sea floor and came back bent double. The easiest and quickest way to sound and sample from the *Challenger* was to use the Baillie sounding machine, which brought up some bottom material in a tube attached to it for that purpose, and gave the depth as well. This machine, like the one designed by Brooke in 1853, had a short tube which projected below the sounding weights and which might bring back close to a cupful of ooze.

The commander of the *Challenger*, Captain Nares, modified the Baillie sounding machine still further, and a tube, two inches in diameter, was made to project as much as 18 inches below the weights.

Dredge used on the *Challenger*. (*From the* Challenger *Report*, Deep-Sea Deposits)

With this arrangement, a quart of sediment could be retrieved, thus enabling Murray, Buchanan, and others to run many different tests on each sample.*

Still larger samples of sediment were acquired by dragging a dredge or a trawl along the bottom. A fine-meshed "bread bag" was sewn inside the rough outer bag when muds and oozes were to be collected, whereas the outer dredge bag, made of heavy twine netting, was used alone for collecting rocks and nodules. Even with these relatively sophisticated means, the oldest method of sediment collection was not ignored, and every time the *Challenger* weighed anchor the heavy iron flukes were examined for traces of mud and ooze, which, if found, were bottled and added to the collection.

---

* It has been noted that another type of sediment sampler, the gravity corer, had been designed in 1866, but was not used on, or even considered for, the *Challenger*. The corer was first made and used by Henry Marc Brunel for a geological survey of the floor of the Strait of Dover. Wishing to ascertain whether a proposed tunnel beween Dover and Calais would pass through a continuous bed of limestone, he designed a corer with steel-tipped lead pipes which fit beneath an 80-pound weight. The corer was dropped onto the bottom of the English Channel in 207 places, and in 111 cases the punch leads brought up samples of chalk. Brunel's gravity corer was not as different from a regular sounding machine as its name might imply, and samples that it could retrieve could probably be brought up as well by the Baillie machine, which could be weighted with 400 pounds of lead.

After using these methods of sediment collection for three and a half years, the scientists on the *Challenger* were able to describe in an admirably accurate, albeit general, way the composition and distribution of deep-sea sediments. In intermediate depths—"except in the neighbourhood of coasts, where the deposit at the bottom consists chiefly of the debris washed down by rivers . . ."♆—the most common sediments sampled were organic oozes largely composed of planktonic shells and skeletons. These sediments were usually named for the kind of plant or animal whose remains made up the greater part of the ooze, such as globigerina ooze (composed of the limy tests of globigerina), pteropod ooze (full of mollusc—pteropod—remains) and diatomaceous ooze (made from the siliceous skeletons of diatoms).

The first of these, globigerina ooze, was the supposed precursor of chalk. Thomson and Murray, politely disagreeing for the first two or three years of the expedition, studied hundreds of samples and compared these to pieces of chalk chipped from cliffs or dug from the ground. By the time the *Challenger* had returned, Thomson had been shown the error of his hypothesis, and admitted reluctantly that "there is great difficulty in pointing out rocks . . . which correspond entirely . . . with the beds now in process of formation at the bottom of the ocean."♆ The subsequent discovery of wave-worn pebbles and shallow-water fossils in land chalk deposits pointed definitely to their shallow-water origin and finally closed the debate on the continuity of chalk.*

Another question which arose from the study of oozes concerned the theoretically simple but apparently more complicated relationship which sediments bore to the plankton living above them. Murray, who had already singled out marine sediments as his specialty, collected plankton from the surface and from depths of 10, 20, 50, and 100 fathoms and compared them to his oozes. While Thomson saw in these surface collections "a number of things beautiful in their brilliancy of colouring and . . . strangely interesting for the way in which their glassy transparency exposed the working of the most subtle parts of their internal machinery,"♆ Murray saw the ingredients of deep-sea ooze. After a year or more he found that "hundreds of observations of this kind, repeated day after day, led to a very accurate conception of

* A similar reevaluation was made in the case of red clay, which Thomson had first believed was "very like one of the paleozoic schists" (a continental rock).♆

the part played by surface organisms in determining the nature of the deposits now forming on the floor of the ocean at different depths and in different latitudes throughout all parts of the world."* Oozes, he said, referring to those forming at intermediate depths, were mostly made up of surface plankton and contained only scattered plates and spicules from bottom-dwelling animals. (Thomson had originally believed that oozes were almost entirely composed of organisms that had lived on the bottom.)

So adept did Murray and his colleagues aboard the *Challenger* become in predicting the nature of the surface fauna from the oozes on the bottom and vice versa, that toward the end of the cruise games were played with unmarked sediments, the object being to guess from a look at a sediment where it had been taken and from what depth.

Depth was an important consideration in the study of sediments, for it was found that portions of the planktonic rainfall that constantly fell toward the bottom of the sea were dissolved below a depth of approximately 2,500 fathoms, and that the character of the deepest sediments reflected this loss of organic litter.

This unsuspected process of dissolution was discovered as the *Challenger* first made her way across the deep eastern basin of the Atlantic. A pasty, dark brown clay was brought up from below 2,500 fathoms, and in it were found few of the shells or spicules which made up the bulk of the more common oozes. But at the same time the naturalists on board reported that life in the surface layers was just as abundant as it had been over areas carpeted with ooze.

Buchanan and Murray explained the apparent contradiction by correctly surmising that the corrosive power of sea water increased with pressure (which essentially meant depth), and that the calcareous portion of the planktonic debris falling through three or more miles of water was largely dissolved before reaching the bottom. In some areas this process could prevent the deep-sea floor from receiving all but the smallest amounts of debris, and in these regions there was little to mask the incredibly slow accumulation of a fine, brown clay.

Red clay, as Murray insisted upon calling this brown deposit, was the first of several nonorganic deep-sea sediments to be discovered aboard the *Challenger*, and when it was first brought in, it commanded the attention of all. As the first samples were taken from the sounding tube and dredges, the naturalists gathered on the dredging platform, and Murray, squeezing a lump between his fingers, pronounced it "soft, plastic and greasy."^ψ Passed around for inspection the clay was molded into various shapes, rubbed on the tongue to detect gritty particles of sand or shell, peered at through hand lenses, smeared on slides for microscopic inspection, sniffed, squeezed, and generally examined in a scientific manner. A portion of the sediment was baked until thoroughly dry—a process which took much longer than expected—then the testing was begun all over again. The small brown brick was scraped, poked with a knife, and finally broken with a hammer into fragments that were found to "stick to the tongue when dry."^ψ

As the *Challenger* proceeded on her journey, samples of red clay were brought up from almost every deep station, and when large amounts were hauled in, a quart or more could be spared for another experiment. This quantity of clay was added to a sinkful of sea water, and through the muddy mixture was drawn a magnet wrapped in iron paper. After sufficient swirlings and stirrings, the magnet and paper were withdrawn and the latter examined for magnetic particles which might have been pulled from the clay. From each quart of clay thus examined Murray found between 20 and 30 cosmic spherules and some half-dozen chondres (metallic globules with a crystalline structure) which he speculated were thrown off by meteorites passing through the earth's atmosphere. (It is currently estimated that from 35,000 to a million tons of cosmic matter plummet into the ocean annually.)

Although red clay was dredged only from deep basins, usually thousands of miles from land, it was found to be largely composed of a mixture of volcanic debris and terrestrial minerals. Murray explained the puzzling occurrence of the latter by assuming that dust was blown by winds far out to sea. Other scientists doubted that wind could transport appropriate quantities of dust so many thousands of miles, but Murray, with characteristic industry and ingenuity, was ready to answer their arguments with more than speculation. First, he found in ships' logs accounts of dust storms far from land. Most of these reports could be linked to the infamous Harmattan, a trade wind that each winter drives roiling clouds of dust across the Sahara and out over the South Atlantic, but mention was made too of similar storms that blew seaward from the desert regions of western Australia. Next, Murray managed to collect, or have collected for him, large samples of water from mid-ocean, and even in the clearest of these he found small quantities of dust and sand. To further clarify the origins of red clay,

in other words to show that its other constituent, finely ground particles of pumice, would have had time to drift from the site of a volcanic
eruption to a bed of red clay perhaps a thousand miles away before
becoming water-logged, Murray let bits of the spongy volcanic rock
that he had dredged from the sea floor dry out for seven years. He
then put these fragments into beakers of sea water, where, he noted
with satisfaction, some of them remained afloat for 20 months.

In the very deepest portions of the oceans, still another deposit was
found—radiolarian ooze. As the *Challenger* sailed through the eastern
Pacific, near the island of Guam, a sounding happened to be made in
what is now known as the Mariana trench. Before reaching this trench,
or "deep," soundings had shown that the ship was sailing over a broad
plateau that lay some 2,300 fathoms beneath the surface. Not expecting
any sudden change in depth, the sailors swung a sounding machine
over the side with the usual 300 pounds of lead weights. The sounding
line spun off the huge reels, rattled through the blocks, and continued
to disappear beneath the surface. The 3,000-fathom mark was passed,
the 4,000, the 4,500, and just as the officer on watch concluded that a
strong current was stripping off the line, the hemp slowed its pace.
The depth indicated was 4,575 fathoms, almost 5.3 miles. This depth
seemed so extraordinary that a second sounding was made using
heavier weights. The depth then registered was 4,475 fathoms, and this
time three inches of oozy clay, red on top and fading to the color of
straw below, was brought up in the sounding tube. (The one thermometer to return intact from the bottom registered 26.4° F.)*

When the new deposit, the deepest that had ever been retrieved by
any ship, was examined under a microscope, the glassy shells of the
small, planktonic animals called radiolarians could be seen scattered
within a matrix of red clay. Radiolarians had been captured in the
surface waters, although their delicate, transparent bodies, shaped like
stars with a hundred arms, or clusters of minute glass bubbles, were
often badly mutilated. The skeletons of these animals, made of silica,
did not dissolve as readily as the limy tests of some other types of
plankton, and consequently, the radiolarians could complete the long

---

* The *Challenger*'s sounding, in what is now known as the Challenger Deep,
had been exceeded the year before by a measurement of 4,655 fathoms made
from the U.S.S. *Tuscarora* off Japan. No deposit was brought up with the cast,
which led some men (but not those aboard the *Challenger*) to doubt its validity.
Recent surveys have recorded depths as great as 6,000 fathoms in the same general area.

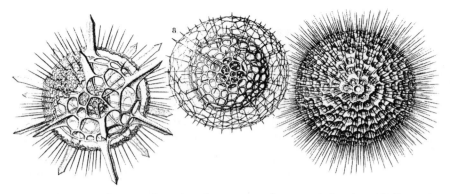

The glassy skeletons of radiolarians were discovered in the *Challenger*'s deepest bottom samples. More than 3,500 different species of this simple microscopic animal were collected on the expedition. (*From the* Challenger *Report,* Zoology, *Vol. 18, Plates*)

descent to the deep-sea floor and accumulate there. (An incredible number of new species of radiolarians were discovered on the expedition—over 3,500.)

In the same deep ocean basin where the dredge returned from day-long journeys bearing traces of red clay or radiolarian ooze, it was common to bring up bushel upon bushel of what appeared to be irregular potatoes. The bulging dredge was winched clear of the water and, with a great thumping and bumping—very different from the clattering fall of shells—a hundred or more manganese nodules would fall out upon the deck.

The *Challenger* had first encountered manganese deposits in the Atlantic. There pieces of deep-sea coral coated with a glossy black veneer were dredged from the Mid-Atlantic Ridge. Buchanan had examined the curious crust and pronounced it an almost pure oxide of manganese. Nodules, on the other hand, were most frequently found in the Pacific and were not nearly so pure.* As described by Murray, the nodules ranged from the size of a mustard seed to that of a cricket ball, and most of the larger ones had a bit of volcanic glass, a shark's tooth, or the ear bone of a whale for a nucleus. The sharks' teeth were particularly interesting, for they had not worked loose from the jaws of contemporary animals but were the fossilized teeth of Miocene sharks which had lived several millions of years before. That these teeth, overlain by a coating of manganese, were still lying on the

* Nodules may contain varying proportions of iron, nickel, tin, copper, cobalt and other metals in combination with variable amounts of manganese.

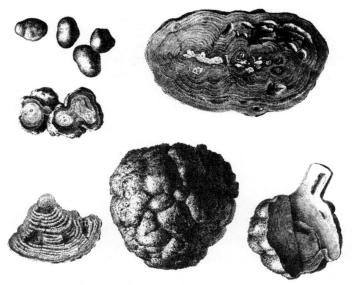

Manganese-iron nodules dredged by the *Challenger* expedition. (*From the* Challenger *Report*, Deep-Sea Deposits)

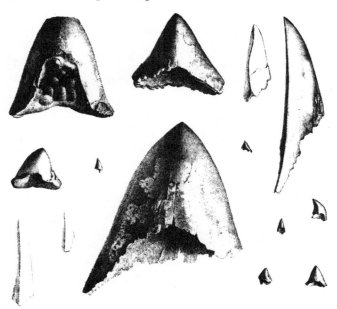

Fossilized shark's teeth, many millions of years old, dredged by the *Challenger* expedition. (*From the* Challenger *Report*, Deep-Sea Deposits)

surface of the sea floor led the *Challenger*'s scientists to the justifiable assumption that such deep-sea deposits were collecting with unimaginable slowness.

Such discoveries as manganese nodules, red clay, or unsuspected submarine ridges were described by Thomson, Murray, and others while still on the ship, and their preliminary reports were mailed back to England, where they were read before scientific societies or translated into more vigorous, less technical language for an occasional newspaper account. Having thus appeared, the information was fair game for all users, and men of disparate interests and reputations wasted little time in fitting the *Challenger*'s discoveries into their pictures of the world.

Submarine mountain ranges were particularly susceptible to this treatment, especially in the hands of naturalists who wished to see in the records of deep-sea soundings the dim outlines of a drowned continent or the shadowy form of a submerged land bridge. These ancient connectors between existing continents were needed by zoologists, paleontologists, botanists, and geologists (many of whom were studying the distribution of past and present life) to explain the patterns of closely related rocks and animals that they were finding in widely separated lands. When, for example, the fossil mammals of the Wasatch (an area including parts of New Mexico, Utah, Colorado, and Wyoming) were found to resemble the flightless birds and shuffling tapirs of western Europe, then a path was made—at least in minds and textbooks—which allowed the ancient animals to pass between the two continents. More evidence of prior connections came from the studies of rocks and plant fossils, and, armed with such suggestive data, many geologists fearlessly crisscrossed the oceans of past geological ages with meandering land bridges and island stepping stones, or plugged the entire gap with a slowly sinking continent.

When news arrived from the *Challenger* that a long and sinuous mid-Atlantic ridge had been discovered, the northern portion of which lay in the path of the wandering Eocene mammals, articles such as "The 'Lost Atlantis' and the 'Challenger' Soundings" appeared which announced that "not only a 'sunken island' but a ridge does lie in the mid-Atlantic between the Old and New World,"[*] and hinted that this ridge just might be the lost land of Atlantis.

As other rises were discovered, they too became the preferred locations of the imaginary land bridges and ancient island chains. One of these, "Lemuria," named for a primitive monkeylike primate of Madagascar, was thought to thread its way for some distance along the Carlsberg Ridge in the Indian Ocean in order to connect India with Madagascar and the Seychelles and thereby explain the similarities in fauna, flora, and geological structure in those areas. ("Lemuria" then supposedly continued even without the support of a submarine ridge —from the Seychelles to southern Africa, while the ridge continued southward toward Antarctica.)

Scientists who did not agree with this interpretation of submarine ridges—and Murray was among them—insisted that composition, not configuration, was the most important characteristic to study. If a ridge truly marked the location of a former continent, they reasoned, then from it could be dredged continental rocks (granite, gneiss, schist, etc.). But if the rise were an oceanic structure, then only volcanic materials such as basalts would be scraped from its slopes. This simple criterion for distinguishing between terrestrial and oceanic formations was complicated to some extent by the fact that glaciers and icebergs were known to float continental rocks far out to sea and, upon melting, dump them there. Sealers and surveyors had reported seeing great boulders embedded in icebergs, and Darwin had understood that if "but one iceberg in a thousand or ten thousand transports its fragment, the bottom of the Antarctic Sea, and the shores of its islands, must already be scattered with masses of foreign rocks."¥

The *Challenger* began bringing up these foreign rocks and gravels, many with glacial scratches on them, as she sailed north from Bermuda to Halifax, and, at one station, in 1,340 fathoms of water, the dredge came up with a 500-pound boulder wedged in its mouth. The origin of these rocks, and similar ones found in the Antarctic, was not difficult to imagine, but when similar debris was dredged from the iceless waters between Bermuda and the Azores and between Tahiti and Valparaiso, Chile, then Murray and Alphonse Renard, co-authors of the *Challenger* Report on marine deposits, admitted "that these fragments have been found some distance beyond the known limits of floating icebergs," and they called upon a changing climate to explain the anomalous distribution.

"It is evident that floating ice must have had a wider extension formerly than at the present time," they stated, and, to explain the occurrence of continental debris just a little bit further, added that "both seals and penguins carry to sea large numbers of stones . . . in their stomachs, to which the sealers give the name of 'ballast.' These animals may . . . distribute rock fragments to great distances from the land."✲

With these few exceptions, however, the *Challenger* expedition, and other voyages that followed, found neither continental rocks in deep ocean basins nor marine sediments identical to deposits found on land. Oceanographic data could not be construed to support the theories of drowned continents or impermanent ocean basins. But a few soundings and sediment samples did not weigh heavily against the more numerous observations made on land, and scientists continued their debate on the history of the earth with little notice of oceanographic contributions.

There were, of course, other expeditions besides the *Challenger* whose oceanographic investigations included geological studies of the sea floor, but among them the *Challenger* expedition led the way—in time, and in discoveries—and its Report on deep-sea deposits, published in 1891, was the most complete statement of what was then known. Before Murray and Renard so arranged and augmented the existing information on marine sediments, knowledge of the sea floor was scattered piecemeal through books and journals and was expressed in terminologies so diverse as to defy comparison or compilation. It was Murray's intention to correct this situation and, with the Abbé Renard, a professor of geology and mineralogy at the University of Ghent, he embarked upon a 15-year study of "all available samples of . . . Deep-Sea Deposits, whether collected by the 'Challenger' or otherwise. . . ."✲

Murray had requested, and for years continued to receive, thousands of rocks, clays, and oozes collected by nearly 100 ships over tens of years. British Survey Ships in every sea and ocean sent parcels of sediment to the *Challenger* Office*; The U.S. Hydrographic Office

---

* A sample of phosphatic rock from Christmas Island, sent to Murray from the commander of Her Majesty's Surveying Ship *Egeria*, indirectly helped British oceanography by adding a considerable sum to Murray's fortune which he then largely spent on research. From the sample, Murray realized that the rock would be easy to mine for use in fertilizers. He encouraged the British government to annex the uninhabited volcanic island, then formed a company to work the deposit. By World War II (as the story goes) the British Treasury had received in royalties and taxes from the operation more than the cost of the *Challenger* expedition. Like the Treasury, Murray received a large share of the profits, but unlike the government, he channeled much of his earnings back into oceanography. He helped pay for the last volumes of the *Challenger* Re-

offered samples taken by the *Tuscarora, Gettysburg,* and *Enterprise* on their cable surveys; Alexander Agassiz sent the sediments collected aboard the *Blake;* the India Rubber, Gutta-percha and Telegraph Works Company forwarded oozes from their ships working off the coast of western Africa; and from Nares' north polar expedition and the old collections of Sir James' Antarctic Expedition came more muds and oozes from the opposite ends of the earth.

Each sample of sediment arrived with information concerning its location and depth, and Murray first used this data, in combination with many ships' soundings, to prepare topographical maps of the Atlantic, Indian, and Pacific oceans. These maps, printed in 1886, were based on approximately 6,000 deep-sea soundings.[*]

The sediments sent in from other expeditions, as well as the ones in the *Challenger's* own extensive collection, were thoroughly studied and described, but by the time Murray and Renard were ready to put this mountain of material into print, funds for the report were nearly gone, and only a single volume, describing the deep-sea deposits collected aboard the *Challenger,* was published. Still, Murray could write without exaggeration that the 525-page volume with 22 diagrams, 29 plates, and 43 charts, "is the first attempt to deal systematically with Deep-Sea Deposits, and the Geology of the sea-bed throughout the whole extent of the ocean."[Ψ]

The influx of sediment samples and Murray's concern for deep-sea deposits did not end with the publication of the report, and by 1912, when his book *The Depths of the Ocean* appeared, he had examined close to 12,000 samples of sediments. "Notwithstanding the large

---

port, paid the operating costs of the *Michael Sars* expedition in 1910, and endowed the Alexander Agassiz Medal for distinguished work in oceanography.

[*] In the last years of the nineteenth century, maps of the ocean basins appeared with increasing frequency—James Dana produced one in 1890, Alexander Supan's "Chart of the Ocean Depths" appeared in 1899, and several smaller maps were printed in the intervening years, all based by and large on the same soundings. An attempt at coordinating undersea mapping was made in 1899 when the International Geographical Congress organized a commission to draw up a general map of the ocean basins and to standardize the terminology of undersea features. Prince Albert I of Monaco undertook the task, and expense, of consolidating the world's deep-sea soundings and by 1904 had completed charts of all the ocean basins, based on 18,400 soundings. The job of keeping these charts up to date has devolved upon the International Hydrographic Bureau and a group of 16 nations who pool their data to produce the General Bathymetric Chart of the Oceans.

Bridle Path, Tahiti, by John LaFarge. One of the luxuriant, reef-fringed islands that so captivated European scientists. (*Courtesy Fogg Art Museum, Harvard University. Gift—Edward D. Bettens to the Louise E. Bettens Fund*)

amount of sounding-work carried on since that Report [*Deep-Sea Deposits*] was published," he wrote, "the general results, the classification, and the nomenclature given therein have been fully substantiated and found adequate in every respect. . . ."[*] It is to Murray's credit that the classification and nomenclature of deep marine sediments—even his dark brown "red" clay—remain today essentially as he set them down nearly 100 years ago.

As described by one of their number, oceanographers—like other sea birds—return to the land to breed and raise their young and, generally speaking, the coasts to which the first of them returned were cold and wet. These drafty conditions inspired among them a certain vulnerability to warm and sunny climes, and if there were scientific problems to be found in such pleasant regions, then these questions would not be ignored for long by seagoing scientists. Such an enticing problem was presented by coral reefs. Like other aspects of oceanography which grew from the common stock of the previous centuries' multipurposed voyages, early studies of coral islands were but minor parts of some surgeon-generals' broader duties. Naturalists sailing with Captain Cook, for example, had no choice but to give the corals of Australia's Great Barrier Reef a close look, for in 1770 the *Endeavor* "struck and stuck fast . . . upon the south-east Edge of a reef of Coral Rocks."⸸ Even without such intimate contact, other expeditions sailed home with vivid descriptions of the beautiful reefs and atolls.*

One of the first distinctions to be made among coral islands (and one that still impresses visitors to the tropical seas) was the difference between high and low islands. The high ones were correctly assumed to be the towering peaks of volcanoes which often rose thousands of feet above the sparkling sea. Their irregular flanks were thickly forested, and from their indented shores extended wide flats of brightly colored corals. "On the other hand, we have the atoll [or low island]; a thing of problematic origin and history, the reputed creature of an insect apparently unidentified; rudely annular in shape; enclosing a lagoon; rarely extending beyond a quarter of a mile at its chief width; often rising at its highest point to less than the stature of a man—man himself, the rat and the land crab, its chief inhabitants; not more variously supplied with plants; and offering to the eye, even when perfect, only a ring of glittering beach and verdant foliage, enclosing and enclosed by the blue sea."⸸

This, the low island, the coral atoll, posed several problems. At first men believed that atolls grew from the bottom of the sea, but this straightforward explanation was soon contradicted by naturalists who observed that living corals, of the reef-building variety, were rarely found below one or two hundred feet. Allowing corals such a restricted building zone, it became necessary to supply the thousands of known atolls with pedestals of some other but their own construction. Naturalists pored over charts of doubtful accuracy and soon saw in the low islands' roughly circular clusters the suggestion of volcanic craters. Atolls, then, were supposedly built along the crests of barely submerged craters, while a longer, more strung out pattern of islands that could not be made to fit the crater shape was apparently built upon the crests of a submarine ridge. This hypothesis gave geologists the task of explaining the improbable existence of range upon range of undersea mountains, the peaks of which all rose to within 100 feet of the sea's surface. Geologists boggled at the thought. Some other explanation was clearly needed, and Charles Darwin, sometime between his twenty-second and twenty-seventh years, came upon it.

---

* An early and particularly thorough study of reef structures was made between 1815 and 1818 by Adelbert von Chamisso, the naturalist on the Russian expedition led by Otto von Kotzebue.

Darwin, who lived from 1809 to 1882, sailed around the world in the *Beagle* in the years 1831 to 1836, and in the course of the voyage visited several islands with fringing reefs and barrier reefs and landed on a single atoll. In spite of this rather limited association with coral structures, he worked out a simple and elegant theory of coral-reef formation. (The theory was, in fact, worked out in his mind before he saw any coral islands at all.)

Darwin believed that the three prominent forms of coral reefs—the fringing reef growing out from a coast; the barrier reef, separated from the coast by a fairly deep channel or lagoon; and the roughly hoop-shaped coral atoll—were progressive stages of a single process. The process itself rested on two assumptions: first, that great portions of the earth's crust were slowly sinking, and second, that reef-building corals grew only in warm, shallow water. Assuming these to be true, Darwin's theory supposed that coral larvae drifted into the shallow waters around a volcanic island and established a colony which, in time, extended like a fringe from the perimeter of the land. Since the island was slowly sinking, the corals were forced to grow upward to keep close to the surface, and since the volcanic island was roughly conical in shape, its circumference diminished as it sank, creating a widening channel between its coast and the inward edge of the coral reef. (Corals did not usually fill in this channel, for the communities situated on the outer reef took most of the food and oxygen from the incoming water. This left the inner groups with a poor diet, on which they barely grew.) As the process continued over thousands of years, the island finally disappeared beneath the sea, leaving only a rough coral outline of its earlier dimensions—an atoll.

These views Darwin expressed most completely in his book *The Structure and Distribution of Coral Reefs* (1842), and those whom his theory interested included land geologists as well as oceangoing ones. (The former used Darwin's explanation to account for certain isolated stacks of limestone found in Switzerland, Belgium, and Austria called "fossil atolls.") Darwin's most respected and knowledgeable supporter in the matter of reef formation was the American geologist James Dwight Dana. As an assistant professor of geology and mineralogy at Yale College, Dana had been appointed to accompany the U.S. Exploring Expedition of 1838, and he had sailed aboard the *Peacock* among the coral islands of the South Pacific. When his ship put into Sydney, Australia, in 1839, Dana read there a brief newspaper account of

Darwin's coral-reef theory, and the ideas seemed to fit perfectly with things he had already seen. After a three-month visit to the Fiji Islands (or rather to the reef flats of the Fijis, for "the character of the natives" prevented exploration of the islands), Dana was able to add an important piece of geological evidence to Darwin's argument. Dana recognized that the deep, sometimes fjordlike fingers of water that poked into the sides of volcanic islands were drowned valleys that had been flooded as the mountainous islands subsided. His observations strengthened Darwin's case for general subsidence, which seemed to later geologists the most speculative portion of his coral-reef theory.

Dana's thoughts on coral reefs were included as a chapter of the Exploring Expedition's geological report and were expressed more fully in his book *Corals and Coral Islands*, published in 1872.

The Yale professor was certainly in Darwin's mind when the latter, in 1876, described his own book on coral reefs as being "thought highly of by scientific men, and the theory given is, I think, now well established."⸶

Four years later (two years before Darwin died) John Murray of the *Challenger* expedition read his paper "On the Structure of Coral Reefs and Islands" before the Royal Society of Edinburgh and with it bluntly challenged Darwin's theory.*

Murray was bothered by Darwin's contention of widespread subsidence, and he felt that the observations made aboard the *Challenger* suggested an alternative. "There are other agencies [than subsidence] at work in the tropical oceanic regions," read Murray, in the presence of Sir Wyville Thomson, who wished he were reading something else, "by which submarine elevations can be built up from very great depths so as to form a foundation for coral reefs; [and] . . . all the chief features of coral reefs and islands can be accounted for without calling in the aid of great and general subsidences."⸶

The "agencies" that Murray had in mind were the billions of planktonic organisms that lived in the surface waters of the sea and whose remains were constantly accumulated on the floor of the sea. On board the *Challenger* Murray had tried to estimate the average concentration of these tiny creatures in the surface waters by dragging a tow

---

* Murray was not the first to disagree with Darwin on corals. Several German geologists, among them Carl Semper and J. J. Rein, had previously disputed Darwin's theory and Murray had incorporated some of their ideas.

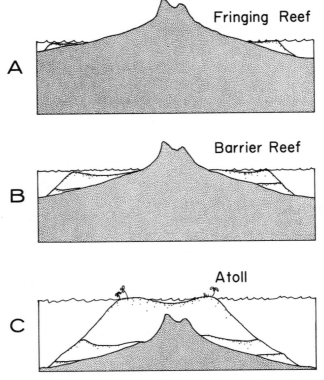

Fringing Reef

Barrier Reef

Atoll

According to Charles Darwin, the development of coral reefs and atolls was a continuous process. Corals, growing on the sides of a slowly sinking volcanic island, grew upward as the island subsided, and when the island had disappeared entirely, only an atoll, or circular ring of coral, remained. (*After Charles Darwin,* The Structure and Distribution of Coral Reefs)

net of specific size for half a mile. The live portion of the collection, the soft tissue, was eliminated by boiling the haul in caustic potash. The shelly remainder was washed, weighed, and multiplied so as to yield a product roughly indicative of the weight of skeletal material existing in a volume of water one mile square by 600 feet deep. (Murray erroneously assumed plankton to be distributed evenly throughout such a space.) His staggering answer of more than 16 tons per square mile seemed to provide ample basement material for thousands of coral islands.

Murray supposed that this wealth of debris rained down upon submarine crests and ridges and accumulated like falling snow until the piles were built high enough to provide a suitable base for corals. Once the corals were established, he believed they grew to the surface,

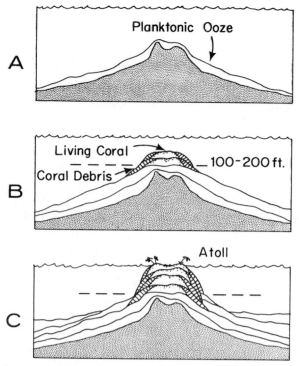

According to John Murray, an atoll developed when layer upon layer of planktonic ooze fell upon a deeply submerged mountain and gradually built upward to within one or two hundred feet of the surface. Coral larvae swimming in the water could then colonize this sediment bank, and their natural pattern of growth produced an atoll. (*According to John Murray, "On the Structure and Origin of Coral Islands,"* Proceedings of the Royal Society of Edinburgh, *Vol. 10*)

whereupon those on the outside of the colony continued to grow and thrive while those toward the center, cut off from necessary nutriments, languished and died. The remains of the latter dissolved, the quiet waters of a lagoon flowed over the central portion of the reef, and there, without the dubious process of subsidence, was the history of an atoll. This more complicated hypothesis of Murray's won many adherents, but not without a curious struggle.

Within a year of the *Challenger*'s return, Murray had felt ready to criticize the Darwin-Dana theory of coral reefs. Sir Wyville Thomson had not agreed. The latter, buttressing the still shaky supports beneath the new *Challenger* Office, was reluctant to risk an emotional squabble with prestigious members of the scientific community whom he felt would surely take offense if a "scientific worthy" of Darwin's stature were contradicted. Murray appears to have taken Thomson's advice, at least for a while, and it was not until 1880 that he read his paper on coral reefs with Thomson, vice-president of Edinburgh's Royal Society, "in the chair."

No controversy erupted. Thomson had underestimated the combined effects of Darwin's reputation and other scientists' lack of experience with reefs and atolls. Unless those scientists, who were meant to choose between Darwin's and Murray's theories, had visited a coral reef, it was very difficult for them to weigh the merits of the two theories, a problem complicated by Murray's use of oceanographic techniques and data which were themselves less familiar than the reefs and therefore hard to evaluate. For many, Murray's theory was an unknown quantity. Only a few accepted it, and only a few others—notably Darwin and Dana—actively rejected it.

This unexciting situation was not allowed to continue indefinitely. In 1887 the Duke of Argyll, a prominent liberal in the House of Lords and a prolific writer on the subject of religious and scientific compatibility, elected to deliver "The Great Lesson." In it the duke charged men of science with idolatry—with blind adherence to the words of Darwin—and the example he chose to illustrate his controversial point was the nonexistent controversy between Murray's and Darwin's theories. It seemed clear to him that Darwin was wrong, Murray right, and that the latter had unjustly met with a "conspiracy of silence."

The duke's article succeeded where Murray's theory had failed in raising the indignation, ire, and interest of the entire scientific community. Huxley, a fast friend of Darwin's, refuted all "The Great Lesson's" teachings, and a few more besides, in his prompt reply, entitled "Science and the Bishops." The fight then moved to the scientific journal *Nature*, where weekly publication allowed for a quicker and more satisfying exchange of insults. The duke charged Huxley with "reading into it ['The Great Lesson'] a great deal that is due to his own excitement and nothing else."[*] To which the accused replied that, "as fast as old misrepresentations are refuted, new ones are evolved out of the inexhaustible inaccuracy of his Grace's imagination."[*]

The debate lasted for half a year, and although it had never concerned corals except in a secondary way, it did make it easier for scientists opposing Darwin's views to get a hearing. Papers began to appear supporting Murray's views, and in 1888 Alexander Agassiz' book *Three Cruises of the U.S.C.G.S. Steamer "Blake"* included a lengthy chapter explaining why Darwin was wrong and Murray right.

For Agassiz the cruises through the Caribbean aboard the *Blake* were the beginning of some 25 years of coral-reef study which soon made him the most experienced observer of reef structures. For others, such studies had been a small part of broader investigations, but for Agassiz reefs became the *raison d'être* for his costly oceanographic expeditions.

Agassiz had first talked to Murray about coral reefs when the two were together in Edinburgh unpacking the *Challenger* collection. Upon his return to the United States Agassiz accepted the Coast Survey's invitation to do some dredging aboard the *Blake,* and during two cruises among the reefs of Florida, the Bahamas, and eastern Yucatán (a third cruise was in colder water), he compared what he saw to the theories he had heard. His observations favored Murray's views.

"All naturalists . . . have felt the difficulty of applying Darwin's theory of reef formation to the peculiar conditions existing along the Straits of Florida,"⁺ he wrote, then curiously went on to acknowledge Darwin's belief that the reefs of the West Indies (and several other regions) were exceptions to the general rule and were formed without subsidence upon platforms lying just beneath the surface. But for Agassiz the reefs of the Caribbean were neither the exception nor the rule. Each group of islands and atolls scattered through the Pacific and to a lesser extent through the Indian and Atlantic oceans might, he felt, have been formed by a different process. Each would have its own history and each must be studied as a unique problem.

With this philosophy, Agassiz determined to visit as many different kinds of coral formations as he could find, and in 1896 left his old collecting grounds in the Caribbean and set out for the Great Barrier Reef. His plan was to spend a month studying the reef, and to that end he chartered a steamer and equipped her with the gear he had developed for the *Blake* and had had shipped, at considerable expense, to Australia. A month of foul weather wreaked havoc with his careful plans. Agassiz was miserably seasick during most of the voyage and on the few only moderately rough days when the dredge could be cast, he emerged, pale and unsteady, to watch the dredge careen over the rail and its contents scatter along the rolling deck.

The following year Agassiz rode on calmer waters among the Fiji Islands. He had chosen this group on the advice of Dana, who, having

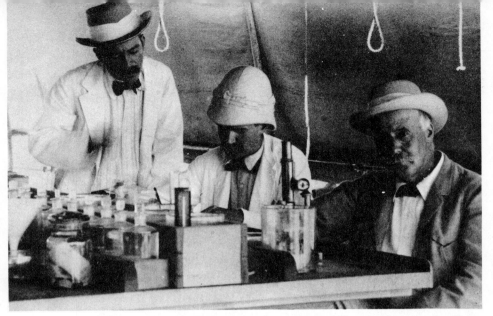

Alexander Agassiz, right, on the *Croyden* during his expedition to the Great Barrier Reef in 1896. Two assistants from the Museum of Comparative Zoology are with him—William McMichael Woodworth, left, and Alfred Goldsborough Mayor. (*Courtesy Museum of Comparative Zoology, Harvard University*)

walked over the same reef flats 57 years before, felt that the Fijis offered good examples of each type of coral structure. Agassiz left the capital city of Suva on the chartered steamer *Yaralla* and visited some of the larger, high islands in the group. Then the *Yaralla* ran down along the line of smaller islands which included barrier and fringing reefs and numerous atolls.

Agassiz planned his work so that most evenings the *Yaralla* could make for the lee of a high island or anchor in the protected waters of a lagoon. By five o'clock the next morning he was up, first to watch the early boat traffic pass out of the lagoon and next to measure distances for the day's run. Each day's investigation included towing for coral larvae, photographing the islands, taking a few shallow soundings, and buying samples of coral from native divers.*

---

* Agassiz was fascinated by the natives whom he encountered on his numerous journeys, and his letters are full of observations and comparisons of the many groups he saw. The people of the Fijis were among his favorites, although he admitted being greatly surprised by their custom of wearing nothing but a skimpy loincloth "in the presence of ladies!" It would be nice to know what the Fiji Islanders thought of Agassiz as he took his morning stroll along the deck of the *Yaralla* dressed in pajamas.

One leg of the Fiji expedition included a visit to the Exploring Isles, a five-island cluster enclosed by a single barrier reef and named by the U.S. Exploring Expedition. The largest of the islands, Vanua Mbalavu, stretched in a backward "S" for 14 miles, and along its length lay row upon intersecting row of harshly weathered volcanic hills. Deposited along the crests of some of these hills, hundreds of feet above the sea, rose a ragged crenellation of ancient corals.

Agassiz could not but interpret this curious phenomenon as evidence of elevation, and arrived at the same conclusions upon seeing other tall islands built entirely of coral. Many of these, he noticed, were flat or dished on top. These mysterious depressions, often quite deep and very wide, Agassiz considered analogous to the "Banana Holes" (solution pits) of the Bahamas. To other observers the elevated bowls were former lagoons; the tall coral islands, ancient atolls; and the process, Darwin's, followed at some later date by erratic elevation.

By the time the rainy season came, forcing Agassiz and his collections into a wing of a local hotel, the self-appointed investigator of coral structures had been studying islands and atolls for over 20 years. Still he refrained from expressing a theory of reef formation that would unite the masses of data he had accumulated and make sense of all its separate parts. He put off this task, which to him was distasteful, and visited instead the two major groups of coral islands remaining for his inspection. At the turn of the century he embarked upon his longest expedition, which took him from San Francisco to Japan by way of hundreds of South Sea Islands, and in 1902 made the last of his coral expeditions among the Maldives in the Indian Ocean.

This will be the end of a most successful expedition [he wrote to Murray], perhaps to me the most interesting visit to a coral-reef group I have made. For certainly I have learned more at the Maldives about atolls than in all my past experience in the Pacific and elsewhere. I should never have forgiven myself had I not seen the Maldives with my own eyes and formed my own opinion of what they mean.—Such a lot of twaddle—it's all wrong what Darwin has said, and the charts ought to have shown him that he was talking nonsense. . . . At any rate I am glad that I always stuck to writing what I saw in each group and explained what I saw as best I could, without trying all the time to have an all-embracing theory.*

---

* Darwin, on the other hand, admitted he had an hypothesis for everything. "My mind seems to have become a kind of machine for grinding general laws out of large collections of facts. . . ."⍦

Now, however, I am ready to have my say on coral reefs and to write a connected account of coral reefs based upon what I have seen. It will be a pleasure to me to write such a book and illustrate it properly by charts and photographs. But it will be quite a job with my other work on hand. I hope to live to 100! or rather I don't hope, but ought to! to finish all.✣

Agassiz never wrote a connected account of coral reefs, nor did he live to be one hundred, but it was not a lack of time that frustrated his efforts. He found it exceedingly difficult to extract general laws from the soundings, photographs, and pieces of coral he had collected. When, in 1903, he was invited to have his say on coral reefs before the Royal Society of London, he "simply described what he had seen . . . ," wrote Murray, "and did not enter into any controversial matters."✣

His plans for his coral-reef book followed the same path. After outlining the work three times, to his continuing dissatisfaction, he determined to drop all attempts to theorize, and state only what he had observed. He left for his son "a vast collection of extracts from the literature . . . marked and scored, and a few rough notes, of no use to anyone but himself."✣

By the turn of the century enough work had been done on corals to make it clear that each of the several conflicting theories was capable of explaining the visible characteristics of reefs and atolls, and each ridge, slope, and depression had as many interpretations as there were theories. Proof for any one of them could not be obtained from additional observations of coral islands but would have to come instead from holes drilled deep into an island or atoll. Drilling was the crucial experiment, for cores, taken from below the zone of possible coral growth, would either be of compacted ooze, proving Murray's theory of ooze-built platforms, or of sunken corals, supporting Darwin's views. (Agassiz expected to find either volcanic rock or compacted ooze, but not drowned corals.)

In a frequently quoted letter to Agassiz, Darwin expressed the wish "that some doubly rich millionaire would take it into his head to have some borings made in some of the Pacific and Indian atolls, and bring home cores for slicing from a depth of 500 to 600 feet."✣

Such experiments were not as easy nor their results as conclusive as Darwin had imagined, but he was addressing his thoughts to the right person. Agassiz, close to "doubly rich" from his copper mines, took to the Fijis "a boring apparatus and the most skillful man of the Diamond Drill Company,"֍ and prepared to drill 350 feet into the island of Wailangilala. On that tiny island, set on the rim of an elliptical reef and tufted with thickets of bay cedar and scraggly palms, Agassiz deposited his drilling rig and a five-man crew. The men and machines were committed to the care of the island's lighthouse keeper and left to drill for six weeks while the *Yaralla* steamed off for elevated islands.

When Agassiz returned to Wailangilala he found that a hole had been drilled for only 80 feet, the first 50 through sand, the lower 30 through reef rock. Comparing this coral rock to deposits he had seen— on the elevated island of Vanua Mbalavu, for instance—he declared that the samples had been formed under similar conditions. He thought both the corals under Wailangilala and those on top of Vanua Mbalavu had been deposited on a slowly rising platform. His arguments were not convincing, and, more important, his drill holes had not penetrated below the depth of possible coral growth and could not, therefore, be considered conclusive. In spite of these drawbacks, he gleefully predicted that "our English friends who are howling for joy at the results of the boring in Funafuti will be greatly surprised when they hear from me."֍

Scientists were even more surprised when they heard from Agassiz' English friends, who were themselves drilling very deep holes into what they hoped was a "typical atoll" in the South Pacific.

Coral-reef investigations, like other oceanographic activities, were becoming too costly for even the doubly rich, and the financial burden was shifting to government agencies and in a few cases to well-established scientific societies. In the case of the British drillings on Funafuti, in the Ellice Islands, the project was jointly organized and financed by the Royal Society and the British Association for the Advancement of Science.

In May of 1896, H.M.S. *Penguin* arrived in the Ellice group, and within a short while her drilling crew had exhausted their supplies and their patience on two sandy holes which repeatedly caved in. Two more expeditions were sent to the atoll, and finally a hole 1,114 feet deep was bored. Below a superficial sandy layer the drill ground its way not through compacted ooze or volcanic rock but through a fairly continuous bed of coral. Surprisingly, this did not win the day for Darwin.

Sections of the core retrieved from the deep hole were sent back to

England, where they were distributed for examination among experts of contradictory points of view. The committee in charge of the project had decided that these experts were to include no mention of coral-reef theories in their reports. It was hoped that this policy would produce unbiased accounts of the corals from which other scientists might freely draw their own conclusions. This they did.

Supporters of the Darwin-Dana theory claimed victory. They were sure that the corals retrieved from several hundred or more feet had once grown near the surface and had subsequently sunk. Murray received this news calmly and contended that the bore hole had been put through a pile of coral debris (the atoll's talus slope) rather than through the reef proper and that the drilling had only sampled shallow-water corals, which, when dead, had tumbled down the reef's steep seaward slope toward the ocean floor. (See diagram p. 161.) Agassiz agreed with Murray but did not choose to finance a second drilling expedition of his own to prove the latter's point. Agassiz' active collecting days were over (his last major expedition had been to the deep and landless regions of the eastern Pacific in 1904–05), and he devoted what time his accumulating ailments allowed him to the continued study and arrangement of his coral collection. Still wrestling with his proposed book, he made a last trip to Europe in 1910 to talk with Murray and to visit more coral collections. With more, if not new, thoughts and plans in mind, he sailed for home aboard the S.S. *Adriatica*, and on Easter morning, 1910, Alexander Agassiz died at sea.

"I have been in all I have undertaken most successful from the world's point of view," he had written his sister-in-law, "but from mine it has lost its charm long ago."ᵛ

The coral-reef controversy took one more turn before it was settled. This last twist was the glacial control theory of coral reefs advanced by the American geologist Reginald A. Daly in 1919. Daly believed that many of the puzzling features of coral reefs could be explained in terms of changes in sea level which were themselves the result of the alternate freezings and thawings of polar ice sheets. The majority of coral reefs were not alive during the last ice age, he argued, for the oceans had been too cold for coral growth except along the equator. Then, as the ice sheets melted, the level of the sea rose and the water became warmer. Corals established themselves on the tops of barely submerged volcanic peaks or on other platforms of sandy and shelly materials and grew upward in time to the sea's advances.

If Daly's theory were correct, the thickness of present reefs should roughly equal the most recent rise in sea level, which was then estimated to be approximately 250 feet (and is now thought to be closer to 400 feet). Several borings had been made in Bermuda shortly before World War I, and the drill had broken through a capping of coral into weathered volcanic rock at a depth of 250 feet—perfect support for Daly's theory.

Other scientists became interested, but just as new plans for drilling were laid, the Japanese Mandate (1918–1944) went into effect and scientists no longer had free access to the islands in the Pacific. A few investigations were carried out by the Japanese while their occidental peers were forced to confine their work to the reorganization and reevaluation of previously collected data.

At the end of World War II the situation, both politically and technologically, was vastly altered. Many of the most interesting coral islands now lay within the United States Trust Territory, and in 1945 the decision was made to test nuclear weapons in the Marshall Islands. Several of the group's five islands and 29 atolls were studied both before and after the weapons tests, and these investigations included deep drilling.

The first holes were drilled into Bikini Island and penetrated more than 2,500 feet of what appeared to be shallow-water corals, but the most conclusive evidence for Darwin's theory came from a pair of holes drilled on either side of Eniwetok. These holes finally went through the coral cap and into volcanic rock at depths of 4,158 feet and 4,610 feet.✤ In deep sections of the cores were found fossilized land snails, pollen, and weathered coral, all indications that the island's downward trip of almost a mile had not progressed continuously but had included some periods of relative elevation. Darwin's theory had not encompassed the complexities of reef building encountered by scientists working after him, but his theory had provided, and still provides, the basic explanation of coral-reef formation.

CHAPTER V

The Sea in Motion

> —all the deep
> Is restless change; the waves so swell'd and steep,
> Breaking and sinking, and the sunken swells,
> Nor one, one moment, in its station dwells:
> —George Crabbe, "The Borough"

In the last quarter of the nineteenth century a group of Scandinavian scientists began to investigate the complexities of the sea in motion, and their concern for movement and for temporal change led them first to make more accurate descriptions of the ocean's circulation and then to devise a new, theoretical method of computing currents. Before making this attempt to unscramble the ocean's currents, tides, and waves, oceanographers had seemed content to define the limits and describe the contents of the ocean. This approach had necessarily led them to construct static pictures of the ocean's average conditions. For some branches of the young science such an approach worked well enough. Murray had shown, for example, that deposits of red clay in the deep ocean were not likely to change perceptibly for hundreds of thousands of years, and Agassiz believed that many populations of deep-sea creatures had struck an unexciting balance with their environments and continued with few changes for year after year. But for other aspects of the ocean, to ignore change was to disregard the very essence of the phenomena. The currents that flowed through all parts and strata of the sea, mixing and moving the ocean's quintillions of gallons, could not be studied in the same way that a submarine mountain or a colony of sea urchins could. It was being discovered that Maury's "river in the ocean" was reversing its flow, jumping its banks, swelling, and shrinking. Information collected on a long cruise could neither indicate the changing patterns of such vagrant currents nor suggest the reasons for their variations. A new approach to the problem was needed, one that would permit oceanographers to see in detail how small portions of the sea moved, from top to bottom, from side to side, and from season to season.*

By the late 1800's several ways of obtaining information about currents had already been developed. There were, first of all, direct methods, and the simplest of these were observations made of drifting objects. The very ship that the observer was aboard was set by currents, and so were icebergs, wreckage, drift bottles, driftwood, and drogues. All of these could be studied, but their meandering paths could not be followed all the time, and except for drogues, their drift told nothing of subsurface currents. In spite of these drawbacks, drift experiments were popular in the late 1800's and, being easy and inexpensive, were undertaken by the U.S. Coast and Geodetic Survey, the Danish Meteorological Institute, the prince of Monaco, and others. In this way, the major surface current systems of the world, most of which had been discovered by fifteenth- and sixteenth-century navigators,† were traced with at least enough precision to suggest a host of unsuspected complexities.

---

* The study of motion in the sea is usually divided into two main parts, ocean currents and the periodic phenomena of waves and tides. The first of these is the topic of this chapter.

† Early navigators, whose ships were set by these powerful currents, could neither estimate the speed nor determine the precise direction of the currents in which they rode. Without chronometers or some other way of determining longitude (the former were not regularly used until the nineteenth century), they were not able to fix the position of their ships, and so could not know just where or how they were drifting. But drift they did. On his third voyage, Christopher Columbus noted the existence of the North Equatorial Current, which pushed his ship westward from Africa toward the West Indies; in the same years, Portuguese explorers were trying to stem the Agulhas Current which flows south along much of Africa's east coast; and on the opposite side of the world, fishermen had already recorded their encounters with the Peru Current (also called El Niño or later the Humboldt Current), which flows north past Chile and Peru, and with the Japan Current, or Kuroshio, which flows north, then northeast along the coasts of the Philippines and Japan.

When more precise measurements were desired, especially of sub-surface flows, then current meters were used. In 1876 Lieutenant Pillsbury designed a meter which he used in his investigations off the East Coast of the United States, and, just after the turn of the century, propeller meters, pendulum meters, and even a photographic recording meter designed to be left hanging beneath a moored buoy for two weeks at a time, were built and tried. Most of these were meant to be used from a ship at anchor, and although the procedure of anchoring and metering was difficult and time-consuming, it did yield the most accurate measurements then available.

There were also indirect methods of studying currents. For example, certain patterns of ocean temperatures implied the existence of cur-rents, and this had been known and used since the late 1700's. At first, temperature measurements had been used to delineate surface currents such as the Gulf Stream and the Peru Current, but soon curious scholars and navigators had begun making subsurface measurements as well. One result of these experiments was the discovery of extremely cold water in deep ocean basins, even in the tropics, and this encour-aged several German scientists to suggest that a much grander, albeit slower, system of oceanic circulation must exist throughout the entire ocean. They believed that cold water at the poles was continually sinking and flowing slowly along the ocean floor toward the equator. To replace the sinking polar waters, masses of warm surface water from the tropics floated toward the poles. Thus two cycles, or endless belts, of water were imagined—one in each hemisphere—and their endless turnings explained the occurrence of cold water at the bottom of the equatorial seas. This concept was clearly set forth in 1847 by the German-Russian physicist Emil von Lenz, who in a short paper on ocean temperatures verbally drew a picture of the two symmetrical cycles of moving water. His simple model influenced Maury, Ferrel, Carpenter, and later workers, even though temperature (and salinity) measurements made on the *Challenger* and on later expeditions showed that the masses of water in the oceans were not arranged symmetri-cally but were interlocked in a much more complex pattern.

The water masses, whose properties were being studied and whose existence implied oceanic circulation, were large volumes of water that had approximately the same temperature and salinity throughout. It was understood that these masses were formed at the surface of the sea where the water attained its temperature and salinity and also acquired a certain proportion of dissolved gases. Subsequently, this water might sink to a particular depth according to its density, and flow hundreds or even thousands of miles from its source, but all along the way the mass would still retain its characteristic properties and could be traced by means of them. (Dissolved gases were not as good indicators as temperature and salinity, since the amount of oxygen, for example, could be affected by animals living within the water mass.)

Widely spaced temperature and salinity measurements, then, could be used to construct a picture of what might be called the average arrangement of water masses (which in turn implied an average pattern of circulation), but such observations could not show how the masses, and the currents that moved them, swelled and diminished nor how they changed their positions from one season or one year to the next. To follow these changes, detailed and continuous investigations were required and, understandably, these costly studies were not begun until a practical reason for their execution arose. In Scandinavia the desire to explain the erratic comings and goings of food fish furnished such a reason, and the belief that currents influenced the migrations of fish enabled scientists to gain government support for a wide variety of hydrographic studies.

The idea that the movement of fish might be related to the movement of water was suggested as early as the 1860's by the Norwegian zoologist Georg Ossian Sars.* In 1864 Sars received a grant from the Norwegian Parliament to study the life cycle and migration of cod. To do so, he packed his violin, which he took everywhere with him, and headed north with the cod fishermen to the cold Lofoten fishing grounds. There, to his surprise, he found cod eggs floating in the water, although the assumption had been that these eggs sank, as did those of herring, plaice, and other food fish. During incubation, then, cod eggs were at the mercy of surface currents—reason enough for Sars to become interested in ocean circulation. But another reason was soon discovered, for when Sars turned his prodigious talents to the study of herring, he ascertained that the fish's summer migration was not for spawning, but for feeding, and that the fish fed on plankton that drifted with the currents.

Sars and others continued to study the life cycle and behavior of

---

* G. O. Sars was the son of Michael Sars, the theologian and zoologist whose collection of deep-water fauna so impressed Sir Wyville Thomson.

fish, but before they could explain the complicated relationships among fish, migrations, currents, and water masses, more had to be learned about the chemical and physical arrangement of the waters. This half of the picture was first drawn by a few chemists and physicists who set out to study the temperatures and salinities of coastal waters. Luckily for these men, the water layers and masses that they were trying to map were more easily discernible than those which moved through the open ocean. In the relatively shallow bays and fjords, where the water warmed and cooled more rapidly and where light, fresh water from the land flowed out over the surface, water masses with distinctive characteristics were formed, and the differences among these could be detected even with the relatively imprecise instruments of the time.

In the late 1870's a Swedish chemist, Gustav Ekman, began studying the physical conditions of the Skagerack, a portion of the waterway connecting the Baltic and the North seas. Like Sars, Ekman was motivated by fisheries' problems, and he sought to explain why great shoals of herring had suddenly reappeared in the Skagerack after an absence of 70 years. These fish had been known to come and go with frightening irregularity over the past thousand years. When they came, under clouds of screaming gulls, new fisheries sprang up in their path, and when they left, the ruin was disastrous. Ekman believed that he could relate the appearance of the herring to certain changes in the temperature and salinity of the waters he studied. In the Skagerack he found several layers of brackish water from the Baltic floating out over the heavier, more saline North Sea water, and the latter extended like a wedge beneath the fresher layers. He found too that herring preferred a particular water layer of intermediate salinity. This shelf or bank water, as it was called, moved in and out of the inland seas, and with it went the fish.

Ekman knew that his finding would not be of much use to fishermen unless the bank water and other layers could be mapped and followed with precision, and he knew that his solitary investigations were incapable of yielding a comprehensive picture of the mixing, moving waters even within the Skagerack. So, in 1890, when the opportunity arose, Ekman joined forces with another chemist, Otto Pettersson, and together they helped organize a new and impressively thorough series of hydrographic investigations. (In Scandinavian countries, "hydrography" is approximately synonymous with "physical oceanography.") At this time, Pettersson, who was soon to emerge as one of the most

Otto Pettersson, a Swedish chemist and physical oceanographer, organized many of the earliest cooperative cruises in Scandinavian waters and promoted the idea of an International Council for the Exploration of the Sea. (*Courtesy Jöran Hult*)

influential spokesmen and supporters of physical oceanography, had not yet begun his career as an oceanographer, but was working as a chemist with a particular interest in warming and cooling. His work on ice conditions in the Baltic Sea had come to the attention of A. E. Nordenskjold, leader of an expedition on the *Vega* (1878–1879), and the latter had asked Pettersson to use the *Vega*'s data to write a report on the hydrography of the Siberian Sea. This Pettersson did, and in consequence, he not only met John Murray, who was then working on the temperature measurements brought back by the *Challenger*, but also became involved in problems of hydrography which he never tired of for the rest of his long lifetime.

When Pettersson joined forces with Ekman, both men understood that to obtain a useful picture of circulation a series of expeditions involving several vessels that could work together at many times throughout each year would have to be organized, and this in itself was a new approach to the study of the sea. In the name of fisheries' research such a series of cruises was begun in the early 1890's, and Pettersson and Ekman managed to involve as many as six or seven steamers at a time. The methods and objectives of these systematic investigations were described by Pettersson in a "Review of Swedish

Hydrographic Research in the Baltic and North Seas." (A reviewer noted that the article, written in English, was the first on the subject to be "published in any language usually read by scientific men. . . .")

The system adopted by the Swedish naturalists at the present time, in their exploration of the sea which surrounds the Scandinavian peninsula [wrote Pettersson], is to despatch a number of ships simultaneously from different ports across that part of the sea which is to be explored, each ship being provided with a complete set of hydrographic instruments worked by assistants specially trained for their task in the laboratories of Stockholm's Högskola, the Polytechnic Institute, etc. The route of each ship, and the position of every sounding station, are determined beforehand, according to previous experience. In every successive expedition the same sounding-places are chosen, in order to ascertain the alterations which have occurred in the arrangement of the water strata. The advantages of this method of research are obvious. All the observations being taken within a few days, the hydrographic state of a certain part of the sea at all points is exhibited practically simultaneously, unaffected by changes of wind and weather. In February 1890 we explored the Kattegat and the Skagerack with five steamers. In less than a week we collected at about seventy sounding-stations more than 1000 water samples, and took the same number of temperature observations, besides 200 samples for gas analysis. . . .

For the past two years the same method of simultaneous hydrographic research has been adopted by the Danish Government for the Kattegat and parts of the western Baltic and Danish Sounds. Four times a year, on the first of February, May, August, and November, thirteen hydrographic sections [lines of stations] in different parts of the Kattegat and the Sounds are made by Danish gunboats. We have proposed to the Swedish and Norwegian, as well as to the British, Governments a plan for a common international hydrographic exploration, according to which Swedish and Norwegian research should proceed in the Baltic, the Skagerack, and on the eastern side of the North Sea simultaneously with Danish soundings in the Kattegat and British observations in the northern parts of the North Sea. [The Pettersson-Ekman proposal was adopted and in 1894 ships from Sweden, Denmark, Germany, Scotland, and Norway made four joint hydrographic surveys.]*

On all of these cruises the routine was approximately the same. At each of hundreds of stations the depth was first measured with a sounding machine, for no pattern of circulation could be worked out within a basin of unknown dimensions, and soundings were also needed so that the water samples could be taken at proper intervals.

For measuring the temperature of the water, insulated water bottles were usually used. The English firm Negretti and Zambra had manufactured, since 1874, a reversing thermometer which in theory ought to have been more accurate (see p. ooo *et. seq.*), but in practice these instruments were difficult to use and prone to disorders. The water bottles were more reliable, and Pettersson favored one of his own construction in which concentric layers of water were used as the insulating material instead of the more usual gutta percha. After the depth had been sounded, the water bottles were lowered one at a time to depths of 5, 10, 20, 30, 40, 60, 80, 100, 150, 200, 300, 400 meters, etc., and each bottle was tripped shut at the desired depth by a propeller mechanism which was built onto the bottle.* The bottle was then hauled to the surface, where a thermometer was put in through a special slit in the lid and read, in the belief, or at least in the hope, that the temperature of the water had not changed. Typical summer readings in Gullmar Fjord in western Sweden, for example, were 17.50° C. (63.5° F.) at the surface, 5.04° C. (41.1° F.) at the bottom, 400 feet below, and a minimum reading of 4.60° C. (40.3° F.) in the middle.

After the temperatures had been read, the water from the bottles was transferred into green Norwegian soda-water bottles, Danish common white bottles, colorless Swedish bottles, or small medicine bottles (a bottle that would not contaminate the water was difficult to find) and saved for further testing. The determination of the sample's salinity was particularly important, for this was used, with temperature, to identify the water mass.

Pettersson and Ekman determined the salinities of their water samples chemically, by means of chloride titration. They were actually measur-

---

* This propeller mechanism, which was supposed to close the water bottle as it started upward through the water, did not always work properly, and Fridtjof Nansen, who designed at least half a dozen water bottles, including an automatic one which could be used from a moving ship, remarked that "it is also very desirable to have a telephone connected to the sounding-line; by such an arrangement one can hear when the bottles are closed even at great depths. Then there could not easily be room for doubt as to the temperature observations, however astonishing they might be."Ѱ

By 1910, messengers—lead shot or bronze weights that slid down the lines to open and close water bottles, plankton nets and other equipment—had almost completely replaced propeller mechanisms. Messengers had been used sporadically at least since 1843 when Georges Aimé used them to detach the weight from his sounding device.

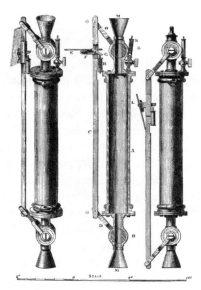

Stop-cock water bottles of the type used aboard the *Challenger* were also popular on many other expeditions. The bottle is shown open, left and center, and closed, right. (*From the* Challenger *Report*, Narrative, *Vol. 1, First Part*)

ing the amount of chloride in each sample, which they then converted to salinity. This could be done, because much earlier in the century chemists had discovered that sea water, whether briny or only brackish, contained almost exactly the same ingredients, and furthermore, that the proportions of the major salts remained constant. These relationships were precisely expressed by the German chemist William Dittmar, who, while teaching in Glasgow, Scotland, made very careful and well-known analyses of 77 of the *Challenger*'s water samples. Dittmar showed that by measuring one ingredient, usually chloride, it was possible to estimate the sample's total salinity.

Using this principle, oceanographers gradually abandoned the older, more complicated method of determining salinities by means of hydrometers and began determining chlorinities instead. Unfortunately, their results, however accurate, could not be compared to all other determinations, for until the turn of the century oceanographers did not agree upon a standard formula for converting chlorinity to salinity.

In the open ocean, the normal salinity of sea water was known to lie in the range between 33 and 37 parts per thousand, but in the North Sea and adjacent waterways, the variation was much greater, and

Pettersson and Ekman found salinities as low as seven parts per thousand, and as high as 35.

In addition to measuring the physical properties of water, plankton studies were soon added to the hydrographic work, because it was found that distinctive communities of drifting organisms lived in different water masses. Consequently, the plankton could be thought of as naturally occurring markers and used to trace the constantly shifting boundaries between masses.

Pettersson began his study of plankton in the summer of 1893 using an apparatus which towed several nets at different levels and which, therefore, could sample plankton from water layers at different depths.

"Our very first trial . . . in the Gullmar fiord," he wrote, "convinced me that the waters of the Skagerack at different depths often contain different and characteristic forms of Plankton. On that day we could distinguish three strata of water in the fiord, which showed marked differences with regard to the temperature and salinity."*

With scientists from five countries thus using all the means at their disposal to trace the elusive water masses, a new concept of coastal and oceanic circulation began to emerge. Actually, the concept was not really new, but its details and complexities were. In the North Sea, for example, where the seasonal surveys were undertaken by ships and men from the Fishery Board for Scotland, it was found that water from two different sources flowed together and mixed in a most intricate manner. Ocean water, cold and saline, flowed in from the north and west, while fresher, and usually warmer, land water poured in from the Baltic Sea and from various rivers. A mixture of the two was considered typical "North Sea water," even when found outside the sea itself. As the distribution of ocean and land water was mapped each season, it became apparent that the size and position of each mass changed much more rapidly than had been expected and that the pattern "may be altogether different at the same season in different years, and practically the same at quite different seasons in the same year."*

Seasonal charts were drawn up to follow these changes, and Pettersson even tried to integrate a great number of observations to show how the whole western arm of the Gulf Stream behaved. He drew sketches to demonstrate how in the summer months warm water ballooned northward almost to the island of Spitzbergen far north of Norway, but that by the next spring the Stream had dwindled and pulled back to the latitudes of Norway. Almost exactly a year later, in the spring of 1897, a very different pattern was observed, and Pettersson's sketch showed that while the main portion of warm water lay exceptionally far south, there was also a great lens of Gulf Stream water which had apparently been left behind north of the Arctic Circle. Although these seasonal sketches and the other charts produced as a result of the joint surveys showed how misleading the idea of an average current was, they could only give a general indication of where a current had flowed in a certain season of a particular year. No attempt was made to indicate the speed of the currents nor to depict the relatively small details, such as vortices and meanders, that scientists later found to be so characteristic of moving water. To study such details Pettersson knew that still more careful and more extensive surveys were needed, and he knew too that these would be of interest to fishermen as well as to physicists. Consequently, he encouraged the establishment of a Swedish Hydrographic Commission which in turn urged the Swedish government to hold a meeting with representatives from Great Britain, Norway, Denmark, and Germany. The objective was to organize an international program of investigations into the chemical, physical, and biological conditions of the water in the North Atlantic for the especial benefit of fisheries.

His Majesty King Oscar II of Sweden lent his royal support to the plan, and in 1899 representatives from those countries named, and from Ireland, the Netherlands, and Russia as well, accepted the king's invitation to meet in Stockholm. There was begun what in 1902 officially became the International Council for the Exploration of the Sea.

At the same time that Pettersson and Ekman were urging that detailed investigations be extended into the Northeast North Atlantic and beyond, a Norwegian explorer, Fridtjof Nansen, was getting ready to lead a scientific expedition through the ice-clogged North Polar Sea. There were several reasons for attempting such a dangerous voyage, and among them was the desire to investigate the circulation of the Arctic Sea. Nansen, like other scientists who had worked in more familiar waters, realized that even the most detailed investigation of coastal seas could observe only the last links in a long chain of events. Another type of investigation—the long, pioneering expedition to distant, little-known regions—was still needed to complement the systematic studies. In particular, Nansen felt that some information should be

gathered from the North Polar Sea as part of the attempt to understand the circulation of the important Norwegian and Greenland
Seas.*

Fridtjof Nansen, born in 1861, studied biology and zoology, then
put to sea on a sealing vessel at the age of twenty-one. Six years later
he led the first expedition successfully to traverse the Greenland ice
cap, and this feat made it somewhat easier for him to gain support for
his Arctic explorations, which many knowledgeable men called fool-
hardy and impossible.

His plan was to have a ship built to withstand the pressures of the
pack ice and with her and a crew of 13 to drift in the ice from the
New Siberian Islands (northwest of the Bering Strait) to the Greenland Sea (see map p. 186). Along the way—his course being determined by the winds and the currents that might move the ice—he
proposed to make meteorological and oceanographic observations. An
attempt was also to be made to reach the North Pole by dog sledge
were the ship to drift within sledging distance of that inhospitable
point.

At the time Nansen laid his plans there was no record of a ship, once
caught in the crushing pack ice, ever leaving it again in one piece,† nor
any direct evidence that currents flowed across the polar basin. In fact,
Nansen's hopes were largely based on the discouraging evidence
provided by the American ship *Jeannette*, which had been crushed by
the ice after drifting, unintentionally, for two years. A list and a few
pieces of gear from the ship had drifted from the East Siberian Sea to
the east coast of Greenland.

Even if Nansen's ship withstood the pressure of the ice, there was
the chance that the unknown currents would not carry the vessel in

---

* Within the Arctic Circle some meteorological and magnetic observations had
already been made as a part of the First International Polar Year, 1882–83. Scientists from several countries had collected data from temporary stations set up
on ice-bound rocks and promontories, but none had worked extensively at sea.

† In 1777 a number of whaling ships were locked in the ice north of the Norwegian island of Jan Mayen and crushed in turn. The last had drifted 1,100
nautical miles in 107 days before foundering. Almost a century later, the German
ship *Hansa* was frozen in and carried southward until crushed. The crew, stranded
on an ice floe, drifted 1,080 nautical miles in 246 days before landing in Greenland. Less spectacular drifts were also recorded, and it was fairly common to
find driftwood from Siberia on the eastern shores of Greenland, and Scottish
golf balls on the island of Lofoten off northern Norway.

Fridtjof Nansen in his cabin on the *Fram*. The etching was made from a photograph taken in February, 1895, just a month before Nansen left the ship to try to reach the North Pole. A picture of Nansen's wife and baby daughter hangs behind him. (*Courtesy H. Aschehoug & Co., Oslo*)

the direction or at the rate of speed anticipated. Uncharted lands might block her way, and this was of considerable concern, for at that time it was believed that the Arctic Sea was a shallow sea, dotted by many shoals and islands. The crew on such a voyage knew too that scurvy, polar bears, bugs, and boredom could further reduce their chances of success. In spite of the hazards, Nansen received two-thirds of the funds he needed from the Norwegian Parliament, and the remainder was oversubscribed within a few days of the expedition's announcement. He immediately set about having a most extraordinary vessel designed and built. He named her the *Fram*.

The *Fram* (meaning "forward" in Norwegian) was a small double-ender approximately 123 feet in length. She was rigged as a three-masted schooner, and the rounded shape of her hull had been designed to let her slip up and out of the crushing ice. Her bow and stern were each built of layers of oak planking four feet thick, and her hull was covered by an outer layer of greenheart (wood so dense it will not float), which the ice might rip away without damaging the rest of the ship. The hold resembled an oak grove after a hurricane, with shoring beams crisscrossed in every direction.

The *Fram* left Vardö, Norway, in July, 1893, and the Norwegian North Polar Expedition, as the voyage was officially titled, was under way. Nansen steered his strange vessel north and east, skirting the pack ice until, late in September, the *Fram* was trapped by the ice.

Fog in the morning, which cleared off as the day went on, [wrote Nansen in his diary for Sunday, September 24], and we discovered that we were closely surrounded on all sides by tolerably thick ice. Between the floes lies slush-ice, which will soon be quite firm. There is an open pool to be seen to the north, but not a large one. From the crow's nest, with the telescope, we can still descry the sea across the ice to the south. It looks as if we were being shut in.⚓ . . .

Friday, October 13th. Now we are in the very midst of what prophets would have had us dread so much. The ice is pressing and packing around us with a noise like thunder. It is piling itself up into long walls, and heaps high enough to reach a good way up the *Fram*'s rigging; in fact it is trying its very utmost to grind the *Fram* into powder.⚓

But the *Fram* would not be ground. The ice pushed in sometimes so hard that the 700-ton vessel was lifted several feet, but each time the *Fram* broke the ice beneath her. Pressure ridges were not the only danger.

Great bug war today. We play the big steam hose on mattresses, sofa-cushions—everything that we think can possibly harbor the enemies. All clothes are put into a barrel, which is hermetically closed, except where the hose is introduced. Then full steam is set on. It whizzes and whistles inside, and a little forces its way through the joints, and we think that the animals must be having a fine hot time of it. But suddenly the barrel cracks, the steam rushes out, and the lid bursts off with a violent explosion and is flung far along the deck. I still hope there has been a great slaughter, for these are horrible enemies. [Bugs were a danger to the food on board, the woolen and fur clothing, and to the health of the men.] Juell tried the old experiment of setting one on a piece of wood to see if it would creep north. It would not move at all, so he took a blubber hook and hit it to make it go; but it would do nothing but wriggle its head—the harder he hit the more it wriggled. "Squash it, then," said Bentzen. And squashed it was.⚓

The *Fram* drifted constantly with the pack ice, but, discouragingly, she seemed just as often blown to the southeast as pushed to the north,

The *Fram* trapped in the ice, exactly one year after she had left Norway. Polar-bear skins being cured are tacked to her hull and hung from lines. (*Courtesy H. Aschehoug & Co., Olso*)

so that by January, 1894 she was no farther north than she had been in September. This was particularly discouraging, as Nansen and his crew had made the startling discovery that the Arctic Sea was not shallow, but lay instead in a deep and landless basin. Soundings showed that the vessel had at least 1,000 fathoms beneath her keel.

> In one point only have my calculations proved incorrect [wrote Nansen], but unfortunately in one of the most important. I presupposed a shallow Polar Sea, the greatest depths known in these regions up till now being 80 fathoms, found by the *Jeannette*. I reasoned that all currents would have a strong influence in the shallow Polar Sea, and that on the Asiatic side the current of the Siberian rivers would be strong enough to drive the ice a good ways north. But here I already find a depth which we cannot measure with all our line, a depth of certainly 1,000 fathoms, and possibly double that. This at once upsets all faith in the operation of a current; we find either none, or an extremely slight one; my only trust now is in the winds.*

The following summer, when the days grew long and the temperature warmed to between minus 22° and minus 40° Fahrenheit, Nansen had one of the ship's steel cables unlaid into separate strands and these twisted together end to end to make a lead line thousands of fathoms long. With this he measured unexpected depths as great as 2,100 fathoms. This was much too deep to uphold the popular assumption that the polar region had, until recently, been covered by an extensive tract of land.

While the men aboard the *Fram* let the winds blow them a mile or so each day across the frozen Arctic, they undertook many kinds of studies.

Those that involved the greatest labor were, of course, the meteorological observations, which were taken every four hours day and night. . . . Also among our scientific pursuits was the determining of the temperature of the water and of its degree of saltness at varying depths; the collection and examination of such animals as are to be found in these northern seas; the ascertaining of the amount of electricity in the air; the observation of the formation of the ice; the investigation of the currents in the water under it, etc., etc. . . . Not an inconsiderable item of our scientific work were the soundings and dredgings. At the greater depths it was such an undertaking that everyone had to assist; and from the way we were obliged to do it later, one sounding sometimes gave occupation for several days.[∀]

Meanwhile, the *Fram* was proceeding in a more northerly direction, and by March, 1895 she had reached 84° North, 360 miles from the Pole. Nansen believed that this was about as far north as the vessel was likely to drift, and so, with Frederik Hjalmar Johansen, dogs, sledges, and a minimal amount of food and gear, he left the ship and set out for the "big nail." Nansen and Johansen drove slowly north over the drifting ice until they were within 225 miles of their goal, but by this time they realized that they had neither the time nor the energy to complete the journey. So, from that point, the farthest north that men had ever been, the two explorers turned south toward Franz Josef Land, where they hoped to encounter a fishing boat during the short summer season. While sledging for three months over rough, icy hummocks, which Nansen described as congealed breakers, the two men lost their way, and were forced to winter on an island which Nansen later named Frederick Jackson Land. The following summer, by a freakish piece of luck, the Norwegians encountered an English expedi-

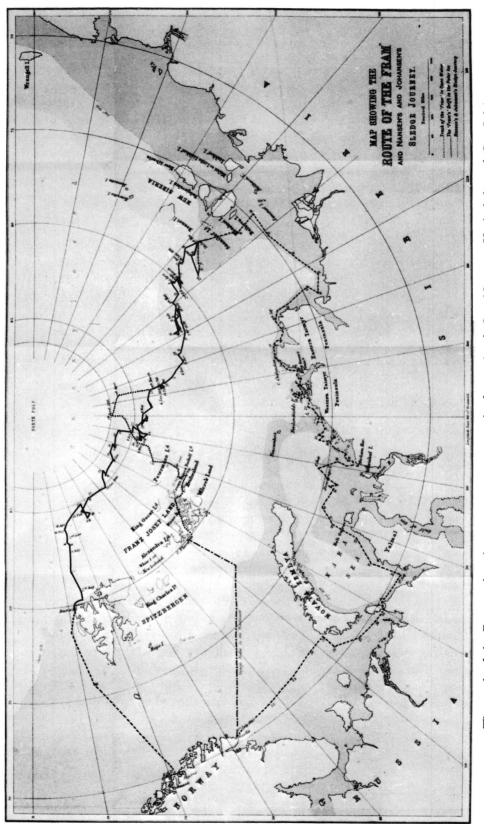

The track of the *Fram* on her three-year voyage across the frozen Arctic Sea. (*Courtesy H. Aschehoug & Co., Oslo*)

tion (led by Frederick Jackson) and late in the summer of 1896 they sailed home aboard the *Windward*. The *Fram* meanwhile had drifted south and west and had finally emerged from the ice north of Spitsbergen. She sailed back to Norway across the Greenland Sea, and arrived just a week after Nansen and Johansen.*

As Nansen had stated, one of his primary objectives was "to form a more complete idea of the circulation of the northern seas," and to do this he had carefully observed the character of the Arctic water by measuring its temperatures and salinities. Using one of Otto Pettersson's insulated water bottles, with a reversing thermometer attached, he had found that a layer of cold but relatively fresh water (derived in part from the rivers of Siberia, Canada, and Alaska) lay above a warmer, more saline layer of ocean water. Below these two was a third mass of water—cold, saline, and of great density. Upon his return to Norway, Nansen compared this arrangement of waters with the more familiar structure of the Norwegian Sea and saw that the bottom layers of water in the two seas were distinctly different.

---

* Curiously, the safe return of Nansen and the *Fram* spoiled a Russian plan to collect money for an ice-breaker which, when finally built, was used for Arctic exploration.

"Just at the time when Dr. Nansen proposed to build his *Fram*," wrote Stephan Osipovich Makaroff, the "restless admiral" of the Imperial Russian Navy and commander of the steamer *Vitiaz*, "I had the idea of adopting offensive tactics against the polar ice. . . ."Ψ His plan entailed the building and the use of an enormous ice-breaker.

"I wrote to Dr. Nansen a letter in which I stated that I was entirely of his opinion, that he would be carried by the currents somewhere in the direction he imagined. . . . I [also] thought it quite possible that he would not complete his voyage in three years; I also thought that, if in four years nothing was heard of him, people would be anxious to send help, and that would be a good pretext for collecting money.

"In my opinion the best way to penetrate into the Arctic regions is by means of a powerful ice-breaker. . . . Fortunately for Dr. Nansen, the current carried him on very well, and . . . I was happy to learn that he and his *Fram* had safely returned home. Of course that deprived me of my excuse for collecting the necessary money for building a large ice-breaker, but I found another motive, this time purely commercial. I proposed to build an ice-breaker which in winter-time might clear the way through the ice to the port of St. Petersburg. . . ."Ψ

Makaroff's commercial motive served as well as his more humanitarian one might have done, and in the late 1890's the 305-foot ice-breaker *Yermak* came off the ways in England. On her maiden voyage she plowed through 160 miles of ice to reach St. Petersburg and in 1899 made two summer cruises into the polar pack ice to study the properties of ice. Makaroff made a film showing the *Yermak* in the ice fields which was used by the Russian academician A. N. Krylov to write a learned paper on ice resistance.

"We are therefore obliged to assume that the basin of the Norwegian Sea is separated from the North Polar Basin by a comparatively shallow suboceanic ridge (or plateau) uniting Spitsbergen and Greenland." (Later, explorers on a Russian ice-breaker—not the *Yermak*—found such a plateau rising to within 1,000 to 1,500 fathoms of the surface but also found a deep, narrow channel running through it that allowed for some exchange of waters.)

This barrier that Nansen had indirectly discovered prevented a vigorous exchange of deep waters between the Arctic and Norwegian basins, but did not hinder surface circulation, and Nansen tried to chart these currents by comparing the daily drift of the *Fram* to the direction of the winds. He worked on this problem with the Swedish hydrodynamicist Vagn Walfrid Ekman, and together they saw that the *Fram*, and the ice in which she was frozen, did not drift exactly with the wind, but bore off some 20 to 40 degrees to the right. Ekman explained this curious deflection as an effect of the earth's rotation, and went on to publish a masterful explanation of the effect of rotation on wind-driven currents.

Ekman also used Nansen's data to explain another puzzling phenomenon known as "dead water." When a ship was caught in dead water—and this happened frequently in the Scandinavian fjords—she quite suddenly lost way, refused to answer the helm, and might even come to a shuddering halt.

Ekman asked the ship owners and naval officers to describe their experiences with dead water.

"Often the phenomenon appears in the manner described below," wrote an admiral.

> A small vessel sails free . . . on the starboard tack. . . . Suddenly a stripe begins to form on the water. . . . Outside . . . this stripe the water, as usual, runs aft by the vessel, but inside it the water forms a kind of wake in which the water seems to follow the ship. . . . On the whole the phenomenon presents itself *to the sailor,* as if part of the water separated from the rest, attached itself to the ship, and was dragged by it through the sea. When the stripe is developed, the vessel runs up into the wind in spite of the helm and lies quite powerless with her sails shivering. By rowing with an oar on the windward side she may be brought off again, and sailing is possible for a short time, but soon a new stripe begins to form and the vessel runs into the wind again. The masters of quite small craft often try to free their vessels from

the dead-water by stirring the water with an oar, and it is said they sometimes succeed.⸙

A skipper might also try to break free by sheering off course, running his whole crew forward and aft on deck, pumping violently, pouring petroleum ahead of the ship, firing guns into the water, or cutting and beating the water alongside with oars and handspikes.

Ekman approached the problem in a less spectacular manner. Since his most precise information on dead water came from Nansen, who had encountered the phenomenon while sailing on the *Fram*, Ekman had a model of the vessel built which he could tow through a glass-sided tank. He filled the tank with layers of colored water, each of a different density, and these were meant to approximate the conditions found in fjords. In the latter, such a layered effect often occurred when fresh stream water or melt water flowed out across the saltier, denser sea water.

As Ekman pulled the wooden model through the stratified water in his tank, he noticed that in addition to the usual waves formed behind the ship on the surface of the water, a second set of normally invisible waves was also forming at the boundary of the first and second layers of water. These unseen waves (called boundary or internal waves) exerted a powerful drag on the ship and were the cause of what seamen called dead water.

Ekman published his findings in 1904 as a part of the scientific results of the Norwegian North Polar Expedition (the *Fram* report). Although in theory the possibility of boundary waves had been discovered by physicists many years earlier, oceanographers had not expected to find these waves in the ocean since the density differences between layers of sea water were much smaller than those artificially produced in a laboratory tank, and conditions at sea were further complicated by winds, currents, and a host of other variables. Once Ekman had related boundary waves to dead water, however, scientists began looking for the invisible waves and soon found them. By 1909 Otto Pettersson had devised an instrument which could automatically record the oscillating boundary between water layers. He installed this device in his own research station on an island in Gullmar Fjord.

"The little place was full of gadgets; water could be pumped from various depths, and all sorts of recording instruments ticked away. A copper vessel floating on the surface of discontinuity [the boundary between water layers] rose and fell with the submarine waves, and transmitted their motion to a recording drum."Ψ Some of the waves so measured were 100 feet high and had crests so far apart that they passed only once each day.*

In spite of all the data that Nansen was able to give Ekman and others and which led to reports on dead water and boundary waves, the Norwegian explorer began to realize that much more could have been done from the *Fram*.

"But even our oceanographical work in the North Polar Basin . . . is not nearly so complete and accurate as could be desired, and as I now, too late, see that it easily might have been; for a strong ship, frozen fast in the solid ice, drifting slowly across the deep sea, offers . . . advantages for the most delicate oceanographical investigations. . . ."Ψ

Nansen wrote this in 1902, six years after he had returned from the Arctic, and what prompted his expression of regret was the introduction of a new way of estimating the circulation of the ocean which, unfortunately for him, was dependent on the "delicate oceanographical investigations" which he had not made.

The new method involved studying the forces that produce motion in the sea, and in doing this, use was made of the theoretical knowledge contained in the laws of fluid dynamics. Based on these laws, hydrodynamic equations could be written which stated how a fluid ought to behave in response to certain forces. To be of use in studying the real ocean, these equations had to include terms for the most important forces, such as the earth's gravity and rotation, and these had to be selected from among a complex multitude of lesser factors, all of which do, in fact, influence the water's behavior. If too many of these factors were ignored, the equations would express relationships that were so oversimplified as to be useless, and if too many factors were included, then the equations themselves would be unwieldy and unworkable. It was a delicate balance to strike, yet if workable and essentially accurate equations could be written, their solution would yield information about fluid motion that was actually occurring in the sea and physical oceanographers would gain a useful new way of investigating currents.

In attempting to adopt this theoretical approach, oceanographers

---

* This type of wave is believed by some to have caused the *Thresher* disaster in 1963. The submarine may have slid suddenly and uncontrollably downward below its collapse depth on the back of a huge internal wave.

Bjørn Helland-Hansen aboard the R/V *Armauer Hansen* in 1923. Twenty years earlier, he and Johan Sandström had developed one of the first sets of useful equations for calculating geostrophic currents. (*Courtesy Håkon Mosby*)

were following an example set by meteorologists, and this they could do, for both air and water are fluids (as far as physics is concerned) and the principles involved in studying the two are the same.*

In 1898 a professor of meteorology at the University of Oslo, Vilhelm Bjerknes, published a circulation theorem—a statement expressing how fluids of different densities ought to circulate—in a paper, "On Fundamental Hydrodynamical Principles and Their Special Application to the Mechanics of the Atmosphere and the Oceans." Using this theorem, two other Scandinavian scientists, Johan Sandström and Bjørn Helland-Hansen, made estimates for each term

---

* As the American meteorologist William Redfield wrote in 1834, "the waters of the ocean are subject to the same impulses as the superincumbent atmosphere, [and] it is probable that the principal movements of both fluids have their origin in the same causes."Ψ

in the theorem, then the two of them went on to work out a set of useful equations for calculating ocean currents. This formula was based on the assumption that in the open ocean the water at any level moves in response to two primary forces. The first of these is a horizontal pressure (in a sense the water's desire to seek its own level) which exists whenever the surface of the sea or the surface of some intermediate layer in the sea is not level. When broad hills of less dense water rise, even slightly, above troughs or basins of denser water, then the highs tend to drift directly inward toward the lows. But the second force, the Coriolis effect that results from the earth's rotation, deflects the moving water, to the right in the Northern Hemisphere and to the left in the Southern, and actually turns the drift so that it moves at right angles to the slope of the water.

Based on this understanding of what were later called geostrophic currents (from the Greek *geo*, "earth," and *strophe*, "turning"), Sandström and Helland-Hansen saw that it would be possible to compute the speed and direction of these currents from an equation that included the Coriolis force, which could be determined from the latitude, and the water's topography which, being the result of unequal densities, could be determined from precise measurements of temperature, salinity, and depth. (Several factors of lesser importance were also included in the formula.*)

Probably the first attempt to use geostrophic equations was made in 1900, nearly a decade before the paper describing it was even published. Helland-Hansen and Fridtjof Nansen had been invited by the director of the Norwegian Board of Sea Fisheries, Johan Hjort, to participate in the first cruise of the fine new research vessel *Michael Sars*. The two men were to take charge of the hydrographic work but first were to help select the best instruments available. This pleased

---

* As early as the 1870's and 1880's isolated attempts had already been made to apply hydrodynamic laws to the sea. In 1878, a German high-school teacher, E. Witte, published a paper on ocean currents in which he explained why, on a rotating earth, the boundary or interface between two water masses of different densities and different velocities cannot be a horizontal surface, but must have a slope. His paper was not well known, and the principle it contained was independently rediscovered some 28 years later by the Austrian meteorologist Max Margules. (See Neumann and Pierson, *Principles of Physical Oceanography*.)

The Norwegian meteorologist Henrik Mohn, who led an expedition into the North Atlantic in 1876–1878, also anticipated some of the theoretical work done at the turn of the century. In 1885 Mohn worked out a formula for computing horizontal velocities from the ocean's dynamic topography, and in 1903 Sandström and Helland-Hansen rediscovered essentially the same one.

Nansen, who was especially interested in new instruments and improved accuracy. By mid-July, when the *Michael Sars* set off toward Iceland, a great variety of gear had been stowed aboard.

"It would be worthless to make observations of the temperature and salinity . . . in this sea [the Norwegian Sea]," wrote Nansen, "with no more accurate methods than have hitherto generally been employed, for the errors of observation are much greater than the natural variations. Nevertheless the slight differences in density, etc., which may exist in the deeper strata of the Ocean are certainly very important for the circulation of the whole hydrosphere, and it is therefore very desirable that they should be exactly determined. . . . I consider it therefore to be one of the first and most important aims of oceanographers to improve materially the methods of oceanographical research, and then with improved methods to collect new materials of observations from all parts of the Ocean."⍦

As before, the measurements that most concerned Nansen and Helland-Hansen were of temperature and salinity, for in addition to their new use as ingredients of the geostrophic equation, they were still primarily used to plot the positions of water masses. To measure the deep-sea temperature, and to bring up samples of water for salinity determinations, five different kinds of water bottles were used aboard the *Michael Sars*. Three of these were insulated in one way or another and two were not but carried Negretti and Zambra reversing thermometers down with them on separate reversing frames. (By 1910, Nansen had attached the water bottle and a pair of thermometers directly to the reversing frame to make what he called a "reversing water bottle." Later this became known as the Nansen bottle, which is still used routinely by oceanographers today.)

The reversing thermometers should have given more accurate readings than the water bottles, but, as Pettersson had found, they did not. Georges Aimé, the French scientist who had designed deep-sea sounding mechanisms and other apparatus years ahead of his time, had introduced what was probably the first model of a reversing thermometer in the 1840's. Its design was too intricate to be practical, however, and it was not until the late 1870's that the British firm Negretti and Zambra patented a more useful model.* Their thermometer, looking

---

* The firm's first reversing thermometer, manufactured in 1874, was nearly as impractical as Aimé's, for to register properly it had to be turned a full 360°— slowly. Several of these thermometers were sent to the *Challenger* and the first

A Nansen bottle is fastened to a hydrographic wire to be sent down in series with five or more bottles to collect water samples and measure temperatures. A frame for both protected and unprotected reversing thermometers is built right onto the bottle. (*Courtesy Woods Hole Oceanographic Institution*)

more or less like the normal household variety, had a narrow constriction and an S-curve in the tube that led from the mercury reservoir to the scale above it. When the thermometer hung, bulb down, in some deep stratum of the sea, the mercury passed slowly through the constricted tube until it registered the temperature of the water. Then, as it was hauled to the surface, a propeller mechanism, or later a messenger, caused the thermometer to turn upside down. At this moment the column of mercury was supposed to break off exactly at the constriction, and the amount left in the scale portion (which registered the water's temperature) could neither be augmented nor diminished regardless of the temperature of the overlying water. In practice, however, "the difficulty with these thermometers is that the mercury

were tried in the Sulu Sea with irregular results. Thomson and Murray saw promise in the principle but realized too that refinements were needed before the instrument would be reliable.

seldom breaks off exactly at the same place of the contraction," wrote Nansen. "Besides this, the glass of the thermometers is not generally of good quality, so that they easily change their zero point, and further their scales are usually so roughly made that accurate readings cannot be taken."[*] (In 1902, a German firm began producing a more workable reversing thermometer, and in later years these Richter models were believed to be accurate to within plus or minus .01° C.)

To measure the salinity of the water through which the *Michael Sars* moved that summer, Nansen and Helland-Hansen took a double set of water samples from each depth. The salinities of one set were determined by titration, while those of the other were measured with a hydrometer.

"We cannot expect to find absolute agreement between Helland-Hansen's determinations by titration and ours with the hydrometer . . . ," wrote Nansen,[*] and the reasons he gave were that the relationships between chlorinity and salinity (and between salinity, temperature, and density) had not been definitely determined.

At exactly this time, however, a Danish physicist, Martin Knudsen, was working on a set of hydrographical tables which would set forth the precise relationships among these properties. At the 1899 meeting for the organization of the International Council for the Exploration of the Sea Knudsen had suggested that such tables be published in order to facilitate the standardization of hydrographic work. For the same reason, he had also suggested that Standard or Normal Water (water

Early models of the Negretti & Zambra deep-sea reversing thermometer had to make one entire revolution in order for the mercury to be trapped in the capillary tube. (*From the* Challenger *Report,* Narrative, *Vol. 1, First Part*)

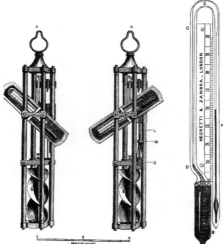

of known and unvarying salinity) be distributed to oceanographic laboratories throughout the world as a standard against which all other salinities could be compared.

Helland-Hansen had received some Standard Water from Knudsen, who was then setting up the Hydrographical Laboratory for ICES in Copenhagen, and the former constantly checked his results against it. Helland-Hansen went on to determine the densities of his samples using Vice Admiral Makaroff's tables, but "for the final report Knudsen's new Tables will be used, when they appear."ᵠ

So, by the end of 1900, some four months after the *Michael Sars* had returned from the last leg of her summer cruise, Nansen and Helland-Hansen had as precise determinations of water densities as was possible to make at that time. They plotted these data on graphs and, along one section—the line of stations run from the north coast of Norway to Bear Island—they noticed that "the water near the Norwegian coast is conspicuously lighter [and therefore higher] than the water to the north, and has consequently a very strong tendency to rise and spread northward. . . ."ᵠ

These observations were taken to Nansen's friend V. Walfrid Ekman, and the latter computed the relative velocities of the surface water flowing away from the Norwegian coast as compared to the deeper water moving toward the coast.* In figuring the velocities, he used Professor Bjerknes' theory and Sandström's equations. Helland-Hansen became fascinated with the method's possibilities and proceeded to work with Sandström to develop the equations further. Their joint paper, "On the Calculation of Ocean Currents," appeared in 1903, and thus, wrote Nansen, the method of computing the velocities of currents was developed into "a very convenient form."

Initially, the greatest use of dynamic computations was made by Scandinavian scientists. The *Michael Sars* made frequent voyages into the Norwegian Sea—that broad arm of the Atlantic that extends north between Norway and Greenland—and there collected the precise measurements needed for geostrophic computations. In 1902 Danish

---

* The geostrophic currents computed in this way (and those determined later from Sandström and Helland-Hansen's formula) are relative motions, but they can be assigned absolute velocities if the motion of any level of the sea can be precisely determined. In practice, current meters were rarely used for such determinations and a level of *no* motion was assumed to exist near the bottom. As more was learned of strong bottom currents, this assumption became less valid.

and Scottish oceanographers began to extend their investigations into the southern parts of the Norwegian Sea, and Nansen and Helland-Hansen used their observations as well in calculating the circulation of this extensive region. Finally, in 1909, the two scientists published a monumental work, "The Norwegian Sea, Its Physical Oceanography Based Upon the Norwegian Researches 1900–1904." Their picture of the currents was the first to be drawn primarily from geostrophic computations and was much more extensive and more detailed than the work done by Pettersson, Ekman, and others only 10 or 20 years before.

> We arrived at the conclusion that there must be many forms of motion of great and far-reaching importance, though hitherto hardly known at all . . . , [wrote Helland-Hansen]. We must picture to ourselves great submarine waves moving through the water-masses . . . , standing waves, and great vortices. We must further conceive of constant fluctuations in the velocity . . . [and] direction of the great ocean currents. . . . There is an interplay of many different forces, producing an extremely variegated picture; the sea in motion is a far more complex thing than has hitherto been supposed.⁴

Nansen and Helland-Hansen's careful study made the Norwegian Sea the most thoroughly investigated and best known body of water in the world, and although no one at that time could be sure how reliable the new method of current calculation was, subsequent observations upheld their conception to a remarkable degree.*

In spite of this study, the use of geostrophic equations was slow to catch on outside Scandinavia, both because the method was complicated and because there was considerable doubt as to its worth. Western European countries belonging to the International Council for the Exploration of the Sea gradually became exposed to the theoretical approach from their Scandinavian neighbors, but in the United States, in spite of a comprehensive report, *Dynamic Meteorology and Hydrography,* written by Bjerknes and others and published by the Carnegie Institution of Washington, D.C., in 1911, the new method was not adopted until the mid-1920's. Before that time few attempts

---

* Nansen continued to work on oceanographic problems, which included directing the Central Laboratory for ICES for six years, until the outbreak of World War I. He then left his colleagues and the science he loved to grapple with problems of refugee resettlement. For this work he received the Nobel Peace Prize in 1923. He died in 1930.

had been made to check the dynamic method (as it was often called) against direct methods to see just how closely the two agreed, and in the early years of the century even Helland-Hansen felt that these "calculations made as to the velocities of currents still rest upon an insecure foundation. . . ."[*]

Then, about 1924, a German oceanographer, Georg Wüst, used dynamic computations to obtain a picture of the currents which flow at different levels through the Straits of Florida. He compared his results to the current profiles that Lieutenant Pillsbury had made in the same area with a current meter in the 1880's. The pattern of currents was essentially the same, and confidence in the dynamic method grew.

A second opportunity for comparing the dynamic method against direct observations arose when the International Ice Patrol began computing the circulation of the northwest Atlantic and at the same time observed the actual drift of icebergs (see Chapter VII, p. 261 *et seq.*). An American Coast Guardsman, Edward H. Smith, was sent to the Geophysical Institute to learn all he could about the dynamic method from Helland-Hansen. He brought back information and instruments, and by 1926 the Ice Patrol (operated by the Coast Guard) was making what Smith called "velocity profiles" through the icy waters. The current patterns thus revealed by dynamic computations agreed well with the visible drift of the bergs and made many converts to the new method.

In spite of its growing popularity, the dynamic method was used only for small, regional explorations until, in 1925, the German Atlantic Expedition set out aboard the steamer *Meteor* to make a systematic investigation of the currents and water masses that moved through an entire ocean. Plans for this extensive exploration were begun shortly after World War I and were encouraged by the Naval Command, since at the time only unarmed surveying or research vessel were allowed to show the flag abroad and keep in touch with the Germans living in foreign countries. Official enthusiasm notwithstanding, funds to pay for the expedition were nearly impossible to obtain. Finally, the Navy agreed to complete the half-finished gunboat *Meteor* as a research vessel and to furnish a crew, while a private group of citizens who had formed an Emergency Organization for German Science raised funds for a scientific staff and for the best and most modern apparatus. The cruise was initially planned as a long pioneering expedition through the Pacific, but when the extra cost of diesel engines was considered, it was decided to complete the vessel as a coal-burning steamer and reorganize the expedition's objectives to fit the smaller, more familiar Atlantic.

The scientific leader of the expedition was Dr. Alfred Merz, a physical oceanographer and a professor at the Oceanographic Institute at the University of Berlin. Merz was particularly interested in the ocean's deep circulation, and in 1921 he and his student and assistant, Dr. Georg Wüst, took temperature and salinity observations from the reports of the *Challenger, National, Fram, Princesse Alice, Michael Sars, Armauer Hansen,* and others, and constructed longitudinal sections of temperature and salinity along the middle meridian of the Atlantic Ocean. The arrangement of water masses thus revealed did not appear to be the simple two-cycle system suggested by Lenz, but a more complicated four-layered structure which included warm, saline surface water, an extensive tongue of Antarctic Intermediate Water, Atlantic Deep Water, and deepest of all an extremely cold layer of Bottom Water. Using this pattern Merz tentatively constructed a new model of Atlantic circulation which he hoped could be verified or modified with the data collected on the *Meteor*.

The Atlantic Ocean, then as now, was the best known of all the oceans, but Merz and Wüst could see that this applied primarily to its northern portion. Consequently, the track of the *Meteor* was laid out to fill in the blanks, and the 14 profiles (lines of stations) that she was to make were arranged at regular intervals from 20° North, the latitude of the Cape Verde Islands, to 55° South, the latitude of the tip of South America.

A longitudinal section of salinities through the Atlantic Ocean at 30° West was made by Alfred Merz in 1925. The arrows indicate the probable direction of flow. (*Courtesy Deutsche Akademie der Wissenschaften zu Berlin*)

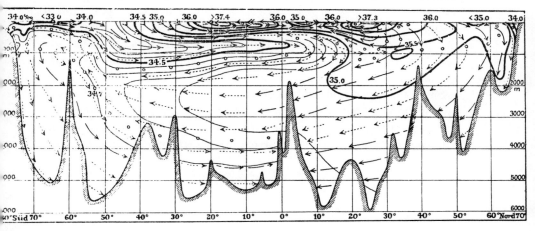

In April 1925 the *Meteor* was finally ready to depart. She was gloriously equipped with two new kinds of echo sounders, protected and unprotected reversing thermometers, large and small water bottles, current meters, plankton nets, deep-sea anchoring gear, and all the instruments and chemicals needed for analyzing water samples; and she was as superbly staffed by five physical oceanographers (Merz and Wüst among them), two meteorologists, one chemist, one geologist, and one biologist. Captain Fritz Spiesz commanded a crew of about 120 men, and each of them, sailor and scientist, knew that a cruise on the *Meteor* might be exciting, but it would not be comfortable. For one thing, the ship, with a length of 216 feet and a beam of only 33, rolled like a log, and for another, she was packed to overflowing with coal. The space that remained for living quarters was roughly comparable to the cramped quarters on a submarine.

On April 16 the *Meteor* left Wilhelmshaven on the North Sea and sailed to Buenos Aires, Argentina, which was to be the starting point of the expedition. On the way, acoustical soundings were made every two or three miles, the deep-sea anchoring gear was tried, and a complete set of meteorological observations was made. These last involved flying kites up to 10,000 feet, releasing pilot balloons that drifted with the winds until lost from sight at 80 to 90,000 feet, and occasionally firing smoke bombs from an antiaircraft gun. (This had to be discontinued because the shock of the explosions disturbed the delicate reversing thermometers.)

Early in June the *Meteor* set out across the Atlantic in the path of the Roaring Forties to make her first profile. As expected, she encountered almost continual foul weather and, when only six days out, with work barely begun, she was forced to roll and pitch back to Buenos Aires to land Dr. Merz, who was suffering a relapse of a lung ailment. With Merz in a hospital, the *Meteor* resumed her stormy voyage, and in spite of a hurricane that drove her close past Gough Island (an isolated peak that rises from the Mid-Atlantic Ridge) she managed to occupy 20 stations, make a near-continuous echo-sounding profile of the ocean floor, and check these depths with numerous wire soundings.

Since the expedition's prime objective was an investigation of circulation, the scientific work at each station was arranged so as to gather the data necessary to plot the arrangement of water masses and to calculate their movements. Most essential were precise measurements of temperature and salinity, and these were made at frequent intervals from the surface all the way to the bottom. From these determinations would come a picture of the ocean's water masses and, when the densities of the layers were calculated as well, geostrophic equations could then be used to estimate the waters' flow. Direct measurements, made with meters while the ship was anchored, were also made (although not in the Roaring Forties) and these served as a check for the dynamic computations. Other factors affecting circulation were also considered, such as the contours of the sea floor which were mapped in duplicate with the two echo sounders, and the weather conditions which were assiduously observed.

After completing the first profile, the *Meteor* put into Cape Town, South Africa, and then, toward the end of July, headed back across the Atlantic, along a more northern parallel. The routines established on the first profile were continued, and in addition, a deep-sea anchor station was made. The ship was successfully anchored in the middle of the Atlantic in 2,000 fathoms of water for over 40 hours while current meters were used at many depths. The *Meteor* then resumed her slow voyage westward, and on August 25 news of Merz' death was received by wireless. The second profile was carefully completed (the scientists felt that Merz would have wished it so), and the ship returned to Buenos Aires. There, it was decided that Captain Spiesz should assume the position of chief scientist for the rest of the voyage.

Before returning to Germany in the spring of 1927, the *Meteor* completed 14 profiles, 13 of which crossed the Atlantic. She had also made several long summer detours to the very edge of the Antarctic ice. In the course of the 67,000-mile voyage she had occupied nine deep-sea anchor stations and 310 regular hydrographic stations and, at the latter, more than 9,000 temperature and salinity measurements had been gathered. In addition, over 33,000 duplicate soundings had been made in an area where only about 3,000 had been made before. The new soundings revealed a rough and irregular topography and also showed that the Mid-Atlantic Rise swung eastward south of Africa and extended toward the Indian Ocean.

It was found too that the basins and swells in the Atlantic affected the ocean's circulation to a much greater extent than had been supposed; and the distributions of temperature and salinity on the east side of the Mid-Atlantic Ridge were significantly different from those on the west. In the main, however, the pattern that Merz and Wüst had

The research vessel *Meteor* was used for the German Atlantic Expedition of 1925–1927. (*Courtesy Günter Dietrich,* from his General Oceanography, an Introduction, *Copyright, 1963, John Wiley & Sons, Inc.*)

envisioned before the expedition set forth proved to be an accurate approximation. As they had predicted, the greatest depths in the western basin of the Atlantic were filled by the Antarctic Bottom Current which brought cold water of relatively low salinity from the edge of the Southern Continent north to latitudes above the equator. A second mass, called the Antarctic Intermediate Water, apparently formed at the sea surface several thousand miles north of Antarctica and sank to a depth of between 2,000 and 3,000 feet while spreading north, again past the equator. Between these tongues of water existed a huge mass of North Atlantic Deep Water, which seemed to form between the latitudes of Newfoundland and Southern Greenland and drift southward at depths greater than 3,000 feet. In the eastern basin of the Atlantic, the same water masses were present, but their arrangement was quite different due to barriers such as the Walvis Ridge that runs from the mid-ocean swell to the coast of Africa and blocks the northern flow of Antarctic Bottom Water.

The actual velocities of these huge water masses were difficult to determine, and even the measurements made with meters at anchor stations were not precise.

"If the vessel is firmly anchored and the anchor holds, it [the ship] is not necessarily a fixed reference-point from which current measurements can be made directly without more ado," wrote the physical oceanographer Albert Defant, who joined the *Meteor* near the end of her voyage. "Any ship anchored with a long cable will be subject to movements due to the changes in the wind and the current. . . ."⚓

Defant identified three kinds of ship motion—swinging round, swinging, and yawing—and the last type of movement, he said, "may be unpleasantly noticeable in the current recordings, even when the yawing movements are unnoticed in the open ocean."⚓

In spite of these complications, the direct measurements were considered accurate enough to be used as a check on the dynamic computations, and it was found that the two methods produced results that differed anywhere from 5 to 25 percent. The usual difference was only about 15 percent, and this was thought to be encouragingly close agreement.

When the *Meteor* had returned to Wilhelmshaven, her records and samples had been efficiently removed to the Oceanographic Institute in Berlin, where scientists of many disciplines began to work on the reports. A total of 16 volumes were written (although scientists continued to use the *Meteor* data for miscellaneous projects long after the official reports were published), and these dealt primarily with physical oceanography, meteorology, and planktonology, and, to a lesser extent, with bathymetry, geology, and chemistry.

A most interesting discovery was made when the three- and four-foot deep-sea cores collected from the *Meteor* with a glass-lined coring device were examined by the geologist Wolfgang Schott. Schott had not himself accompanied the expedition, but he had been given the cores to examine and in some of them he found two distinct layers of planktonic ooze. From cores taken beneath equatorial waters, for example, he found that the topmost layer of ooze was composed of bits and pieces of the warm-water plankton that was known to inhabit the overlying waters. But the lower layer was filled with the remains of cold-water plankton that did not live in equatorial waters and could only have done so during an ice age when the temperature of the surface water, even in the tropics, was much colder. The cores, then, contained a record of the earth's climatic changes and by the nature of the planktonic debris they contained, scientists could determine when the great ice sheets advanced and receded, as they were known to have done several times.*

Equally interesting and valuable reports were prepared by chemists, biologists, meteorologists, and, of course, physical oceanographers, for not only were their data more accurate than those collected on most previous expeditions, but also they were collected in a more systematic way and from a much wider area. Other voyages had been longer than the German Atlantic Expedition—but these had wandered almost haphazardly through all the oceans—and other expeditions had been more thorough—but they had covered only a few hundred square miles in a single arm of the sea. For the cruise of the *Meteor* the best of both approaches had been combined, and her voyage was the first to extend the detailed coverage of a systematic investigation to an extensive area of the sea. The expedition, like that of the *Challenger*, is often used to mark the beginning of a new era or phase in oceanography.†

So, with the voyage of the *Meteor*, the new methods evolved for the study of physical oceanography began to be applied to great areas of the sea, but that is not to say that the methods themselves continued to

---

* From Schott's exciting work, the idea soon developed that an uninterrupted record of the earth's climatic changes over the past million years could be obtained if only cores some 50 feet long could be extracted from the sea floor. A variety of coring devices was devised to meet the need. By 1935, Charles Piggot, an American working at the Carnegie Institution in Washington, D.C., had built a gun corer which shot a pipe about 10 feet into the sediment, and 10 years later a safer, more effective, piston corer (still used today) was designed by the Swedish scientists Börje Kullenberg. With the piston corer cores as long as 70 feet could be extracted, but these did not contain an uninterrupted record. Instead, they showed evidence of slumps and current action on the deep-sea floor and this unexpected activity had badly confused the orderly pattern of deposition that had been anticipated.

† Another well-known example of an extensive, systematic survey is the work of the British research vessel *Discovery*. The ship, a Dundee whaler, had been built for Robert Falcon Scott's first trip to Antarctica. (On a subsequent voyage, Scott and his four companions died of exposure after reaching the pole a month after the Norwegian explorer Roald Amundsen.) In 1924 the *Discovery* was acquired by a group of British scientists who intended to survey the marine resources of the Falkland Island Dependencies. The survey was begun in 1925 and, in response to new methods of whaling which threatened to kill off several species of the valuable marine mammals, the survey was soon broadened to include a detailed biological investigation of all parts of the Southern Ocean. (The ship, now restored as it was when Scott used it, is moored on the Thames at London.)

be developed at a rapid rate. Although Helland-Hansen had written in 1912 that "physical oceanography is confronted with a host of new problems, the solution of which will be a matter of the highest interest,"⍢ the science did not continue to grow at the enthusiastic pace set in the early 1900's. Even as World War I began, concerns outside the field of geophysics were attracting young physicists away from a study of the oceans. Radioactivity was discovered in 1896, quantum mechanics was developed under Max Planck and Paul Dirac at the turn of the century, and before 1920 Albert Einstein had formulated both the special and general theories of relativity. By comparison, the problems within the field of physical oceanography seemed lacking in glamour, and studies of the ocean's physical properties were gradually returned to the care of naturalists and biologists who used them as part of their studies of the life in the sea. This meant that to an increasing extent hydrographic investigations of a rather routine nature were undertaken as a part of fisheries research.

In the first decades of the twentieth century, biological oceanography, and especially fisheries research, grew tremendously in size and sophistication. The International Council for the Exploration of the Sea, founded in 1902, was the focal point for much of this activity, and the organization's aim was to gain an understanding of the sea sufficient to make possible effective legislation regulating major fisheries. Fish, after all, could not outlast man's insatiable pursuit forever, and the aim of fisheries research, as expressed by one ICES member, was "to tell the commercial world how far greed might safely go."⍢

Fisheries' Problems and the International Council for the Exploration of the Sea

So it was also with the herring fishery between Stavanger and Bergen that they did draw so many herring on shore that they lay in great piles, and because they could not make out to salt them, or sell them, they . . . did allow them to lie and rot. . . . And after that the herring came no longer. . . .

And God did also let the herring fishery fall away in the following year, and has since taken it from us; so that now in this year 1599 no herring has come.

—Peder Claussøn Friis, "Of the Animals, Fish, Birds and Trees of Norway"

By 1898, when it was suggested to King Oscar II of Sweden that he initiate an international program of marine investigations, it was generally accepted that the impersonal mysteries of nature, rather than the wrath of God, moved the herring and the cod. Furthermore, men with less than a "perfect faith in the marvellous ways of Nature . . . which enable her to cope . . . with all the wonderful advances in apparatus for the capture [of fish] . . ."[Ψ] were beginning to suspect that man, too, had an effect on the fisheries. It is doubtful that these more modern assumptions made the task of explaining the great fluctuations of fisheries any easier, but they did lay the problem on scientific doorsteps and suggested to governments the need for what was inevitably referred to as "rational legislation."

The concern for fisheries that developed in the second half of the nineteenth century resulted from that industry's sudden and dramatic expansion which itself reflected the needs of a rapidly growing population whose more inventive members had applied steam power to boats and winches, designed new trawls and nets, and found efficient ways to refrigerate and transport mountainous loads of fish. With all this, the fishing industry had been jostled from its thousand-year slumber and forcibly encouraged to change to faster, noisier, dirtier—and more profitable—methods which it had done in a remarkably short period of time.* Eight hundred yawl-rigged sailing smacks had ridden at their moorings in the British fishing port of Grimsby in the mid-1880's, for example, and by 1902 all but 29 had been crowded out by a new fleet of 500 steam trawlers. The fish, as well as the people, were bound to feel the change.

But with all the changes that mechanization brought about, the traditional questions and problems remained. Why was the fishing so good some years and so poor in others? The long-line fisherman who sat alone in his dory on the stormy North Sea nursing an implacable hatred for the new steam trawlers knew the answer. It was the noisy trawlers. They came in increasing numbers with their wide, weighted nets to scour his fishing grounds clean, and they were responsible for each poor trip and each poor season. The trawlermen, of course, did not agree. They blamed storms and calms, jellyfish and the Gulf Stream, for, like the liners, they too were being forced to spend more time covering more ground to fill their holds with the same catch that had once been gathered within sight of land.

Whatever the causes, the good, and more noticeably the bad, fishing years had begun to affect a very large number of people. Their grievances were real, their complaints loud and numerous, and by the 1860's and 1870's the governments of seafaring nations had responded by appointing committees and commissions to study the fishing in their territorial waters. The gesture was a fine but in many ways empty one. It required no study at all to know that fish respected no political boundaries, and the deficiencies of a program arbitrarily confined to a small segment of the sea was frustratingly apparent.

Governments, then, were receptive to King Oscar II's plan, which promised them the benefits of a larger, more comprehensive program of fisheries research at modest cost, and accordingly they sent their best, most energetic and inventive oceanographers to the king's international conference in Stockholm in the spring of 1899.

The gathering was impressive. The chief British delegate was Sir

---

* Paradoxically, some ways of catching and handling fish—the setting of drift nets for herring, the backbreaking salting of cod, and the slow unloading of fish from a hold basket by basketful—did not change, and *still* have not changed.

A great fleet of sailing trawlers and steam trawlers moored in the British fishing port of Lowestoft. The photograph was taken in the late 1930's. (*Courtesy Ministry of Agriculture, Fisheries and Food, Lowestoft*)

John Murray, knighted the year before for his work on the *Challenger* expedition; from Germany there was Victor Hensen, the controversial advocate of quantitative plankton studies, Friedrich Heincke, an expert on herring, and Otto Krümmel, author of a *Handbuch der Ozeanographie*. (A review of the book in *Nature* noted his use of the bizarre term "oceanography.") Otto Pettersson, Gustav Ekman, and Theodor Cleve—who together had formulated the content of the king's plan—represented Sweden, and from Norway came their friends Fridtjof Nansen and sharp-eyed Johan Hjort. C. G. Johannes Petersen left his floating biological laboratory in Denmark to attend the conference, and from Copenhagen, the future seat of the International Council, came the future author of hydrographic tables, Martin Knudsen.

In all, several dozen scientists and statesmen were present, and dressed in their formal attire gay with ribbons and orders, they gathered in the palace halls to discuss in French, German, English, and other languages their plans and hopes for a council. Many ideas were brought forward, for each delegate had come with a slightly different understanding of the sea, and each had his own approach to its investigation.

To this animated group was presented a plan for the study of the Northeast Atlantic and its adjacent seas drawn up by Pettersson and his friends and based in the main on the experience they had gained in organizing the multinational cruises of the 1890's. Predictably, their concern for physical oceanography was apparent, and the plan had at its center a program of hydrographic investigations aimed at understanding the circulation of the sea. Although most scientists agreed that circulation was of central importance to fisheries' problems, the Scandinavian approach did not exactly suit the tastes or conjectured needs of every government.

"You should make it clear," read the Foreign Office instructions to British delegates, "that the principal object which Her Majesty's Government has in view is to secure a careful inquiry into the effect of present methods of fishing in the North Sea . . . for determining whether protection against overfishing is needed. . . ."⸸

It was the squabble between the liners and the trawlermen that the Foreign Office wished to settle, and it was the biology and behavior of fish rather than the salinity of a water mass that appealed most to British scientists.

It is surprising, to say the least, that only three years were required to reconcile the scientific, political, and financial haggles and start the council on its illustrious career. Not that there was a great deal of quibbling among men all anxious to have the ocean more thoroughly explored, but there was such a diversity of training, philosophy, and goals among them. In the scientific ranks alone there were several clearly definable groups. There were hydrographers. Educated as chemists or physicists, these men were interested in the physical conditions of the sea, inclined toward the use of mathematics and theories, insistent on the need for basic research, and willing to combine their interests with a study of fish if that was where government support lay, and it did. Proposing to work with them in the council was an even larger number of biologists and zoologists. They were interested in the life of the sea, and their approach to it was largely, though not exclusively, descriptive and nonmathematical.

The methods preferred by each group differed too, and while hydrographers wanted their oceanographic stations set out across currents or drifts, the biologists wanted theirs laid out in a grid as if the ocean were an extension of fields and forests. It was as difficult to get hydrographers and biologists really working together as it was to combine their dissimilar observations.

It might be argued that there was nothing new in the council's proposal to bring these groups together. The U.S. Coast Survey had hired both hydrographers and zoologists, the two worked side by side in some marine laboratories, and many oceanographic expeditions, like the *Challenger*, had sent a chemist to sea among their naturalists. But in these instances one man or group was usually subordinate to the other. Buchanan on the *Challenger* was taken along to analyze sea water and sediment samples and to provide other bits of information helpful to the naturalists whose task of describing the contents of the sea was the major one. Conversely (and examples of this sort are rare outside Scandinavia), Cleve's plankton studies were not primarily biological in character but were undertaken to provide physical oceanographers with another way of identifying and following water masses.

In ICES the groups were to be more finely balanced. Neither hydrographers nor biologists could claim that the answer to fisheries' problems lay within their field, and neither was supposed to enlist the other as subordinate.

In spite of the differences, each of which influenced the character of ICES, a program including hydrographic, biological, statistical, and practical investigations was finally worked out to the satisfaction of eight nations, and in the summer of 1902 Germany, Denmark, Norway, Sweden, Finland, Britain, Holland, and Russia became charter members of the International Council for the Exploration of the Sea.

To implement such an ambitious program, the area to be studied was divided among participating nations, and each was given responsibility for investigating the regions closest to its shores. (This made a rather complicated pattern of areas for a country like Denmark, whose territorial waters circled the Faroe Islands, Iceland, and Greenland as well as the Jutland peninsula.) The very reason for starting ICES, of course, was to study the sea as an *undivided* system, and so, in spite of the partitions, made for convenience, the oceanographic cruises were carefully coordinated in order that the data gathered might be fit together to give a single picture of the entire Northeast Atlantic. The hydrographic portion of the council's program needed the strictest coordination, for it was important that temperature, salinity, and current measurements be made at the same time (at least within the same week) over the entire area. Quarterly cruises were launched each August, November, February, and May to collect hydrographic data and some biological information as well.

The other parts of ICES' program—studies of fish, statistics, and so forth—required less coordination and were pursued by each country at its own pace, although still in accordance with an overall plan. To complete the administrative picture, each participating nation was allowed two official delegates and a flexible number of "experts" or advisers, all of whom attended the annual council meetings. Committees were set up among these delegates and experts to cope with specific problems such as overfishing, eels, the biology of a particular sea, or fish tagging.*

When the newly elected delegates of ICES met officially for the first time in Copenhagen in 1902, the question that seemed the most important to answer was the same one that had been asked with weary repetition for the past three or four hundred years—why does the abundance of fish fluctuate? Why, for example, did dense and flashing shoals of herring swim into the Baltic Sea year after year and then, quite suddenly, fail to appear? The traditional answer, one that the fisherman admitted gave him no satisfaction, was that when fish failed to appear it was because they had chosen to swim someplace else instead. In other words, fishery fluctuations were caused by some kind of migrations. This was a logical, if misleadingly oversimplified assumption. It was logical because fishermen knew that fish did migrate, first to their spawning grounds, then back to their feeding grounds, and fishermen also believed, but here they were mistaken, that all fish traveled throughout the entire area where the species was found. Putting fact and fiction together, the belief emerged that there were huge stocks of fish which moved more or less in unison along a favored route that took them over a huge area of the sea. Along this route the fisheries had been established, but if for some reason the fish changed their route, then the old fisheries would fail, while new ones would gradually get started along the fishes' new path.

To use herring again as an example, there was a fishery in Scotland that began each June on the grounds near the Shetland Islands. From Stornoway and Castlebay, Stronsay, Wick and Lerwick, more than 1,000 herring luggers with red-brown sails worked over these grounds all summer. Then, in early autumn, they moved south and shot their

---

* In subsequent years other international organizations for the study of the sea were established, and most followed ICES' program both in form and content. Examples are the International Commission for the Scientific Exploration of the Mediterranean Sea, suggested in 1908 and begun after World War I, and the International Commission for the Exploration of the Adriatic Sea, formed in 1910.

nets off the coast of England, and in good years sailed farther still to catch winter herring in the English Channel. From this annual procession grew the belief that the Scottish fleet was following a veritable army of herring. These fish were believed to live in the Arctic Sea, and each spring they moved south, splitting into at least two groups (a division needed to explain the Norwegian herring fishery), and each group prescribed a fairly regular path before returning north. In the years when the herring fishery in Scotland or Norway failed, it was naturally assumed that the fish had swum to more favorable shores to spawn and feed. Similar explanations were put forward for other fisheries.

Of course fishermen and scientists realized that to give this explanation of fluctuations any meaning, it would be necessary to ascertain why fish migrate now to familiar grounds and now to some unknown bank or shore, but at this point in the reasoning both logic and agreement fell apart, and a multitude of "causes" were advanced, each of which, according to the frustrated director of Scotland's Fishery Board, "appears to be equally valueless." Like the fishermen, scientists had their "causes" for these migrations, and although few observers would be so disrespectful as to call them valueless, the explanations did involve an awesomely convoluted mixture of scattered facts and haphazard statistics.

ICES, then, inherited the suspicion, if not the outright assumption, that fishery fluctuations were caused by migrations, and that these in turn were influenced both by changes in the sea—temperature, salinity, winds, etc.—and by changes in the fish themselves—their age, feeding and breeding habits, and so forth. The first set of factors interested hydrographers, the second, biologists, and in spite of the council's efforts to sponsor "hydro-biological investigations," the two groups resisted hyphenation, and each proceeded to work out its own story without being greatly influenced by the other.

ICES' hydrographic program was centered around a schedule of quarterly cruises which were designed to gather the information necessary to plot the physical parameters of the sea throughout all seasons of the years. That these factors were important to the distribution and behavior of marine life had been demonstrated at least since the 1860's, when Thomson, Murray, and others had explored the Wyville Thomson Ridge and found that the curious difference in water temperature on the two sides of the barrier dictated the kinds of animals that could live on either slope. (The few animals that were found on both sides of the ridge lived deep on the warm-water side and fairly close to the surface on the cold side; in other words, in water of about the same temperature.) Temperature was also known to affect the hatching time of fish eggs and the growth of plankton on which fish fed. Currents seemed of still greater importance to the fisheries, for they moved the fish eggs, the bug-eyed larvae that hatched from them, and the plankton that the larvae ate. Currents also affected the location of water masses, and soon it seemed that they were responsible for the location of fish as well.

In the 1870's Gustav Ekman had explained the movement of fish as a result of their preference for water masses of certain temperatures and salinities, and in the 1890's Cleve had suggested that "the still obscure causes of the migration of fishes may be found to be intimately connected with the change of water containing different kinds of plankton."[*] Later, Otto Pettersson offered the most purely hydrographical and mechanical explanation of migrations—namely that fish did not actively swim from place to place but were passively sucked off the banks and into the fjords by the movement of huge internal waves. Could all these hypotheses be true? ICES set out to learn how the environment worked and how fish responded to its changes.

August, 1902, was chosen for the first of the quarterly cruises, and although the council was barely a month old, many of her members were already equipped with research vessels. The beautiful, white Norwegian steamer *Michael Sars* had come off the ways in 1900, Finnish scientists had modified a heavy ice-breaker, the Swedes had borrowed a ship from their navy, the Germans used the royal steamship *Poseidon*, the Danes the *Thor*, and the Russians the *Andrei Perwoswanny*. The plan for the first cruise—as for each of the quarterly cruises—was for a steamer from each participating nation to leave port on a specified day for a cruise over some portion of its territory. A line of stations was assigned to each vessel, and at each station routine collections of water and plankton were to be made. These samples were to be returned for analysis to the laboratories in each country which had been established for ICES' work and the results sent to the Central Bureau for integration, publication, and distribution.[*]

---

* In addition to being a data center for ICES, the Central Bureau soon took over the work assigned to the council's short-lived Central Laboratory. Martin Knudsen was director of the bureau's hydrographic service from 1902 until 1948, and from him came the well-known hydrographic tables used until very recently to determine salinity from chlorinity and density from temperature, salinity, and depth. Knudsen also made and distributed Standard Sea Water.

The research vessel *Michael Sars*, built in 1900, was used by the Norwegian Board of Sea Fisheries for the cooperative cruises sponsored by the International Council. (*Courtesy Institute of Marine Research, Bergen, Norway*)

From the beginning these quarterly cruises were strenuous, rushed affairs. The time was short, the work extensive, the weather often bad, and the boats small. During a March cruise, for example, the *Michael Sars* put out from Bergen for the Shetland Islands, ran 15 stations between Norway and Lerwick, then headed north. At 6:10 A.M. the ship hove to for station 16, by lunchtime all hands were back on deck for station 17, and just as supper ended they were roused again for the eighteenth and deepest station of the cruise. As usual, the ship's captain, Thor Iversen, headed the *Michael Sars* into the weather and went forward to measure the depth of the water with a Lucas sounding machine. Marks on the thin wire showed the water at station 18 to be 1,100 meters deep, and with that information Hjort and others manned the small winch and began lowering and raising insulated water bottles to depths of 20, 50, 75, 100, 150, 200, 300, 400, and 500 meters. As each bottle returned, the temperature of its water was measured either by reading a thermometer that had ridden within the bottle or by sticking one in through a special slit in the top.* Once the temperature was recorded and the water poured into a number of smaller bottles, the sampling bottle was cocked open for its next trip, reattached to the small winch cable, and sent down again. The man fastening and unfastening the bottles from the wire needed both hands for his work and was unable to follow the sailor's maxim—one hand for yourself and

---

* On the first of these quarterly cruises most ships were equipped with insulated water bottles, but by about 1910 most countries had changed to thermometers, which by that time had been made more reliable.

one hand for the ship. In rough weather he was put on a leash tied to a davit to avoid being washed overboard.

But on station 18, weather was not a problem, and while water samples were collected other men used the ship's large winch to make a vertical plankton haul from a depth of 400 meters. Because the water was so deep at this station (most of ICES' work was done in relatively shallow coastal waters), there were more than the usual number of samples to be collected, and the routine work was carried on beneath the ship's swaying deck lights well into the night. When the station work was completed, fry trawls, for catching young fish, were fastened to the port and starboard wires, egg nets were attached farther up the same wires, and a surface tow net was hung from the ship's boom. With all this gear streaming out behind, the *Michael Sars* left station 18 and steamed through the early-morning darkness until at dawn her navigator announced that station 19 lay just ahead. The sound of the ship's engine throttling down was enough to roll most men from their warm bunks. The deck lights came on, the overcast sky paled slightly, and the process of sampling began all over again.

On some ICES cruises, when the weather permitted and when the steamers rode in fairly shallow water, current meters of various designs were let over the side or sent down from a dory to measure the surface and subsurface flow of water. In poor weather, especially on the February cruises when the sea was almost invariably rough and the days so short that almost all the sampling was done in darkness, few current measurements were made, and even the stations themselves were occupied on a less rigorous schedule.

"About midday we had to look sharp and get out again [from the ice near Jan Mayen]," wrote Captain Iversen from the *Michael Sars* one February, "as the wind increased to a gale, accompanied by severe frost and remarkable shrouds of mist, which assumed the most fantastic shapes and were constantly in motion. I have never seen anything like them before. We shaped our course for Vesteraalen, and got sail on her to steady her a bit. [The ship carried small stay sails for stability.] The whole of the afternoon we were pretty well cased with ice—hull, spars and standing rigging—and on running suddenly into the middle of an ice-floe about nine o'clock that evening we had a hard job to get the ship round against the wind, her sails being so stiff with ice that it was impossible to take them in. However, we managed gradually to get her bows up against a large cake of ice and brought her round with the help of her engines. . . .

"The stations we took up during the severe frost were the reverse of easy, as the metre-wheels froze up, and we had to keep them warm with thick, red-hot iron bars that were brought from the engine-room and held close to the wheel-axles.

"On the night of the 17th we ran into another storm, which lasted till we arrived in port."✻

The information won from the sea under such difficult conditions was published first in lengthy tables and then, in later years, in atlases which showed the seasonal salinities, temperatures, etc.; for the entire North Sea area. Scientists were still looking for the perfect (almost magic) relationship between the abundance of fish and some physical factor which would enable them to predict from one year's hydrographic information the success of the next year's fishery. Since it was not known what this factor might be—or if a way would ever be found to shortcut the awesome task of considering everything that affected the life and behavior of fish—every kind of data that could be collected, was, and every one of these measurements was recorded and filed away. This made for considerable bulk and awkwardness, but it wasn't a totally unmanageable system, and in the council's early years, as scientists cast among the effects of man and nature for an explanation of fisheries' fluctuation, the data from ICES' voluminous data lists were pored over, recombined, and astutely pushed about like so many pieces of a fiendishly complex puzzle.

While hydrographers within the council pursued elusive water masses, other ICES scientists spent their time with fish, and one of the first tasks they set for themselves was marking. Fish marking or tagging, if done well on a large sample of fish, could yield extremely valuable results, for not only could recaptured fish help indicate a pattern of migration and a rate of growth, but they could also be used to test certain hypotheses concerning the age determination of fish and the division of fish into separate stocks. C. G. Johannes Petersen of Denmark was among the first to mark large numbers of fish, and the method he introduced to the International Council was adopted and used extensively in many countries.

Petersen was a zoologist, a fine-looking man who began his career as a fisheries scientist even as he was completing his doctorate, and continued his work until he died in 1928 at the age of sixty-nine. In 1889 Petersen was appointed director of Denmark's small but mobile bio-

The Danish zoologist C. G. Johannes Petersen successfully implemented many of the first tagging experiments on plaice. He designed the Petersen disks, which are still used to tag many species of flat fish. (*From ICES, Journal du Conseil, Vol. 3*)

logical station. "The Station," as it was commonly called, was a beamy wooden vessel which could be towed from place to place through the country's numerous sounds and waterways. Petersen carried out all manner of investigations from the station. He hung plankton nets over the stern of the vessel and relied on sluggish currents to drift a few plankton into their meshes, he dove on oyster beds to count their numbers from season to season, and, with an eye for utility—a characteristic he shared with Johan Hjort of Norway—he harvested quantities of wild, stringy mussels, which, although too tough for humans, were enthusiastically gobbled up by the chickens that lived on the bank by the station. In the station's official report, Petersen wrote that in spite of a powerful fishy odor from the dying mussels, "fowls eat them eagerly, and produce good eggs, with no disagreeable taste. . . ."♥

In the 1890's Petersen turned his attention to the study of a flounder-like fish called plaice, and began marking hundreds of these fish, using a method he'd first tried in 1887. He chose plaice for his experiments both because of their importance to Danish fisheries—their catch being more valuable than either herring or cod—and because plaice were hardy fish, easy to keep alive, and apparently indifferent to the pair of bone buttons which Petersen fastened like cufflinks (joined with a silver wire), to either side of their backs. (Black plastic disks have

A flat fish marked with Petersen disks lies on a scale which is used to measure its length. At right are the disks or buttons with the silver wire that connects them. Pliers are used to attach the disks to the fish. (*Courtesy National Marine Fisheries Service*)

replaced Petersen's bone buttons, but the method of marking flatfish remains essentially the same.)

To tag plaice, Petersen borrowed fishing boats and went out into the sounds and bays along the Danish coasts, where he made short, half-hour hauls with a trawl. The fish thus caught were hauled on board and dumped into tubs of water, and from there transferred to rows of tanks fitted with stacks of zinc trays, each large enough to hold a single fish. Fresh sea water came sloshing through these tanks from the deck wash, while Petersen and his helpers, in thigh boots and oilskin aprons, stood for hour upon hour stringing buttons on each healthy plaice, measuring the fish, and recording its vital statistics in spattered note-books. When the weather was calm, the marked plaice were returned to a large holding tank for observation—the object being to return only healthy, active fish to the sea—but when the boat, pitching and yawing, slopped fish and water from the tanks onto the deck, over boot-tops and up sleeves, then the plaice, once marked, were "tipped into the sea with diddles" or poured, in a gentler manner, down wooden chutes.

Petersen marked several hundred plaice each year in this fashion, and fishermen, working in the shallow Danish waters, caught and returned at least 30 to 40 percent of each lot within 14 months. (Fishing for plaice was so intense that in one experiment 80 percent of the marked fish were returned.) Most of the plaice returned to Petersen were caught only a few miles from where they'd been marked, while those marked and recaptured in the North Sea by Petersen's opposite number in England, William Garstang, seemed to travel farther, probably making inshore-offshore migrations of 40 or 50 miles.*

These marking experiments were of practical as well as scientific import. In the first category, the relatively conservative wanderings of the plaice suggested to both Petersen and Garstang the possibility of improving the growth rate of this important fish by transplanting young plaice from overcrowded grounds to other suitable but underpopulated waters. A plan for moving plaice to the Dogger Bank (a shoal the size of Wales situated in the North Sea halfway between Britain and Denmark) was put forward in 1894 by a Grimsby trawler captain, and the following year a less ambitious scheme for the transportation of young fish to Danish sounds was advocated by Petersen. Funds for the projects were not raised immediately, but by the spring of 1904, hundreds of small plaice were caught in the Baltic Sea, dumped aboard shallow, flooded barges, and towed through canals for release in Denmark's inland waters. Thousands of other fish from the same overcrowded grounds were transported in holding tanks to Dogger Bank. As Petersen and Garstang had predicted, the transplanted plaice grew abnormally fast, and the fall fishery on the new plaice grounds brought in ten times the money expended on the project—and almost all the plaice as well. Like fish hatching and stocking, however, transplanting didn't work well in the open sea, where vessels from any nation could fish out the new recruits. Annual transports of plaice were made in Danish waters, but the experiment was not continued on Dogger Bank.

In addition to suggesting plans for transplanting, the marking of plaice also indicated that fish from different areas did not mix freely with each other. Plaice tagged in the Baltic stayed there, and English plaice rarely if ever crossed the North Sea to take up residence on the other side. Furthermore, each of these ill-defined groups of fish seemed to have slightly different characteristics. This observation was not new. Generations of fishermen had known that fish from one ground looked a little different from others of the same species caught somewhere else, but they didn't know why. Fish changed in appearance

---

* Still more extensive migrations were discovered by Johannes Schmidt, a Dane better known for his work on the migrations of eels than for his marking of plaice. He tagged plaice in Iceland and found them to travel hundreds of miles annually. Similar observations led to the somewhat unwelcome realization that the same species of fish was likely to have a number of very different life styles, depending on its environment.

according to the seasons, and fishermen had not differentiated these seasonal differences from what might be termed geographical differences. Consequently, they had not concerned themselves with the possibility that separate races of fish might exist, each inhabiting a more or less restricted region and each restricting its migrations to this region.

In Germany Dr. Friedrich Heincke had been working on just this problem of races. Adopting methods used by anthropologists who were then measuring men's skulls to sort them into round-headed or long-headed races, Heincke used his calipers on herring. He measured their length and their thickness, he took down the dimensions of their heads, he studied their scales, and he laboriously counted their vertebrae and fin rays. He found no single measurement to distinguish Dogger Bank herring from the Norse or Scottish varieties, but he did find that the relationship between several measurements was so characteristic that he could use a group of them to divide herring into several races. There were, for example, Icelandic herring which had remarkably large eyes, a short skull, and a fairly long tail, and then there were Norwegian herring, which had small eyes and a short tail. These races, he claimed, did not interbreed, but lived in more or less separate regions of the sea—quite a contradiction to the still-popular belief that all herring lived in the Arctic and swam now to Norway, now to Scotland, and so on.

Heincke knew that he needed to mark herring to prove his hypothesis, but marking was extremely difficult, for herring were not nearly so hardy as plaice, nor could Petersen's bone buttons be adapted to fit the fish's round shape. Consequently, most early attempts to mark herring failed, and Heincke, while still sure that his measurements were sufficient to show the existence of races, could not estimate with any certainty how these stocks moved. (It wasn't until after World War II that a herring marker was designed which neither hurt the fish nor pulled off. When the tubular plastic tags had been returned from thousands of fish, the results of the tagging program confirmed Heincke's belief in the existence of races but revealed a more complex relationship among stocks than he had anticipated.)

Heincke's study of herring stocks and Petersen's marking of plaice both suggested that there were many groups of fish in the Northeast Atlantic which confined their travels to relatively small areas, and one consequence of their findings was that it became increasingly difficult for scientists to use extensive migrations as an explanation of fisheries' fluctuations. This left them with overfishing as a cause, or forced them to consider the possibility of some natural but still unsuspected factor which regulated the abundance of fish. The question—why do fisheries fluctuate?—was the same, but the answer was suspected by some to lie outside the realm of migrations.

One key piece of the puzzle remained to be discovered before an explanation of fluctuations could be put together. A method was needed to determine the ages of fish, and scientists began to examine several possibilities, not so much with the problem of fluctuations on their minds as with the hope that determining ages would allow them to reckon (in a rather circuitous manner) whether or not a stock was being overfished. The reasoning went like this. Dating fish, as the determination of their age was termed, would enable scientists to compute the age composition of any catch (the age composition being an expression of the proportions of younger, middle-aged, and older fish in any catch), and a continuous record of age compositions would enable them to spot overfishing. If, as was expected, the age composition of catches was found to remain about the same from year to year, this would be taken to mean that a fishery was in balance with the stock's ability to repropagate. If, on the other hand, the record of age compositions showed that each year the catch included a diminishing percentage of older fish, it would be understood that the fishing was so intense that almost all the fish were being caught before they had a chance to become perhaps five, then four, then three years old. This was overfishing, and if the trend were not reversed, the fishery would soon wipe out the mature spawning fish and the stock would decline drastically. Of course, no one could prove that a fishery was hauling in too many members of the older generation until a way could be devised of accurately dating fish. So, motivated by the need to recognize instances of overfishing, men tried various ways of telling a fish's age.

By the 1890's Petersen had already hit upon a tedious method of determining ages, but one that was applicable only to immature fish. He had plotted the length of blennies (a small, oddly shaped fish commonly found along Europe's rocky shores) and found that their measurements fell within three ranges, which he believed represented three ages. The smallest, he said, were the fry of the year, and so were less than a year old. The intermediate fish were the fry of the previous year, more than one but less than two years old, and the largest fish were mature individuals over two years old. Among these full-grown fish, whose size did not change much as they grew older, Petersen's method made no distinction. A more comprehensive method was needed.

In about the same years, Heincke too had been casting about for a way to determine the age of his herring. Initially he had approached the problem as Petersen had done, expecting to find that the age of a fish could be determined by the relationship of its thickness to its length or by some more complicated set of physical proportions, but by the end of the century he had turned his attention from shapes to scales, and thereby rediscovered the method of scale dating. This method, originally used to determine the ages of carp, involved counting the rings or bands on a fish's scale, markings which result from the fish's uneven rate of growth. Heincke knew that in summer fish grew more quickly, and he found that during this time each of their scales increased by a wide, translucent margin. Then, in winter, they grew slowly, and each scale increased only by a narrow, opaque band. Heincke found similar markings on the vertebrae and otoliths (ear bones) of some fish, and these were particularly helpful in the case of plaice and a few other fish whose scales failed to show the alternating bands.

Dating fish by growth rings had its complications and limitations, but in spite of these the method was adapted for use on most commercially important fish, and confidence in the procedure grew as scale-dated and marked fish were recaptured, reexamined, and found to have grown the proper number of additional rings. By 1904 the method was recognized by the International Council as a possible aid to research, though there were still many scientists who felt the method to be unreliable. For those who did choose to count growth rings, however, a new degree of precision in the computation of age composition was possible. In Norway young Johan Hjort decisively embarked upon such investigations, and his results were revolutionary.

Johan Hjort was born in 1869 and grew up in Oslo (then Christiania), where his father was a professor of ophthalmology. After briefly attempting to study medicine, the younger Hjort changed to zoology, which he studied both in Munich and at the famous Zoological Station in Naples. In the early 1890's he returned to Oslo with his Ph.D. to succeed Georg Ossian Sars as lecturer in zoology at the university and, like his predecessor, to turn his attention to fisheries' problems. His interests were broad and ecological, and his enthusiasm for plankton, fish, bottom animals, currents, sea ice, temperatures, and salinities soon put him in contact with Pettersson, Ekman, and Cleve in Sweden, Petersen in Denmark, and many others. In 1897 Hjort was appointed director of the biological station at Drobak and three years later moved to a position of still greater influence as director of the Board of Sea Fisheries in Bergen. Here he remained for 16 years, and during this time accomplished his most important fisheries work. Hjort was a member of ICES from 1902 until his death in 1948, and within that group was almost as well known for his volcanic temper as for his fine work.

Like many of his friends, Hjort was interested in solving the stubborn problem of fisheries' fluctuations, but unlike his peers he attacked the problem both on its most general and particular levels and, what was rarer still, on its biological, hydrographic, and statistical fronts. Predictably, Hjort's struggle with fluctuations was a long and complicated one.

Hjort's first task was to convince himself that fluctuations were a real problem and that greater and smaller catches did not merely reflect man's variable fishing effort. Digging back into Norway's fishing statistics, he compared the number of fishermen working the northern cod grounds with the number of fish caught in the years 1868 to 1900. He found no connection between the ups and downs of the two groups, and satisfied that fluctuations had some natural cause, went on to compare the location of good fishing grounds (which changed from year to year) to annual ice conditions. Here he found a better correla-

Johan Hjort, director of the Board of Fisheries in Norway, originated the concept of the "year class" among fish. (*Courtesy International Council for the Exploration of the Sea*)

tion, which, he said, led him to seek the cause of the fluctuations in the sea itself. If, in the next several years, the cod had cooperated by appearing along the northern coasts of Norway in something approaching their usual numbers, Hjort might have been free to concentrate his energies on studies of ice, weather, and other hydrographic factors which seemed to many the most likely cause of fisheries' fluctuations. But the cod did not appear, and the first years of the new century were lean ones for the cod fishermen. Naturally the government was concerned, and Hjort and the Sea Fisheries Board were directed to study cod, not currents. This they did with all the means at hand.

One of their aims was to determine the age composition of previous years' catches to see if overfishing were a factor in the cod's decline. Using information laboriously extracted from loosely worded market statistics—the week's catch was "mainly Mark 2 fish" or "Merchant's fish" (fish of a certain weight)—Hjort and his assistants tried to construct the age compositions of cod and herring catches.* Their results were based on the weights of fish, for that was how fishery statistics were kept, even though Hjort knew that weight was not a reliable indication of age. His erroneous age compositions fit well with current expectations. Each year's catch seemed composed mostly of middle-aged fish. A small percentage of younger and older fish were caught too. This was the familiar pattern of age composition that had long been known to characterize animal populations and human ones too, and it was assumed that such a pattern remained constant or changed only gradually over the years.

Then, at the 1904 meeting of ICES in Amsterdam, Heincke presented the results of his scale-dating studies, and Hjort was immediately struck with the advantages to be gained in using this method to determine the age composition of his cod and herring catches. After the meeting Hjort visited Heincke at his laboratory on the island of Heligoland to learn all he could of rings and bands, scales and bones, and upon his return to Norway, mobilized the considerable talent in his office to look for the best way of dating cod, herring, haddock, sprat, and other fish.

Within a year Hjort had decided that scale dating, rather than the use of otoliths or vertebrae, was best suited to his purposes, and he and his colleagues began tweezing a scale or two from thousands upon thousands of fish. Each of these bony petals was examined microscopically, and Hjort counted rings and bands until his head ached and his eyes crossed. The first surprises were the longevity of many fish and the broad range of ages in a single catch. Instead of finding that most of the catch were three-, four-, and five-year-old fish, Hjort sometimes found fish from two to twenty years old in a single haul. But the real surprises were that the age compositions of catches were highly irregular and that they varied from year to year. Many hauls, for instance, had a small percentage of three- and four-year-olds, a mass of five-year-olds, almost no six's or seven's, and then a considerable number of eight- and nine-year-olds. Other hauls of the same fish, taken a hundred miles up the coast, were likely to show the same irregular pattern. But the following year the pattern for both places had changed.

Hjort began to suspect that the population dynamics of fish were very different from those of other animals. The explanation of fisheries' fluctuations, he believed, might come with an understanding of how and why the age composition of fish stocks was so unstable.

"It seems at first sight a bold suggestion to propose studying the fish supply on lines like these . . .," said Hjort to his fellow council members at the ICES meeting in 1907. "And yet," he continued, in his lilting and curiously composed English, "it appears to me a project big with possibility, to regard the discoveries of fisheries research from a similar standpoint to what has been adopted in the science of vital statistics."✻

The members of ICES, suitably surprised and impressed, set up a commission to study the vital statistics of fish, and this allowed Hjort to make use of several thousand more age determinations. The council's interest waned rapidly, however, and in 1909 the commission delivered its report, dropped the study, and left an annoyed and disappointed Hjort to continue the work. This he did, and early in the spring of 1914 he presented his *chef d'oeuvre*, the "Fluctuations in the Great Fisheries of Northern Europe." In this report Hjort fully explained his new concept of fish populations and presented it as the explanation of fisheries' fluctuations.

Of all the fish that Hjort had studied, the spring herring afforded the best example of a fish population's precarious arrangement. In 1907 Hjort found that the catch of spring herring (these are spawning fish, as distinguished from immature ones or "spent," recently spawned, ones) consisted of a small number of precocious three-year-olds and an unremarkable range of four-, five-, and six-year-olds, plus a few older ones. The following year, scale dating showed that more than 34 per-

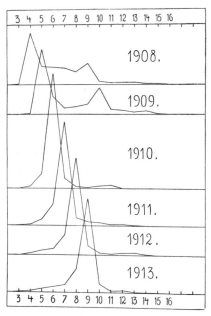

Hjort's diagram shows the age composition of spring herring catches. A tremendous number of herring were spawned in 1904, and members of this year class began to dominate the catch in 1908 as four-year-olds. The same year class made up the bulk of the catch in 1909 as five-year-olds, in 1910 as six-year-olds, and so forth. (*From the* Rapports et Proces-Verbaux, *Vol. 20*)

cent of the year's catch consisted of four-year-olds, that is, fish born in 1904. In 1909, the same brood of fish, which were now five years old, made up 43 percent of the catch, and in 1910 the same brood, then six-year-olds, made up 77 percent of the total catch. Hjort called the fish born in a single year a "year class" and clearly the year class of 1904 had many, many more members than did the broods of preceding or following years. Hjort followed the 1904 year class and found that in 1911 it still contributed more than 70 percent of the year's catch, and in the following years, as eight- and nine-year-old herring, the 1904 year class still furnished more than 60 percent of the catch. (Later studies followed the 1904 year class into the early 1920's.)

Hjort believed that the success or failure of these year classes determined the abundance of fish, or, stated in reverse, the fluctuations of fisheries were caused not by trawlers, storms, or sea ice, but by the varying success of year classes. But what caused good and bad year classes? What allowed so many fish to survive one season, then killed off the young of the next several years? Hjort was not sure, but he suspected that the very early stages of a fish's life were the critical ones, and these he set out to study.

In attempting to explain the occurrence of strong and weak year classes, Hjort first considered the possibility that a successful brood was the result of an abnormally large spawning.* He compared the number of barrels of cod roe salted each year in the Lofoten fishery with the strength of each year class but found no connection. Roe had not been particularly plentiful in 1904 but had been abundant in years when the year classes were weak. Hjort abandoned spawning as the cause of good and bad years, and after checking through several other possibilities, such as storms, that might affect the viable number of fish eggs, he turned to a consideration of fish larvae. Perhaps it was their rate of survival that determined the strength of a year class.

Fish larvae, he knew, hatched from their translucent eggs after several weeks of careless incubation, and began life looking like pieces of carpet thread a quarter-inch long with two round, black eyes at one end and a balloon of a yolk sac in the middle. The sacs provided the larvae with at least part of the food they needed for their first week or 10 days of life. Plankton constituted the rest of their nourishment, and plankton, thought Hjort, must be most important to their survival. He began to think that if plankton were available when the larvae hatched, the latter would feed, grow strong, and survive in great numbers. If plankton were not available, a tremendous proportion of the larvae would die of starvation in the first week or two of their precarious lives.

Hjort made a special trip to the Lofoten fishing grounds early in the spring of 1913 to see if he could observe the effects of a good or poor crop of plankton on a brood of larvae. In February he selected a number of stations on the spawning grounds, and at each of these he

---

* Nansen and Helland-Hansen had tried to show that the temperature of the Atlantic Current determined the quantities of roe, and liver, taken from Lofoten cod, and had found that a high average temperature of the water corresponded to a low percentage of roe and liver taken by the Lofoten fisheries. When Hjort first introduced the concept of year classes, and suggested that fluctuations in the fishery might be affected by the quantity of roe, it was natural for scientists to try to connect the two lines of reasoning. This would allow them to hypothesize that a year class was determined by the amount of spawn, which depended on the temperature of the water, which was controlled by the variations of currents, which some believed were governed by sunspots. Hjort did not think that so pat and so profitless a relationship could exist between such "complicated biological . . . phenomena . . ." and "the simplest physical causes."⚥

cast his fine silk nets week after week. At first he caught almost nothing but fish eggs, and he knew that if the larvae hatched under such conditions they would starve. But then, "later on in the spring . . . enormous quantities of microscopical plant organisms suddenly [made] their appearance . . . in the form of a thick, slimy, odoriferous layer on the silk of the net. . . ."⸎ Here was the food for the larvae, but, reported Hjort somewhat obliquely, anyone who had studied the earliest stages of the fish would know how difficult it frequently was to find the young larvae in the same water where the eggs, a short time before, had been abundant. Hjort had seen the eggs and the food, but not the larvae.* Still, the connection among the three seemed clear enough. If the eggs hatched when the plankton bloomed, a wriggling horde of well-fed larvae would develop, and three or four years later fishermen would begin to catch mature fish from a strong year class. Later, Hjort added a second factor to his suspected cause of good year classes. This was the location of the larvae, for if they were drifted over deep water, he supposed they would run out of plankton before gaining sufficient strength to swim to the bottom and start feeding there. In shallower water, the small fish could more readily augment their planktonic food supply. Both of Hjort's causes were, and still are, almost impossible to test at sea, and both continue to be respectfully regarded as probabilities or good guesses. Scientists are still trying to answer the questions he raised—which of the early stages of a fish's life is most critical, and which physical factors affect it most?

Hjort's concept of year classes was received with enthusiasm and acclaim. In 1913 he received the Alexander Agassiz Medal for distinguished work in oceanography (an award established by Sir John Murray in 1911), and in the same year the International Council reactivated its commission on vital statistics and considered the problem anew. By 1914, when Hjort officially offered year classes as the explanation for "the fluctuations in the great fisheries of northern Europe," scientists were already paying him the compliment of adopting his methods. Utilizing the voluminous statistics gathered and

---

* Forty years later, scientists from the Fishery Laboratory of Lowestoft, England, found two patches of plaice larvae, one of which was naturally supplied with suitable planktonic food, the other not. The larvae in both groups were followed for 12 days, and it was observed that none of the larvae in the poorly fed group survived the yolk-sac stage, while the larvae in the other group appeared strong and active.

preserved by the International Council, results from new studies were quick to appear, and year classes were soon found among sprat, haddock, plaice, and other fish.*

It might appear that Hjort's explanation of a successful year class—the coincidence of larvae with a rich crop of plankton—did little more than throw the question back a step to "what causes fluctuations in plankton?" It did more than that, and one reason that it did was that while Hjort had been working on fish, Victor Hensen and the Kiel School of Planktonologists had been working on plankton. Hjort's hypothesis that fisheries' fluctuations were intimately related to plankton brought two exceptionally productive areas of research together.

It is one of the best-known facts in the history of oceanography that in 1887 Victor Hensen coined the term "plankton," which he defined as meaning all the animals and plants in the sea which are passively drifted about. Almost all classes of marine life are represented in this broad category, for many nonplanktonic animals such as fish, crabs, barnacles, and oysters begin their lives as eggs and larvae which are part of the plankton. A feathery mat of sargasso weed is plankton, and so is a huge, undulating jellyfish, but most members of Hensen's community are microscopic, and as if designed to make their study as difficult as possible, many are transparent as well. Being all but invisible, it is not surprising that plankton was not studied until well into the eighteenth century, by which time most other communities of plants and animals had been quite thoroughly described.

It is impossible to say how plankton was first discovered, but it is certain that long before scientists began straining the sea for plankton, sailors had seen the oceans sparkle luminously at night and, curious as to the cause, had scooped up buckets of silvery water. In it they discerned tiny, flashing specks which some believed were scattered bits of lightning fallen into the sea during thunderstorms, but which others insisted were the smallest of animals. Somehow these minute specks passed from sailor to scientist and from a bucket to the stage of a microscope. The transition had been made by 1761, and in that year Peter Forskal, a Dane, was using a fine dip net to capture what he could from the surface waters of the Mediterranean. Ten or more

---

* The Canadians were so impressed with Hjort's work that they invited him to study their herring populations. Hjort accepted, and found to his surprise that Canadian herring had the same strong and weak year classes as Norwegian fish.

years later his countryman Otto Müller rowed out on the Oslo Fjord with a net and dredge to "make acquaintance with the invisible inhabitants of the world,"* and soon was introducing his new friends to a growing circle of fascinated humans.

In the next century, interest in plankton, like interest in the rest of marine life, grew tremendously. New kinds of organisms were discovered,* and familiar ones rediscovered in new places. By the time the *Challenger* expedition returned in 1876 it was recognized that plankton existed in all oceans, and in the next decades the use of closing nets suggested that plankton existed at all levels of the oceans as well. As scientists labored to identify the myriad forms taken from the sea, they became increasingly impressed with the abundance of plant plankton, which they termed "phytoplankton." Gradually the idea developed among them that this phytoplankton, small as each individual plant might be, was so abundant that it might form the same basis for life in the sea as land plants formed for animal life on the earth.

An early proponent of this view was Victor Hensen. Professor Hensen had come to the University of Kiel on the Baltic Sea in the 1860's to teach physiology, and his interest in the workings of the body had led, in some inexplicable manner, to the study of plankton. The German Minister of Agriculture had also come to Kiel at about the same time, not for physiology, but to establish a Commission for the Scientific Study of the German Seas. Hensen and the Commission soon found each other and together formed a nucleus for a group of zoologists, chemists, and botanists, all of whom were involved with plankton and most of whom were directly inspired by Hensen's controversial methods and results.

When Hensen began to study plankton, the common approach to marine biology was descriptive. Plants and animals were described, their habitats were described, and described too were their food, their predators, and, occasionally, their relative numbers. (This last usually meant saying that this organism thrives in spring and dies off in the winter or that there are more of them here than there.) The descriptive approach, valuable as it was, gave no information as to the actual numbers of plants and animals existing in the sea, and these numbers could be very useful. For one thing, they could be used to estimate the

---

* An unsuspected community of extremely small organisms was found in the intestinal canals of salpas and other marine creatures whose filtering devices are more effective than the finest nets.

Victor Hensen, the German physiologist and planktonologist, introduced the use of quantitative methods for the studying of plankton in the 1880's. Hensen is credited with coining the term "plankton" in one of his early papers. (*Courtesy Institut für Meereskunde an der Universität Keil*)

productivity of the sea, just as a census of peas or potatoes had long been used to indicate the productivity of a field. Counting a crop in a peaceful autumn field, however, was hardly comparable to harvesting a boundless area of heaving ocean, and it was generally understood that quantitative studies could not be pursued at sea.

Hensen did not agree, and consequently he began to devise methods for determining the total quantity of living matter contained in the oceans. To do this, he said, he must first ascertain the quantities of "primitive food for marine animals," by which he meant plant plankton.✹

The vast communities of single-celled plants that interested Hensen grew in the top 200 meters of the sea—the sunlit zone—and to sample them in an orderly fashion he constructed a large hoop net which could be hauled vertically through this zone. The mouth of the fine silk net filtered exactly a square meter of water, which enabled Hensen to express each haul in numbers of organisms per square meter of sea surface. When the unwary prisoners of a single haul were multiplied by the number of square meters in the area under scrutiny, Hensen believed that he had an accurate estimate of the plankton population of that region.

To test his new methods, Hensen organized two expeditions aboard the *Holsatia*, the first, in 1885, to the North Sea, and the second, two years later, through the southern Baltic. On both voyages the big hoop net was repeatedly lowered to a depth of 200 meters, then drawn slowly to the surface. Each time it cleared the water it was left dangling at the rail while sailors sluiced its sides to wash every minute speck into the collecting jar at the apex of the net. The gear was then hauled in, the jar disconnected, and the plankton allowed to settle until its volume could be roughly estimated by sight. Then came the tedious job of counting. Obviously there would never be enough time to count all of the millions of single-celled organisms that came up in almost every haul, so Hensen adapted a sampling technique used by his colleagues to count red blood cells. The plankton in the collecting jar were thoroughly stirred, and a small subsample, hopefully representative of the whole catch, was withdrawn and spread on a special microscope slide which had a fine grid engraved upon it. Then Hensen or his shipmates bent to the microscope and carefully counted and identified all of the organisms in each little square. The counts were totaled, then properly multiplied, first to calculate the total number of organisms in the haul, and then to estimate the astronomical populations of a large area of the sea.*

On the first voyage aboard the *Holsatia* Hensen was primarily interested in perfecting his techniques. He lowered nets to 200 meters, then at the same station to 2,000 meters, compared the hauls, and concluded that there were so few plants below 200 meters that he needn't sample below that depth. On the second cruise he was ready to measure productivity, and from the incredible numbers of plants captured that September in the Baltic he estimated that that sea was as productive as an equal area of prairie land. (Later, plankton hauls made in Kiel bay were analyzed and their proportions of proteins, fats, and carbohydrates compared to those of "ordinary meadow hay, good meadow hay, rye, peas, and potatoes."⁴ The mixed hauls of plant and animal plankton closely resembled peas, but in composition only, not taste. Members of the Kiel School did not comment on this discrepancy, but the British, who seemed particularly fond of eating what they studied, tried plankton in tea sandwiches. A zoologist, Sir William Herdman, served plankton sandwiches to members of the British Association for the Advancement of Science, who, after a few moments of polite

---

* Typical numbers of organisms, taken from the North Sea and calculated to exist in a column of water one meter square by 40 meters deep, are as follows: Diatoms, 3,091,280; Protozoa, 372,000; Coelenterates, 80; Planariae, 160; Actinotrocha, 240; Polychaete larvae, 400; Sagitta, 3,040; Copepods, 198,000; Pteropods, 36,000; Lamellibranch larvae, 32,000; Gastropod larvae, 80; and Oikopleura, 2,400.⁴

reserve, pronounced the dainties to be pasty, bland, and only slightly shrimpy.)

Hensen's plankton hauls involved weightier problems than flavor.

"I am convinced," wrote Ernst Haeckel, who had recently finished a ten-year study of radiolarians for the *Challenger* Report, "that the whole method employed by Hensen for determining the plankton is utterly worthless."[Ψ]

Haeckel objected to all the assumptions on which Hensen's method was based, and consequently was skeptical of all his results. He knew, as did Hensen, that the success of a quantitative plankton sampling technique depended first on the net being able to trap *all* the plankton from the water it filtered, second, on the subsample truly reflecting the total haul, and finally on the haul itself representing the plankton content of surrounding waters. Hensen's method was vulnerable on each count, especially the first and last. Hensen's net could trap neither the smallest forms of plankton nor the even more minute bacteria, and the latter, being newly discovered, were suspected of being enormously important. Even if these two categories of minuscule organisms were dismissed as insignificant, as Hensen thought they could be, the net missed some of the larger, more agile animals and often refused to filter anything at all when its meshes became clogged with "slimy little algae."

Hensen's assumption that plankton was distributed so uniformly in the sea that a single haul could be used to indicate the conditions that prevailed over several square miles near shore and over several hundred square miles in the open sea was increasingly challenged. Although it was not until the twentieth century that sustained investigations began to show just how changeable and patchy plankton distribution could be, scientists of the 1880's already had a feeling that plankton must at times be blown into dense windrows or again dispersed by winds and currents.

In spite of these problems, which were never really solved, Hensen and five others set off aboard the *National* in 1889 on what later became known as the "Plankton Expedition." With unprecedented support raised by the German government, the emperor, and the Academy of Science, the scientists from Kiel equipped themselves with the latest models of Hensen's vertical nets, plus double nets, wagon nets, and other unwieldy paraphernalia. The *National* steamed out of Kiel Bay on the morning of July 15, and, with frequent stops

The steamer *National* was used for the Plankton Expedition of 1889. (*From the* Reisebeschreibung der Plankton-Expedition, *Vol. 1A*)

for the collection of plankton and, for the first time, of bacteria as well, the ship crossed the North Atlantic to the tip of Greenland, proceeded southwest to Bermuda, and from there inscribed a large "X" over the tropical Atlantic, then returned to Kiel in November. The expedition had traveled 15,000 miles.

When the samples collected on the *National* had been counted, Hensen announced that, contrary to expectations, the cold and temperate seas of the North Atlantic supported a much greater mass of planktonic life than did warm, tropical waters. Previously it had been assumed that the reverse was true and that the sea, like the land, produced its most luxuriant growth in the tropics. The assumption was logical, for any biologist knew that the warmer the water the faster its inhabitants' rate of growth, and travelers of any profession knew that the color and profusion of life on a tropical reef surpassed the drab aspect of northern sea wrack and rock fish on European shores. It had been easy to confuse brilliance and variety with abundance.*

---

* When scientists began to suspect that the dreary northern waters might support more life than clear, blue tropical seas, quite a collection of bizarre experiments were tried. One of these involved putting a shark's head in a basket and lowering it to 75 fathoms. Two hours later it was hauled in amid clouds of tiny shrimplike amphipods—relatives of sand fleas and beach hoppers. Six quarts of these amphipods were collected from the shark's head. (Lowering skeletons for natural cleaning was a common practice among marine biologists.)

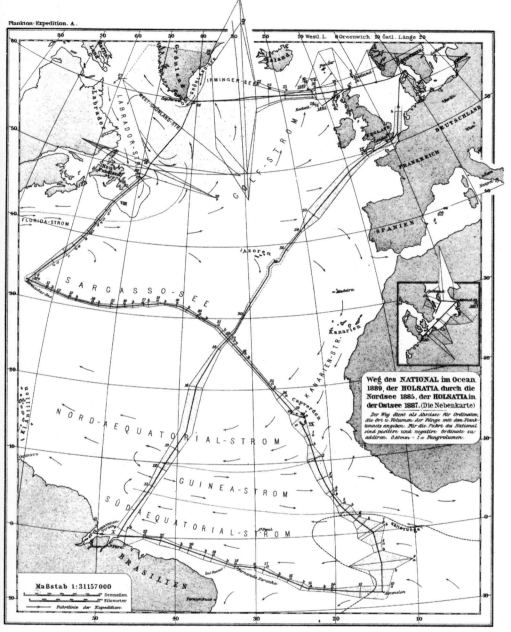

On this track of the Plankton Expedition the width of the band represents the amount of plankton caught. By far the largest hauls were made in the cold waters east of Greenland, but relatively little was found in the tropics. The inset shows the track and catch of the *Holsatia*'s trial cruise in 1887. (*From the* Reisebeschreibung der Plankton-Expedition, *Vol. 1A*)

Hensen believed that this confusion stemmed from scientists' pre-
occupation with coastal waters, which, he admitted, were rich in
plankton and other forms of life at any latitude. This helped explain
the misconception of tropical abundance but did not answer the
questions raised by the puzzling productivity of northern waters. If
light and temperature were not the primary controls upon the abundance of life in the sea, what were? Further studies were made on the
plankton in the bay at Kiel, but the results only made the puzzle more
complex. The greatest plankton blooms did not seem to occur when
expected, during the summer, when the days were long and sunny and
the waters warm, but came instead in the early spring and again in fall.
Hensen's colleague Karl Brandt tried to discover why.

Brandt, who was a professor of zoology, borrowed some of his ideas
from chemists and plant physiologists, who in the 1830's and 1840's had
found that certain chemicals must be present in the soil if plants are to
grow continuously. From such observations, the German chemist
Baron Justus von Liebig had formulated his "minimum law." This
states, in essence, that when any chemical necessary for the growth of
a plant is in short supply, the plant will all but stop growing, even
though all other requisites are present in adequate amounts. Brandt
suspected that the same was true of plankton, and he performed experiments to show that without sufficient nitrates or phosphates, plant
plankton ceased to grow. His hypothesis, then, was that the availability
of nutrients, rather than the amount of light and level of temperature,
controlled the productivity of plankton, and this implied that northern
waters were richer in nutrients than southern seas, and also that nutrients were for some reason more abundant in spring and fall than at
other times of the year.

Brandt's ideas began to explain the curious pattern of plankton distribution that Hensen had noticed aboard the *National*. The prevalence
of plankton in all coastal waters, for example, could be the result of
rivers that washed quantities of phosphates, nitrates, and other materials from the land to the sea, and so fertilized the inshore waters. But
how did the crops of plankton far from shore receive their nutrients?

At first, Brandt suggested that the availability of nutrients in the
open sea might be dependent on the abundance of certain bacteria
whose function it was to break down the tissues of dead plankton into
simpler organic compounds which plant plankton could then use. This
was a step in the right direction, but Brandt was still faced with the

difficult problem of explaining how such simple compounds could be gotten back to the plants in the upper sunlit layer of the ocean once the dead plankton (with its complement of hard-working bacteria) had sunk toward the cold, dark ocean floor. Gradually information was pieced together that indicated that the vertical mixing of ocean waters —itself dependent on temperature, winds, and currents—was the key to the distribution of nutrients and the productivity of the sea. In cold and temperate climates where there is generally more mixing of waters, especially in spring and fall, more nutrients are stirred from bottom waters, and plankton colors the sea in its profusion. In the tropics, on the other hand, where a layer of perennially warm, light water floats on a colder, denser layer, little mixing takes place, and without renewal from below, not enough nutrients are available to support a heavy crop of plankton. The water remains clear, blue, and unproductive.

So the questions involved with the productivity of the sea, which Hensen had set out to answer in the 1880's with his strange nets, had been passed on to chemists who measured nutrients, to zoologists who discovered bacteria, and finally to physical oceanographers who added still another reason to their list of motives for studying the circulation of the sea. Few of the scientists inspired by Hensen—and as many of these were intent on proving him wrong as right—chose to climb as far out on the limb of quantitative studies as he had done, but caution did not preclude the use of this methods. Quantitative sampling became an accepted part of biological oceanography, and within Hensen's lifetime attempts were made to count not only plankton, but whales, fish, and a great variety of bottom animals.*

Although Hensen's methods had been in use for more than a decade before the establishment of the International Council, that body was not anxious to adopt quantitative sampling as part of its quarterly routine.

"The reason . . . why the counting method has not been introduced generally into the international investigations, is probably first

---

* In 1896 C. G. Joh. Petersen designed a grab sampler which he used to collect worms, sand dollars, sea urchins, snails, oysters, mussels, barnacles, starfish—everything, in fact, that lived on the bottom or burrowed beneath it. The sampler differed from a dredge in that it was designed to bring up a precise quantity of the bottom (usually one-tenth of a cubic meter) which allowed Petersen to estimate the productivity of the sea floor in different locations. The grab sampler was a hinged metal box which was lowered into the sea with its two halves open (like the jaws of a steam shovel). When it sank into the bottom, a strong spring snapped the two sides together, and a precise amount of sea floor, with inhabitants, was gouged out. Variations of the device are still in use today.

and foremost this, that the method makes too great demands on the time and energy of the investigator, demands which are out of all proportion to the results we can expect from a general, statistical investigation. . . ."ᵠ

In spite of official reluctance, many scientists belonging to ICES used and further developed Hensen's quantitative methods, and the council's publications, from their very inception, carried numerous articles on Hensen nets, egg nets, the preservation of plankton samples, and alternate methods of counting plankton.

In 1929 an American oceanographer, Henry Bigelow, expressed the opinion that the International Council "may fairly be ranked with the . . . *Challenger* expedition in its importance . . . because the policies of the council have controlled the lines along which oceanography has since advanced in Northern Europe. . . ."ᵠ

His statement implies that ICES, as a scientific organization, was productive and successful, and indeed it was. It had gathered European oceanographers from every field into a fraternity where ideas were swapped and data exchanged. It had pulled the incredibly complex problems of fisheries research together in such a way that scientists from different countries could focus their attention on common goals and build upon each other's work toward their objectives. It had good publications and with them distributed the ideas of ICES' scientists far beyond their own immediate circle, and, perhaps most important, it had gotten countries to build ships, establish laboratories, and commit funds for the study of the sea.

But the council had set out to do still more. Its members wanted to apply the knowledge they'd so patiently gained both to the framing of fisheries legislation and to the development of practical fishing aids. In these endeavors the council was only marginally successful.

The council's difficulties in its role of the fishermen's helpmate stemmed mainly from the organization's small size and modest budget. In early years, when steam power was rapidly enlarging the area of the sea available to commercial fishermen, it occasionally happened that a research vessel came upon a new fishing ground. Hjort, aboard the *Michael Sars*, and Prince Albert of Monaco, on the *Princesse Alice*, found new grounds, but many more such discoveries were made by the fishermen themselves, whose greater numbers gave them far better coverage of the North Atlantic waters.

The same held true for the design of new fishing gear and fishing aids. ICES' scientists built a few novel devices—Petersen's prawn trawl and Alister Hardy's "herring finder"—but more often, new styles of seines and trawls were designed by the fishermen themselves.

The exceptions to the generalization that the council was not able to offer much direct help to fishermen are more interesting than the rule itself, and there is, for example, the story of how Hjort and Johan Ruud accidentally discovered a new fishery when the two were collecting deep-water animals from the soft, muddy bottom of Oslo Fjord. They had begun their collecting in the summer of 1897, and with a borrowed naval vessel had trawled back and forth across the oozy bottom bringing up haul after bristling haul of leggy, long-antennaed prawns. Hjort had suspected that these shrimplike creatures, which had been found in other places along the Norwegian coast and which were haphazardly collected by a few fishermen, might exist in sufficient numbers to support a fishery. The following summer he invited Petersen to come to Norway and asked him to bring along the eel seine which he had adapted for the collection of bottom animals. The seine served admirably as a prawn trawl and nets full of the pinkish-gray animals were taken from depths of 100 to 400 meters in Gullmar, Larvik, and Langesund fjords.

The next step, Hjort knew, was to interest the fishermen in prawns, and interested they would have to be if they were to undertake the trouble and expense of modifying their gear, fishing for an uncommon animal, and selling in an untried market. Knowing this—and knowing fishermen—Hjort began to fish for the deep-sea prawn at night. Along the dark wharves, lanterns in hand, Hjort and Ruud made their way, not too silently, to their boat. They put out into the fjord to fish until dawn, and returned in the first light to unload rich catches of prawn and take them to market. With some reluctance, it seemed, they shared the secrets of their new gear and grounds with other fishermen, and by the end of the summer ten boats were dragging for prawns.* By 1901

---

* In the first years of this fishery, small sailing craft manned by father-and-son teams were generally used. Petersen's trawl, copied with endless variations, was usually towed astern and hauled in amidships. The catch was transferred below decks, where the prawns were boiled in an iron soup kettle over a wood or coal fire.

By the 1920's motor launches had replaced the sailing vessels, and oil-fueled boilers were used instead of kettles. Still later the prawns were frozen rather than cooked.

the number had doubled, and by 1902 the deep-sea prawn fishery was so important that the catch was included in Norway's fisheries' statistics. By this time too the fishery had spread to Sweden, and some 30 years later, after a fortuitous exchange of information between Hjort and Bigelow, the fishery was even introduced to America.

The council's difficulties in its other chosen role, that of fisheries' legislator, were more complex and vexing than those which blocked its efforts to develop new grounds and gear. Fishing restrictions of any sort not only involved fishermen, who jealously protected their right to pull anything and everything from the sea, but also politicians, who rarely dared contradict the immediate needs of their constituents, nations intent on guarding their waters, and scientists, who never seemed able to agree on what should be done to protect the fish.

The first requisite for effective legislation, and one which involved all of these groups, was a firm belief that a stock of fish was actually declining due to overfishing or malpractice, and only when the decline was catastrophic, could everyone agree that something must be done. At that point, a knowledge of the endangered species was essential, and a lack of information produced poor legislation. In 1852, for example, the Norwegian government had been persuaded to believe that the fisheries in certain fjords were declining at an alarming rate. A committee had been appointed to investigate the matter, and this well-intentioned group had stumbled its way through hunches and half-truths to the conclusion that the seiners, who dragged their nets along the bottom, were tearing up masses of fish eggs, ripping seaweed from the rocks, and sweeping away the small crabs and shellfish which bottom fish ate, were, in other words, responsible for the declining fisheries. On the basis of these speculations a law had been passed barring seiners from a number of fjords, and the Danes, who had been watching these proceedings with great interest, passed similar restrictions. The fisheries continued to decline, but the laws remained in effect even after it was clearly shown that most fish eggs float, seiners don't risk dragging their nets over the shallow, rocky areas where seaweeds grow, and shellfish pass easily through the meshes of nets.*

---

* Dragging the bottom actually helped fisheries in some ways, for seine and trawl nets caught some of the larger bottom animals such as starfish which competed with bottom-feeding fish (and Irish lassies) for cockles and mussels. Petersen noted that "in America the starfishes caught are boiled on board, in order to kill them and prevent their causing further damage."Ψ

The experience of the Norwegians and the Danes (and the Scotch, who had closed the broad Moray Firth to trawlers for ten years, with no measurable results) had made it clear that legislation passed with insufficient information was useless. ICES moved in to fill the gap, and as early as 1907 had gathered enough information to formulate measures designed to protect plaice. But now the motivation seemed to be lacking, or rather a growing rivalry among nations proved stronger than their concern for fish. When Petersen recommended that the nursery grounds where the young plaice grew to maturity be closed in certain seasons, there was no agreement as to how this might be fairly done, so the suggestion was made that a minimum size for marketable plaice be established and that fishermen be penalized for landing undersized fish. Hopefully, such a measure would discourage fishermen from fishing on the nursery grounds where a large percentage of their catch would be very small fish. There was no agreement among nations on this measure either, so Petersen and others argued for the use of nets with a standard mesh size, one with holes large enough to let young fish escape. Still there was no accord, and Petersen and the council retreated to the politically expedient measures of transplanting fish and hatching eggs.

The council found the passage of legislation to be particularly difficult at this time, because there were many people who were not convinced that the stocks of the North Sea were being overfished, and there seemed to be no way of proving that man was at least partially responsible for declining catches and for the steadily decreasing size of the fish caught. Paradoxically, World War I provided the perfect experiment. For four and a half years mines, submarines, and enemy cruisers kept fishermen off the grounds, and when the hostilities were over in 1918, the fishing was stupendous. There were more fish, and there were larger fish, and within a decade the renewed supply was gone. By the late 1920's there was again a clamor for fishery protection, and this time efforts were more successful. In 1927 Denmark, Norway, and Sweden agreed on a closed season for plaice and dab, and two years later Germany, Poland, and Danzig joined to protect the plaice and flounder of the Baltic.

While the need for fisheries legislation was dramatized by the long closed season enforced by the Great War, the conflict worked more ill effects than good on the council. As the war began, Germany, Russia, and the United States withdrew (the United States had been a member only since 1912), work at sea was suspended, and many research vessels and laboratories were taken over—some permanently—by various navies and defense ministries.

The plans that had been made before the war for cooperative explorations, independent cruises, floating laboratories, and other projects had to be given up too, and whereas some of these, like Johannes Schmidt's study of eel migrations, could be resumed in the 1920's, others, such as Otto Pettersson's ambitious International Expedition through the Panama Canal, could not.*

In the years immediately following the war, ICES went through a period of reorganization and readjustment. The area of the Northeast Atlantic being studied, which before had been divided according to national boundaries, was redivided along natural geographic boundaries. Committees were formed for the study of the continental shelf, the Baltic Sea, the Belt Seas, and so forth. The council's great data lists were pruned and her administrative machinery modernized. Some new concerns were developed during the twenty-odd years between the two world wars, and new committees were formed for the investigation of whale stocks and for the study of parts of the South Atlantic bordering French holdings on that continent. But the majority of the council's work continued along the lines of investigations set up in

---

* Pettersson the tireless organizer, constantly on the lookout for a way to expand the council's coverage of the ocean, tried to turn a diplomatic ceremony into a scientific outing when, in 1913, he suggested that the nations invited by the United States to open the Panama Canal equip their ships for oceanographic work. The U.S. government had asked each country to provide one or two naval vessels for a convoy that was to be set off from Hampton Roads in 1915, cross the Atlantic, pass through the new canal, and steam as far north as San Francisco. The ships were then scheduled to return along the same route several months later. What a perfect chance to study the general circulation of the North Atlantic and a little of the unfamiliar Pacific as well.

Prince Albert of Monaco agreed to lead the expedition and, with Pettersson, persuaded seven nations to agree to take a series of hydrographic measurements along the route of the proposed voyage. In July of 1914, "His Highness Prince Albert of Monaco . . . assembled the members of the [expedition's] staff on board the 'Hirondelle' then anchored in the harbour of Kiel for a final conference," wrote Pettersson.

"Then came the war."ᵠ The plans were scrapped.

In the early 1920's Pettersson tried unsuccessfully to reassemble the expedition, but Prince Albert had died, and his oceanographic yacht, *Hirondelle II,* was up for sale. Pettersson urged European nations, and the United States and Canada, to purchase the vessel jointly for oceanographic work, but funds were not available. A certain nineteenth-century faith in the hand-in-hand progress of science and man was gone too, and without that the organization of large-scale international ventures became extremely difficult.

previous years. There were refinements, of course, but the topics themselves were familiar. There were "Contributions to the Life History of the Icelandic Plaice" and many other fish, "The Determination of Nitrates in Sea Water," papers "On the Validity of the 'Race Characters' of the Food Fishes," a wealth of year class investigations, and the perennial suggestions for "a proposed closed season for plaice. . . ."

The council's persistent attempts to protect plaice and other fish finally led 14 nations to gather at an International Fishery Conference in London in 1937. There it was agreed both to regulate mesh sizes and to establish minimum legal sizes for several species of fish. Sadly, these measures remained in effect only until the beginning of World War II.

In 1939 ICES' operations were again suspended, and not resumed until 1946. Twenty-five of the council's delegates died in the war, and by the 1950's, when the council was again operating at full strength, it occupied a different position within Europe's oceanographic community than it had in earlier years. ICES no longer "controlled the lines along which oceanography . . . advanced,"ᵠ as Bigelow had admiringly observed in the 1920's, for the science of the sea had grown to include many new interests and organizations. ICES, with its budget of less than $100,000 (for administration and publications), supported only a small portion of marine research.

At present, 17 countries* belong to the International Council for the Exploration of the Sea, and 27 national and international organizations collaborate with it. At a recent meeting in Copenhagen (1970), "working groups" met to discuss the varied problems related to fish tagging, shellfish transplants, fishing gear, plankton counts, larval surveys, statistical records, flat fish, round fish, marine mammals, and many other topics. While adapting to meet the changing needs of European fisheries, the International Council continues "to promote and encourage research and investigations for the study of the sea. . . ."ᵠ

---

* Present members are Belgium, Canada, Denmark, Finland, France, Federal Republic of Germany, Iceland, Ireland, Italy, Netherlands, Norway, Poland, Portugal, Spain, Sweden, United Kingdom, and Union of Soviet Socialist Republics.

CHAPTER VII

American Oceanography in the
Early Twentieth Century

The pursuit of science solely for its own sake, however com-
mendable it may be, is not the spirit that animates our government
in its support of scientific research. In its aims and ambitions this
is a practical age.

—Henry W. Henshaw
U.S. Biological Survey

The practical reasons for the government's pursuit of marine science,
which had evolved in the 75 years between the ascendancy of Maury,
Bache, and Baird and the beginning of World War I, were generated
by a peaceful nation's needs for safe navigation and a good supply of
fish. As long as these goals seemed worth pursuing, the government
had supplied varying amounts of money for the study of both the
physical aspects of the sea—its winds, currents, depths, and shoals—
and its biological ones—namely, edible fish. When the country's use of
the sea had changed, as for instance when steamships began replacing
sailing vessels, the government's policy of support had changed accordingly. With fewer sailing vessels there were fewer dollars for studying
winds and currents. But these changes, both in the use of the sea and in
the government's support of marine science, were minor ones, and
once the Coast Survey, the Depot, and the Fish Commission were
established, the government found no new reasons to appropriate funds
for oceanographic investigations.

Then, in 1914, World War I began, and this was indeed a change.
Quite suddenly the United States government found it had some new

reasons for supporting marine science, and primary among these was the desire to develop methods of submarine detection. But what the war did not give and what the country had not developed were the means of getting military need and scientific knowledge together. Military laboratories or research centers hardly existed in the early 1900's, and in the case of oceanography, private laboratories that could handle government research were lacking as well.

The onset of the war in Europe did not immediately send American military men and civilian scientists scurrying together to correct these deficiencies, and almost a full year of fighting had passed before reluctant steps were taken to prepare for the growing probability of war. It was then, in July, 1915, the the Naval Consulting Board was established (with Thomas Edison as its director) to screen the proffered inventions of civilian gadgeteers. Two years later, this board expanded its powers and began to organize special committees, one of which was a Committee on Submarine Detection by Sound.

Running in some ways parallel to the board's enthusiastic creation of committees were the activities of the new National Research Council (NRC), which had been set up by the National Academy of Sciences in 1916. The NRC also established committees, and these related to an incredible variety of scientific disciplines. There were committees on chemistry, mathematics, astronomy, geology, paleontology, botany, medicine, zoology and animal morphology, *and* submarine detection.

Almost all these committees were supposed to encourage a kind of scientific research which, in the simplest terms, would help win the war. In many instances this meant borrowing ideas from the Allies, who had been turning their scientific thoughts to military problems for a much longer time but, in the case of antisubmarine research, the French and British efforts were in about the same state of nondevelopment as was work in the United States. It appeared that all three powers had taken the advice offered a Prime Minister by a First Lord of the Admiralty in 1804, on the occasion of seeing Robert Fulton's plans for a submarine. "Don't look at it, and don't touch it. If we take it up other nations will, and it will be the greatest blow at our supremacy on the sea that can be imagined."✷

Whatever the reasons, the British, French, and American navies had confined their submarine programs to sporadic tinkerings, and none was prepared for the threat sprung upon it by Germany's extensive use of U-boats.

In spite of this general state of unpreparedness, there was in the United States a single company which knew a considerable amount about underwater sound. This was the Submarine Signal Company of Boston. The company had been established in 1901 by a bold group of men who believed that safety at sea—namely, the prevention of running aground, not protection from enemy submarines—could be much improved by the development of an underwater warning system. What these men had in mind was a network of underwater bells whose sonorous gongs would carry through the water for great distances, much farther, for example, than either a foghorn's airborne groan or a beacon's light. For ships steaming rapidly toward a rocky, fog-bound coast, the sounds from an underwater bell, picked up by special receivers, would give ample and reliable warning of the dangers that lay ahead.

Several models of submarine bells were developed by the company, and several types of ship-borne receivers were built to pick up the underwater sounds.* The first of the company's bells was installed on Lightship 54 in Boston harbor, and in the following years more than 100 bells were installed and more than fifteen hundred ships fitted with receiving gear. (The receivers usually consisted of a pair of microphones placed in water-filled tanks and mounted on either side of the ship's hull.)

The company's goal, however, was not solely the development of a warning device. As its name suggested, the Submarine Signal Company wanted to find a way to send submarine signals. Initially, attempts were made to send Morse code with the submarine bells, but the latters' clangings and bangings had neither the power nor the precision to make the code readable. Then, in 1912, two years before the war began, Reginald A. Fessenden, a former assistant of Thomas Edison, came to work for the company. He immediately set about designing an oscillator that he hoped would be capable of producing a most power-

---

* Experiments in sending and receiving sound through water had begun early in the nineteenth century. In 1826, for example, the speed of sound in water (which is about four and a half times faster than it is in air) was first accurately measured by Jean Daniel Colladon and Jacques Charles François Sturm. The two men had submerged a 140-pound church bell in Lake Geneva, and across the lake, some eight miles distant, had set up a listening post which consisted of a long ear trumpet stuck in the water. The bell was struck with a hammer, and the time required for the sound to travel to the ear trumpet was carefully measured. Sturm predicted that underwater sound would someday be used for signaling over great distances.

Reginald A. Fessenden designed an oscillator which he then used in both the Submarine Signal Company's Iceberg Detector and the company's Fathometer. (*Courtesy Raytheon Company*)

ful underwater sound, and by January, 1914, a stormy time on the North Atlantic, his invention was ready for its first sea trial.

A light snow was falling on Boston harbor as two tugboats cast off their heavy docking lines and eased away from the company wharf. On one, the *Susie D*, the new oscillator swung from a cargo boom, and when the tug had run out as far as the Boston Lightship she anchored and let the oscillator over the side. The other tug, equipped with receivers, headed straight out to sea. The oscillator was switched on, and its signals were clearly heard on the *Neponset* as she crossed Massachusetts Bay and were still coming in as the tug hit heavy seas past Race Point at the tip of Cape Cod. On went the *Neponset*, rolling and pitching in the heavy swells, until the worsening storm and gathering darkness made her captain turn for home. She was 34 miles from the Boston Lightship, and the signals from the oscillator were clear and unmistakable.

Fessenden was pleased with the oscillator's performance and felt it had possibilities as a signaling device. Still, he was anxious to try it for the other purpose he had had in mind—the detection of icebergs.

When the *Titanic* had been sunk by an iceberg on her maiden voyage in 1912, with a loss of more than fifteen hundred lives, a variety of inventions had been inspired, all with the intention of preventing similar collisions, and most with warning systems that utilized sounds that could be bounced off unseen objects. In England, Lewis Fry Richardson had patented a device for "detecting the presence of large objects under water by means of the echo . . ."* but had not moved to implement his invention. In Switzerland, Constantin Chilowsky, a Russian convalescing from tuberculosis in a mountain hotel, spent his leisure hours pondering the problems of echo-ranging (sending sound horizontally through the water) with results which circuitously affected American progress in the same field.

All such thoughts and their consequent inventions took a long time to develop and perfect, but by April, 1914, Fessenden had incorporated his oscillator into something he called an "Iceberg Detector and Echo Depth Sounder" and was ready to try it on a real iceberg. The United States Coast Guard, having recently received the responsibility of maintaining the International Ice Patrol, was more than willing to help Fessenden try his detector, and the gear was installed on the cutter *Miami* and sent to the Grand Banks. The trip started poorly, for it stormed incessantly and nearly everyone on board became hopelessly seasick. Then fog shut in and was so thick that the *Miami* could not move for fear of hitting an iceberg. Finally, after four or five miserable days, the rain stopped, the fog lifted, and with a kind of luck they had not had before, there in view was an iceberg. The oscillator, suspended from a lifeboat davit, was quickly lowered into the sea, and its signals, which went out in all directions, were sent ricocheting off the underwater portion of the berg and off the bottom of the sea. The ship was moved farther and farther away from the iceberg, but the two echoes were still received.

"Bottom one mile. Berg two miles. . . . Test stopped by bad weather,"* read the telegram received in Boston that rainy night.

Fessenden had hardly returned from the successful trial of his instrument when mounting pressures from the war in Europe persuaded him to transform his Iceberg Detector and Echo Sounder into a submarine detector. Accordingly, he spent the rest of the year filtering out the undesirable echoes that bounced off the sea floor, for these might be confused with the presence of an enemy submarine. His first detectors were installed on 10 Canadian-built British submarines in 1915, then Fessenden took his device to Europe, where it was installed on British warships and on a few French submarines as well.*

Two years later, early in 1917, the Submarine Signal Company

---

* Because Fessenden was particularly interested in verbal underwater communication, he had tried unsuccessfully to incorporate his oscillator into a submarine broadcasting system. In the Submarine Signal Company's sound laboratory he had set up an experimental system which had a microphone at one end of the laboratory and a receiver, placed on the bottom of a big water-filled tank, at the other end of the lab. An engineer found the possibilities too wonderful to overlook, and one morning Fessenden walked into the laboratory to find 100 goldfish swimming in the receiving tank, all apparently singing "Annie Laurie."

became even more directly involved in the war. The president of the company was called to Washington, D.C., to appear before the Naval Consulting Board's Committee on Submarine Detection by Sound. He described his company's iceberg detector and other devices, and the committee members were so interested that they traveled to Boston the very next week to see the instruments work. The Submarine Signal Company then offered to establish an experimental station in cooperation with two other private companies (Western Electric and General Electric) for the purpose of developing submarine listening gear, which at the time seemed a more reliable method of detection than did the sound-bouncing technique. The committee agreed, and by April, 1917, when the United States declared war on the German empire, work was already in progress at a station in Nahant, Massachusetts.

The other committee on submarine detection, the one sponsored by the National Research Council, had a longer and more complicated period of gestation. It evolved in part from work already started by the French and British.

Constantin Chilowsky, who had since recovered from tuberculosis in Switzerland, had taken his plan for an echo-ranging device to Paris, where he had shown it to the famous physicist Paul Langevin. The two men built the device, tried it in the Seine, and passed the information on to the British, who were soon doing similar experiments on their own. In 1917, a joint commission from France and Britain came to the United States to forward whatever scientific information might be of value to their new ally. Among this information were plans for submarine-detection devices, and these inspired the NRC's committee to gather a group of university physicists together at the U.S. Naval Experiment Station in New London. Before the end of the war both the private station at Nahant and the government one in New London were experimenting with echo ranging and sounding and had, in addition, developed usable listening devices with which submarines could be detected at a distance of several miles.

Although several models of these listening devices were installed on many of the Navy's ships, it is apparent, in retrospect, that the threat posed by German U-boats was met more directly by the Navy's intensive program of shipbuilding rather than by the development of rudimentary antisubmarine equipment. Nevertheless, the committees on submarine sound had not only convinced the Navy of the usefulness of oceanographic knowledge for military purposes but had also established a pattern of military and civilian cooperation which could be followed when similar needs arose during World War II.* Equally important, the experimental stations in the United States, France, Britain, and Germany as well had pushed the development of echo-sounding techniques to the point where the new method could be used to great advantage by oceanographers and marine surveyors.

The first deep-sea echo sounder to be used extensively, albeit temporarily, on American vessels was one designed by a member of the committee on submarine detection at New London, Dr. Harvey Cornelius Hayes. Before the war Hayes had taught physics at Swarthmore College in Pennsylvania, and as a result of the experiments in underwater sound which he had begun in the college swimming pool, he had been asked to join the New London group. During the war he worked on listening gear and related equipment, and by the spring of 1922, the Hayes Sonic Depth Finder (which used a Fessenden oscillator for its sound source) was ready to be tried on a long run across the Atlantic Ocean. Prior to this time the French had begun to use an echo sounder for running cable surveys across the Mediterranean, and in the United States some naval vessels had been equipped with a shallow-water sounding device for use in less than 100 fathoms of water. Until Hayes' device was installed aboard the U.S.S. *Ohio* and the U.S.S. *Stewart*, however, no ship had been equipped for really deep-water sonic sounding. To test the depth finder the *Ohio* ran a line of soundings between New York and Annapolis; then the equipment was transferred to the *Stewart* for her trip across the Atlantic. With Hayes aboard, the *Stewart* put out from Newport, Rhode Island, and on her nine-day passage to Gibraltar, an average of 100 soundings was made each day. The deepest recorded a depth of 3,200 fathoms. Although the Navy was loath to declassify the workings of the depth finder, brief reports indicated that the finder could not be used in less

---

* While most of the cooperative research generated by World War I was dropped as soon as peace returned, one small part was kept alive. This was due to the Navy's desire to continue the work begun on radio signaling and submarine sounds. Since there were no private laboratories equipped to undertake these tasks, a Naval Research Laboratory was built in Anacostia, Maryland, in 1923. The laboratory was commanded by a naval officer, but almost all the scientists and technicians who worked there were civilians. After World War II a much greater proportion of cooperative research was kept going, and this was handled by private institutions such as the Scripps Institution of Oceanography and the Woods Hole Oceanographic Institution as well as by the Naval Research Laboratory.

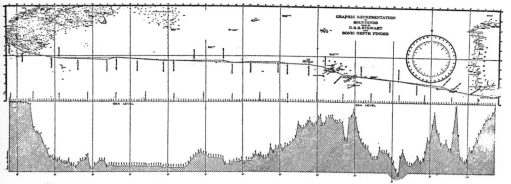

The first continuous profile across an entire ocean basin was made aboard the U.S.S. *Stewart* in June 1922 with a Navy Sonic Depth Finder. The profile extends from Newport, Rhode Island, to Gibraltar. (*Reprinted from* Proceedings *by permission; Copyright* © *1943, U.S. Naval Institute*)

than 40 or 50 fathoms of water, but that in greater depths it could be operated even when the ship was steaming at a full 23 knots.

A continuous profile across the Atlantic was quickly and easily drawn from the *Stewart*'s 900 soundings, a feat so remarkable that a report from the Naval Engineering Laboratory (where the depth finder had been built) stated that "the perfection of the Sonic Depth Finder offers unlimited possibilities to . . . oceanography."⊻

The results were indeed spectacular. The *Challenger* in her entire three-and-a-half-year voyage had taken less than 300 soundings in depths exceeding 1,000 fathoms, and in the same years the Coast Survey considered it a good field season when 100 or so deep soundings were collected.

A second example of the Sonic Depth Finder's usefulness came later in 1922, when the naval vessels *Hull* and *Cory* made sonic soundings for a deep-water contour map off the coast of California. The Coast and Geodetic Survey, not to be left behind by the Naval Hydrographic Office, soon fitted their survey vessel *Guide* with the same depth finder and proceeded from New London, through the Panama Canal, to the West Coast. This line of soundings was particularly valuable, because the sonic soundings, which ranged in depth from 100 to 4,600 fathoms (the deepest being in the Puerto Rican Trench), were consistently checked by conventional wire soundings. In addition, hundreds of water samples and temperature readings were collected at many of the sounding stations.

The results were surprising, and somewhat discouraging. The depths indicated by the sonic recorder did not always agree with the measurements made with the wire sounding machine but they were neither consistently too deep nor too shoal. The scientists on the *Guide* realized that the problem probably lay with their inability to judge just how fast the oscillator's sound pulses were traveling through the water. An approximate speed of 4,800 feet per second was assumed, but it was known that sound would travel faster if the temperature, salinity, or pressure of the water was increased. The discrepancies caused by these natural variations were irregular and affected the *Guide*'s sounding accuracy in unpredictable ways. Tables giving the velocity of sound in sea water of various temperatures and salinities, such as had been worked out by the Norwegian Vilhelm Bjerknes, were needed to correct the depth finder's readings. The Coast and Geodetic Survey soon supplied such tables to users of the Sonic Depth Finder.

With these corrections, and with additional trials, confidence grew in sonic soundings. The Naval Hydrographic Office began using the method to make maps and to run cable surveys, and by 1929 the Office was receiving each year the sonic records made along 10,000 miles of ocean floor. The Coast and Geodetic Survey embarked upon an ambitious coastal mapping project which extended from the Grand Banks south to the Chesapeake Bay, and commercial liners began using the Submarine Signal Company's new Fathometer, brought out in 1924, to find submarine ridges and valleys which helped them pinpoint their position while still far out at sea. In Europe, too, the new method of sounding was becoming popular, and by the mid-1920's all the maritime countries had access to some sort of sonic apparatus, and the Germans were even using theirs on a long oceanographic expedition.*

The development of the echo sounder was certainly a fine example of the kind of benefit that might accrue to the government's support of marine science, but in spite of it and several other devices, the postwar Congress was more inclined to "return to normalcy" than to step too far toward the support of abstract research. In 1924, when the Navy toyed with the possibility of launching a Naval Oceanographic

---

* Probably the first extensive oceanographic expedition to use a sonic recorder was the German Atlantic Expedition on the *Meteor*, 1925–1927. The *Meteor* carried two echo sounders, one modeled after the American Fathometer, the other designed by Alexander Behm, who, motivated by the *Titanic* disaster, had built a workable Echolot by the end of the Great War. The two echo sounders were used to make about 33,000 duplicate soundings, and it was found that the two machines usually agreed to within 40 or 50 feet when operating in deep water.

An artist's conception of the Fathometer in use. The black arrows represent electrical pulses, and the white arrows with rings represent sound waves. (*Courtesy Raytheon Company*)

Expedition—a grandiose plan to do everything for everybody—the project failed "because it demanded a large grant from Congress which was not forthcoming, and which, in fact, there was no reason to expect would be forthcoming."✻

As the government dropped its war research, most of the professors who had worked on special committees returned to their classrooms; the few special contracts that had been given, for example, to Scripps for work on the preservation of edible fish and the harvest of kelp for potash, were terminated; and the consigned ships, borrowed from public agencies, were returned. Private oceanography reflected this readjustment to peacetime practices and took up as before the more abstract elements of the science. A dribble of small, discursive expeditions began again, and chartered vessels and pleasure yachts set out to collect marine curiosities for the American Museum of Natural History, South American shellfish for the Smithsonian, and shallow-water fishes of the West Indies for the new Bingham Oceanographic Laboratory at Yale University. The long-term comprehensive studies that had been interrupted by the war were also resumed, the best-known European example being Johannes Schmidt's 20-year search for the spawning grounds of the European eel.*

In the United States a less popular, but perhaps equally important, series of investigations which had also been stopped by the war was resumed by Henry Bryant Bigelow, who, from 1912 to 1916, had been studying the Gulf of Maine on the fisheries' schooner *Grampus*. The kind of studies that the young Harvard zoologist had chosen to pursue were remarkably comprehensive, including, as they did, repeated trips to a single body of water in all seasons of the year and the collection of physical, chemical, biological, and geological data. This was not the usual kind of oceanographic study being done in the United States, and it certainly was not the kind that Bigelow had been introduced to aboard Alexander Agassiz' long cruises to the Pacific and Indian oceans. The Gulf of Maine studies were, however, similar to the investigations that were being encouraged by the International Council for the Exploration of the Sea in northwestern Europe. There, it was the council that was helping to make the transition from the nineteenth century's great voyages of natural history to the more modern practice of long-term, detailed investigations. In America, it was Bigelow who, with the council's work as his example, began to effect the same transition. It took him a decade to change his own work and another 20 or 30 years to impress the new ways upon others.

Bigelow first put to sea with Alexander Agassiz. In the winter of 1901, when Bigelow was a senior at Harvard, the rumor passed around that Agassiz was again planning to leave the director's room at the university's Museum of Comparative Zoology, this time to lead an

---

* Schmidt had begun his search in 1904 at a time when it was believed that the eels spawned in the Mediterranean. Several eel larvae had been caught there, but no one could find any eel eggs, much less any ripe eels filled with roe or milt. Schmidt began fishing for eel larvae, for he believed that if he could catch smaller and smaller larvae at earlier and earlier times in the year he would eventually find the eggs and then the unknown spawning grounds. He had already ascertained that this area must be somewhere far out in the Atlantic when the war came and his work at sea was curtailed from 1914 to 1920. Finally, in 1922, Schmidt set out again, and this time he found the spawning grounds of both the European and American eels—the former in the Sargasso Sea, the latter slightly farther west, closer to Florida.

expedition to the Maldive Islands in the Indian Ocean. Bigelow didn't know Agassiz, for the latter taught no courses at the university, but being a student of invertebrate zoology, he thought, or at least hoped, that his knowledge of sea cucumbers, jellyfish, and sponges might make him useful on the coming expedition.

"Finally I got up my courage and 'bearded the lion . . .,' " remembered Bigelow. "I told him my name, said I'd heard he was going to the Maldives, and asked him if he would take me with him."⁴

"Yes," was the answer, and Bigelow set off upon one of the most influential careers in American oceanography.

During the few weeks that Bigelow spent aboard the chartered vessel *Amrah* in the Maldives, and later, during the months he passed on Agassiz' Eastern Tropical Pacific Expedition, he became familiar with the policies and practices of a conservative kind of nineteenth-century European oceanography. Agassiz' cruises were modeled for the most part after the successful ventures of the prince of Monaco, Sir John Murray, and other wealthy gentlemen of scientific talent, and were planned as general surveys of large sections of the unknown ocean. The voyages were biological or geological in nature, and Agassiz did not mean to include much physical oceanography or to investigate any area of the sea in great detail.

On these cruises Bigelow watched as sailors, not scientists, operated all the ship's equipment, and when the nets came up, bulging with fish, jellyfish, squid, and other animals, he was given some animals to sort and preserve while the ship got under way and steamed to the next station some two or three hundred miles away. In many respects these were great and glorious collecting trips, and their bounty was expected to keep many a scientist busy for a long, long time.

Bigelow fell into the pattern. When he received his Ph.D. from Harvard in 1906, he immediately went to work in the Museum of Comparative Zoology, where he spent the long New England winters describing and classifying the mountainous supplies of jellyfish he had collected on one or another of Agassiz' expeditions. Then, in the summer of 1908, he was encouraged by the ailing Agassiz to undertake a short cruise on his own. Agassiz hoped that Bigelow would gather some animals from Gulf Stream waters which could be compared to those already collected from the Humboldt Current in the Pacific. It was to be another collecting trip, but at least on this one Bigelow did the work.

The Bureau of Fisheries (the old Fish Commission) was easily persuaded to lend Agassiz' protégé the ninety-foot schooner *Grampus*.* The vessel was made ready in her home port of Gloucester, Massachusetts, and a pile of gear—almost equally divided between fisheries' equipment and European contraptions—was loaded aboard.

To Bigelow's delight, the weather was fair and the winds favoring, and on the afternoon of July 7 the *Grampus* stood out from Gloucester and sailed for Martha's Vineyard. After taking on still more equipment, including a model of the prince of Monaco's fish trap, which had never been tried in the western Atlantic, Bigelow and his scientific staff of two sailed southeast for the deeper water of the continental slope. When the vessel rode in about 300 fathoms of water, the Monacan fish trap, baited with dead fish and luncheon leftovers, was let over the side and allowed to remain on the bottom for an hour. At the end of this time it had to be hauled slowly and carefully to the surface, for it was suspended on a hemp buoy line rather than on wire rope, which had already been found to kink and tangle as the *Grampus* rolled. The trap came up without mishap, and inside were 11 large and ugly hagfish, a long-nosed eel, and several amphipods. A second trial, in deeper water, brought in 23 fish of three different species. The results, confessed Bigelow, were just barely worth the time and trouble inherent in setting the trap from so small and unstable a vessel.

Worse troubles were encountered later in the week when Bigelow tried to use the Tanner closing net. (This net had been designed by the master of the fisheries' steamer *Albatross*, Captain Zera L. Tanner, and was the kind used with consistently negative results by Alexander Agassiz.)

"Even in calm weather a schooner lurches about so violently that it is often impossible to handle any heavy apparatus requiring delicate adjustment," complained Bigelow.[¥] The delicate adjustment he had in mind was the trigger on the Tanner net which was meant to close the net when the haul was over and thus prevent it from collecting animals

---

* The *Grampus* (the name comes from a marine mammal distantly related to whales and dolphins) was a sailing smack with a deep inside well which had been built by the Fish Commission in 1886 both for the collection of live fish and as an example of a safer, more seaworthy fishing schooner. The *Grampus* fulfilled both these functions, and her design was adopted by many New England boatbuilders. As an oceanographic vessel, however, she was none too steady, and the small hoisting engine installed on her deck was unequal to the task of handling heavy equipment.

Henry Bryant Bigelow at the wheel of the *Grampus*. (*Courtesy Museum of Comparative Zoology, Harvard University*)

as it was pulled to the surface. Time and time again Bigelow and the crew gingerly lowered the net over the schooner's side. Up came a wave, the wire rope snapped taut, and the trigger was sprung before the net had even dipped into the water. Bigelow, through his wide reading, was familiar with similar problems encountered by European oceanographers. He knew, for example, that George Murray of the British Museum had led an expedition in 1898 (sponsored in part by the Drapers' Company and the Fishmongers' Company) to study the intermediate fauna off Ireland. He too had struggled with a Tanner closing net that tripped prematurely, and to remedy the defect had lashed the trigger open with a copper wire thin enough to be easily broken at the proper moment by a descending messenger. Bigelow used weak twine instead of wire, which he found worked equally well, but the net still had its difficulties, and nothing was caught in it on the one time it was tried.

The collection of surface fauna was the most successful aspect of Bigelow's short cruise, and in addition to netting a new jellyfish, he brought in several hundred pounds of transparent, barrel-shaped salps (a kind of free-swimming tunicate that often joins with other members of its own species to form long chains of animals). The salps were so abundant, in fact, that the five-foot-wide *Albatross* net could not be used to collect them because the weight of the animals would almost surely have burst the net as it was raised from the water.

The results of Bigelow's first cruise were published a year later in the *Bulletin of the Museum of Comparative Zoology* in a volume that also contained six more lengthy accounts of the animals taken on Agassiz' 1905 expedition. The Pacific collection was by no means exhausted, and after his cruise on the *Grampus* Bigelow returned to the museum to work on the rest of his jellyfish.

Then, in the spring of 1910, Alexander Agassiz died, and Bigelow, although still anxious to continue his study of the sea, was left without direction and without Agassiz' influential support. During the next two years Bigelow went to sea only in a desk chair. Sitting in his office next to the museum's magnificent library, he read about the work being done by members of the International Council and even wrote several articles praising work done by Johan Hjort aboard the *Michael Sars*, although as yet he knew Hjort only by reputation.

In spite of his desire to be out on a ship, Bigelow kept to his museum work (with a docility uncharacteristic of his later years) until the summer of 1912. In the spring of the previous year, just as he finished describing the last of the jellyfish collected by Agassiz in the Pacific, the famous British oceanographer Sir John Murray came to Harvard and convinced Bigelow to launch a modest expedition. Murray was visiting the United States to see the country's oceanographic institutions, or more precisely its marine biological laboratories, and to talk with wealthy individuals who, if properly inspired, might come to the aid of the young and undersupported science.

When Murray had worked his way through teas and receptions as far as Harvard, he received a lengthy reiteration of Bigelow's landlocked complaints and, after ascertaining that the young man could row, borrow a dory, and make tow nets from bobbinet curtains, suggested that he stop asking "any more damn foolish questions!"▼ and get to work.

The encounter was apparently enough to boost Bigelow out of his office. He made arrangements to borrow the *Grampus*, and in the summer of 1912 embarked upon what eventually became a long series of detailed studies of the Gulf of Maine. His areas of major concern were, like the International Council's, fish, plankton, and hydrography, and these he chose to study in the Gulf of Maine not only because that body of water (lying north and northeast of Cape Cod) was near Harvard and handy to Gloucester and the *Grampus*, but also because it was a commercial fishing ground of considerable importance to the

Bureau of Fisheries. A third reason was that the gulf's hydrography appeared intriguingly complex. Warm Gulf Stream water flowed to the east, over Georges Bank, and much colder coastal waters circulated within the gulf itself.

In spite of the small size of the *Grampus* and her penchant for rolling (which repeatedly sent Bigelow to the rail), a most complete battery of observations was carried out during the first summer's cruise. Included were measurements of temperatures and the collection of water samples from the surface to the bottom, the determination of the water's color and transparency,* the circulation as measured by an Ekman current meter, the determination of salinity using the ICES' method of titration and ICES' Standard Water, quantitative hauls of plankton using Hensen's vertical net, and regular plankton hauls using *Albatross* nets, scrim nets, and a hybrid closing net that incorporated all the best features of the Chun, Petersen, and Nansen nets. Dredging and trawling were also done, but because of the difficulty of handling this equipment on the *Grampus*, it had to be considered a minor part of the scientific program.

The *Grampus* occupied 46 long and complicated stations in the summer of 1912, and although the data she brought back were most complete, Bigelow felt that "it is obvious that observations restricted to two months in midsummer cannot afford a picture of the regular series of changes which its [the Gulf of Maine's] waters undergo during the year. . . . Consequently the following report is to be regarded only as the beginning of a survey which, it is hoped, will be continued at other seasons in ensuing years."⩩

The Fates and the Fish Commission cooperated, and Bigelow's hopes for continued investigation were realized over the next dozen years. During winter months, short, bimonthly cruises were made by the fisheries' steamer *Blue Wing*, and in 1913 the *Grampus* sailed on a much longer cruise of more than 2,000 miles to study the waters between Halifax, Nova Scotia, and the Chesapeake Bay. In addition,

---

* The usual way of measuring the transparency of sea water—which was done to get an idea of the amounts of sediment or plankton suspended in the water—was to lower a white plate called a Secchi disk into the water on the end of a marked line and note the depth at which the disk was no longer visible. This particular method had been devised in the 1860's by the Italian astronomer Angelo Secchi while he worked in the Mediterranean aboard the papal vessel *Immacolata Concezione*. (Now the transparency of water is measured with photoelectric devices.)

she occupied the same stations that had been made in the Gulf of Maine during the previous summer. In 1914 Bigelow decided to extend his studies eastward, rather than southward, but when the *Grampus* put in to Halifax he received news of the war and was told in no uncertain terms to get the schooner back into territorial waters. The Gulf of Maine investigations proceeded without enlargement, but without undue impediment, for another two summers, until, in 1918, Bigelow was shipped off as a navigation officer on an army transport. The *Grampus* had already been consigned to the Navy.

By the time the war interrupted the Gulf of Maine studies, Bigelow had become thoroughly familiar with the fishes of the gulf. He knew where the destructive hagfish lived and where the valuable flounder, haddock, and cod were likely to be found. He had learned, too, the general character of the gulf's plankton and had plotted the tiny plants' and animals' seasonal comings and goings. He had found many kinds of tropical plankton and even some floating sargasso weed near the eastern edge of the gulf where the warm Gulf Stream sometimes flowed, while in the colder waters to the west were swarms of copepods, arrow worms, and other forms of cold-water animals. On an early spring cruise Bigelow had finally found some typically Arctic plankton, which indicated that northern waters flowed into the gulf at least at certain brief times in the spring.

The physical or hydrographical studies of the gulf were not as far advanced as the biological ones, and when Bigelow resumed his studies after the war he placed more emphasis on deciphering the water's complicated tidal and nontidal circulation.

In the years between 1922 and 1927, when Bigelow worked from the fisheries' steamers *Halcyon* and *Fish Hawk* (the *Grampus* had been condemned and sold), he released 1,500 drift bottles, and from those recovered deduced that the water in the Gulf of Maine moved in an irregular but basically counterclockwise pattern. In the same years, he experimented with the Scandinavian method of dynamic current computation, which he had learned on his own through reading and through occasional contact with European oceanographers.

Bigelow gathered enough temperature and salinity measurements to construct what he liked to refer to as weather maps or contour charts of gulf waters. These showed the positions of relatively higher, less dense masses of water and of lower, heavier ones as well. Since the higher water could be depended upon to flow downhill (with a twist to the right in deference to the earth's rotation), the dynamic charts showed how the water in the Gulf of Maine should move—at least from theoretical considerations. The circulation they indicated was gratifyingly similar to that shown by the drift bottles. Predictably, Bigelow became a convert to the dynamic method, and encouraged its use where he could.

The results of Bigelow's Gulf of Maine studies were published by the Bureau of Fisheries as three monographs, one each on fish (1915), plankton (1926), and physical oceanography (1927). The 15-year study had sorely tried Bigelow's personal relationship with the sea—he found it increasingly difficult to face the prospect of weeks or months of seasickness as he grew older—but it had also convinced him of the necessity of getting oceanographers out on the water at all times of the year to study all aspects of the sea.

Another project stalled by the Great War (and one subsequently helped by Bigelow's experience in the Gulf of Maine) was the International Ice Patrol. Like the first iceberg detectors, the Patrol was a direct and immediate reaction to the *Titanic* disaster of 1912. Within a month of the ship's sinking two U.S. naval vessels were patrolling northern shipping lanes to keep steamers informed of ice conditions, and the following year the Revenue Cutter Service, precursor of the Coast Guard, took over the job. Also in 1913, an International Conference on the Safety of Life at Sea met in London, and 13 nations agreed to share the expense of establishing an International Ice Patrol. Responsibility for the actual operation of the Patrol was given to the U.S. Coast Guard, but hardly had these arrangements been made when World War I was begun and the work suspended.

In 1919, the Patrol resumed its duties. A special group had been appointed to direct its scientific studies, and Henry Bigelow, an adviser to this group, began to encourage the inclusion of several sorts of oceanographic investigations in the Patrol's program.

The problem facing the Ice Patrol was to find and keep track of the 50 to 1,000 icebergs that broke loose from the glaciers along the west coast of Greenland each spring, drifted down through Davis Strait into the shipping lanes and fishing grounds south and southeast of Newfoundland, and finally melted in the Gulf Stream. This unpredictable procession usually wove its way through a section of navigable waters about the size of the state of Pennsylvania in the three months between April and June.

At first the Patrol relied almost exclusively on direct observation.

Admiral Edward ("Iceberg") Smith, leader of the *Marion* expedition, was third director of the Woods Hole Oceanographic Institution. (*Courtesy Mrs. Edward H. Smith*)

Patrol boats cruised through the dangerous waters, and broadcasts were made each day giving the position and approximate drift of all the icebergs which were sighted. Constant fogs severely hampered this work, and it was obvious that what was needed was some method of predicting the path of icebergs without having to see them all the time. If the circulation of the waters could be plotted with sufficient accuracy, the Patrol would be able to forecast the movement of ice for a week or more, at which time the area could be resurveyed and the changing currents plotted again.

In some ways the Ice Patrol's efforts to understand the complicated eddies and swirls of the Northwest Atlantic recapitulated in a condensed form the history of Scandinavian efforts to develop the dynamic method of current computation and Bigelow's attempts to understand the circulation of the Gulf of Maine. The Patrol cautiously began its work with measurements of surface water temperatures and the collection of plankton, both meant to help define the different water masses that moved through the area. Next, it tried using current meters, but except in shallow water these did not prove practical. Finally the dynamic method of current computation was adopted. The difficult change to Scandinavian methods was made by a young officer in the Ice Patrol who was sent to the Geophysical Institute in Bergen to learn how temperature and salinity measurements could be used to infer the existence of currents.

Edward Hanson Smith, the officer sent to Bergen, had graduated in 1914 from what is now the Coast Guard Academy in New London. He had been assigned to the Ice Patrol, and after the war was sent to the Museum of Comparative Zoology during the fall and winter to learn what he could about oceanography from Bigelow and to work up the reports of the past season's cruises and plan for the ones to come. Under Bigelow's inspiration, Smith became so interested in physical oceanography, and the study of currents in particular, that he began earning his master's degree in the subject (actually in meteorology, which was as close as Harvard could then come to physical oceanography). This he received in 1924, and in the same year set off for the Institute in Norway to study for a year under Bjørn Helland-Hansen who, with Sandström, had adapted Bjerknes' formula for use in the dynamic computation of currents. Smith learned enough in Norway to add a chapter to Bigelow's *Physical Oceanography of the Gulf of Maine* and to introduce the method of making dynamic surveys into the Ice Patrol's routine.

By 1926 the Patrol's two cutters, stationed on the Grand Banks during the ice season, were each equipped with water bottles, reversing thermometers, and an electric device for measuring salinity. With this gear the vessels took turns making four- or five-day cruises over the critical 40,000 square miles. Actually, most of these cruises were ice-scouting or purely observational cruises, but at least three or four times each season, 20 to 25 hydrographic stations were made on a special cruise, and from the data gathered a chart showing the prevailing currents was immediately constructed. The pertinent information from these charts was forwarded to steamers via the Naval Hydrographic Office and was also given to the next Ice Patrol cutter, which began her work as soon as the other ship returned. The accuracy of these charts was checked by the ice-scouting cruises, and if the icebergs actually sighted were located where the previously charted currents should have carried them, then it seemed safe to say that the dynamic method of current computation had successfully indicated the movement of ice and water.

"Here [in the Ice Patrol]," wrote Bigelow, "the most advanced methods of dynamic oceanography are put to practical use . . . and it is here . . . that the soundness of this method of attack has received its most impressive confirmation."[Ψ]

In the summer of 1928 the Ice Patrol extended its oceanographic explorations far to the north of its usual province. The *Marion* expedition was launched to study the waters between Greenland and the northeast coast of America in hopes that the source of the icebergs

The Coast Guard cutter *Marion* made a special oceanographic expedition to the west of Greenland in 1928. (*Courtesy U.S. Coast Guard*)

could be ascertained. The little *Marion*, with Smith in charge, zig-zagged for more than 8,000 miles through fog and rain and sudden gales to visit the fjordlike inlets of western Greenland, where glaciers pushed slowly into the sea and icebergs calved from their advancing edges with thunderous explosions. To study the waters of Baffin Bay and Davis Strait the *Marion* occupied 191 stations, and to improve the bathymetric knowledge of these infrequently visited waters the ship made 1,700 soundings with a Submarine Signal Company Fathometer. The report of the expedition was written in three parts, which included sections on general scientific results, Arctic ice, and physical oceanography. (The last served as Edward Smith's Ph.D. dissertation.)

The *Marion*'s successful voyage further convinced the Ice Patrol of the need for oceanographic investigations, and in 1931 the cutter *General Greene* was equipped with all the necessary hydrographic gear and assigned to northern waters as the Patrol's first "oceanographic vessel." With the new ship at work the number of hydrographic cruises increased, until, at the beginning of World War II, the program was suspended. After the war, much of the Ice Patrol's reconnaissance was done by plane, and the amount of oceanographic research diminished.

The *Marion* expedition also gave Lieutenant Commander Smith some valuable experience, and his reputation as an able physical oceanographer rose with the Patrol's. In 1931 he was lent to the Aeroarctic Society to participate in the remarkable *Graf Zeppelin* expedition. (This was the first successful exploration of the Arctic to be made in a dirigible.) As originally planned, the huge submarine-shaped airship was to fly from Spitsbergen, north of Norway, to Fairbanks, Alaska. This route would take the airship over the pack ice that drifted in the vicinity of Labrador and Newfoundland, and the Patrol believed that this ice directly affected the number of icebergs that came south each spring. The route of the *Graf Zeppelin* was subsequently changed, but Smith was impressed nevertheless by the mapping that could be accomplished from altitudes of 3,000 feet and more.

"In the brief space of three hours an area was mapped [with aerial mapping cameras] equivalent in work to that of four summers by a ground surveying party," wrote Smith.✤ He estimated that the entire glacier front of West Greenland, along which the *Marion* had laboriously traveled, could have been mapped from the air in 10 hours.

In the late 1940's and early 1950's, when the Ice Patrol began replacing much of its sea work with just the sort of air surveys that he imagined, Smith chose to stay near the sea. In 1950 he became the third director of the Woods Hole Oceanographic Institution.

Thus far, American oceanography in the first several decades of the twentieth century appears to have changed only slowly from the pattern that characterized its organization at the end of the preceding century. Private laboratories and wealthy individuals continued to work on basic problems, usually of a biological nature, while the more abstruse realms of physical and chemical oceanography were just beginning to be examined. At the same time, publicly supported offices and laboratories were still trying to deal exclusively with the country's practical needs, which included the old concerns for safer navigation and more fish, and which, temporarily, had embraced a new concern for submarine warfare.

There were some unobtrusive differences between the two periods, however, and one of these was the growing tendency for government agencies to confine their efforts almost entirely to practical tasks such as fish hatching and map making, while privately supported marine scientists moved in the opposite direction toward abstract research. In marine studies government science was becoming synonymous with applied research, private science with basic research.

At this juncture a more dramatic change in American oceanography began to take place. A new benefactor for science had appeared and was beginning to spread its corporate wealth and influence across the country. This was industry, and the boost it gave to science came in two ways: either giant enterprises created very taxable excess incomes which their owners preferred to give to foundations rather than to the government, or the factories and processing plants producing all this wealth needed scientific research for their own improvement, and company laboratories or special studies were set up. Oceanography benefited from both these new practices, notably the first.

One of the first of the grand foundations, organized to give men the benefits of science, was the Carnegie Institution of Washington. It was incorporated in January, 1902, with an initial endowment of 10 million dollars (which was considerably more than the average endowment of Harvard, Yale, Princeton, Columbia, or the University of Pennsylvania, and ten times that of the Smithsonian Institution). With this foundation, Andrew Carnegie hoped "to secure, if possible, for the United States of America, leadership in the domain of discovery and the utilization of new forces for the benefit of man."⸸

Fairly early in his career as a producer of iron and steel, Carnegie had discovered that "nine-tenths of all the uncertainties of pig-iron making were dispelled under the burning sun of chemical knowledge. . . ."⸸ If scientists could do that for pig iron, mused Carnegie, what mightn't they do for the rest of mankind's endeavors? What indeed? To find out, the Carnegie Institution of Washington was begun and within two years was supporting individual research, the publication of reports, and the operation of entire departments and laboratories for collaborative research.* In 1904 the Carnegie Institution established a small marine biological station at Tortugas, Florida, for work on tropical marine life, and in the same year organized the much larger Department of Terrestrial Magnetism. This department was charged with conducting a worldwide magnetic survey which was to concentrate on the more inaccessible portions of the world, namely the oceans, the polar regions, and the deserts.

To pursue its work at sea, the department first chartered the brigantine *Galilee* for a three-year survey of the Pacific. The ship was

---

* The institution's first grant went to the Marine Biological Laboratory in Woods Hole, which, at the time, was in tight financial straits. The Carnegie Institution was actually going to take over the laboratory until prevented by certain legal intricacies in MBL's charter. Luckily, this forced the laboratory to work out its own salvation as an independent institution, a status it very much enjoys today.

The nonmagnetic brigantine *Carnegie* under full sail. (*Courtesy Department of Terrestrial Magnetism, Carnegie Institution of Washington*)

not specifically designed for her task and had a sailing vessel's usual complement of iron nails and fittings, all of which affected the delicate instruments that were kept on board to measure the earth's magnetic field. An additional problem was her lack of auxiliary power. By 1909 the directors of the department had decided that a specially built vessel was needed for ocean surveys, and in the same year work was begun on a nonmagnetic ship which would have almost no iron on it or in it at all. The *Carnegie*, as the 155-foot vessel was named, was framed of white oak, planked with yellow pine, and held together with locust treenails and bolts of copper and bronze. Her auxiliary engine was built almost entirely of bronze, and her anchors were of bronze and manganese. She was rigged as a hermaphrodite brig, and her lines of hemp were rove through wooden blocks and spliced around copper thimbles. (Her crew even wore nonmagnetic belt buckles and of course ate their meals with aluminum flatwear.) An observation room and two circular laboratories were built on the *Carnegie*'s main deck, and the best instruments available were installed.

The first six of the *Carnegie*'s seven cruises had very little to do with oceanography but were most valuable to both the practicians and theoreticians of terrestrial magnetics. On the ship's maiden voyage in 1909, for example, she followed the route taken some 200 years earlier by the British astronomer Edmund Halley in the *Paramour Pink*. Halley had also been collecting data for a magnetic chart and had presented his "magnetic chart and some barnacles which I observed to be of quick growth"[*] to the Royal Society. In the 200 years between the two voyages the magnetic declination (the deviation between true north as shown by the stars and magnetic north as shown by the compass) had changed so much that had the *Carnegie* followed the *Paramour Pink*'s compass courses she would have made landfall, not near Falmouth on the south coast of England as intended, but somewhere along the northwest coast of Scotland. The *Carnegie*'s mission was to measure such variations.

On subsequent cruises the *Carnegie* sailed through all the oceans of the world, until, in 1921, she was laid up for repairs. The ship remained idle for six years, in part because of the rising cost of running such an exceptional vessel. Then, in 1927 the *Carnegie* was towed from her berth on the Potomac River and put into drydock for extensive repairs and for refitting. Plans were being laid for her seventh and most ambitious cruise, and to justify the cost of the projected three-year voyage, it was decided to add a program of oceanographic investigations to her usual complement of magnetic and atmospheric ones. The program had to fit the ship, for there were certain things that the *Carnegie*, without powerful winches and strong steel cables, could not do. Trawling for fish or bottom animals and dredging for rocks could not be considered, but the vessel did have sufficient power to handle water bottles and thermometers, small bottom samplers, and plankton nets. The *Carnegie* also received a radio and a sonic depth finder from the Naval Research Laboratory which enabled her to make a considerable number of accurate and precisely located soundings.

The commander of the *Carnegie* and director of her scientific staff (two hats which are rarely worn by the same person) was James Percy Ault. Ault had studied physics in Kansas, and after graduating in 1904, had joined the crew of the *Galilee* to set out upon what was for him an unfamiliar sea. He liked a scientist-sailor's life, and from the *Galilee* moved to the new *Carnegie* and first commanded that ship on her third cruise. Ault was a pleasant man, by all accounts; he commanded seven scientists and 17 sailors with tact, and he operated much of the deep-sea equipment with considerable skill.

Also on the *Carnegie*'s staff and concerned with the oceanographic program was Floyd Soule, a physicist from the U.S. Bureau of Standards (who later became chief civilian oceanographer for the Ice Patrol); H. R. Seiwell, a chemist who did some biological work as well; and J. Harland Paul, a recent graduate from medical school who doubled as ship's doctor and plankton collector.

On the bright morning of May 1, 1928, the Seventh Street dock in Washington, D.C., was crowded with families, workmen, and dignitaries, all waiting to see the graceful *Carnegie* move off down the Potomac. It was a slow start. First the vessel had to be carefully towed to the mouth of the Chesapeake Bay, where she was swung to make sure that the new oceanographic instruments were not affecting her magnetic instruments and compasses. Satisfied that all was working well, Captain Ault then ordered the new sails raised, and the ship set off for Newport News, Virginia. There, to everyone's frustration, she was put into drydock for readjustment of the oscillator in her sonic sounding gear. At last, toward the middle of May, the *Carnegie* set off across the Atlantic.

The real work of the *Carnegie*, however, did not get under way until the vessel had picked up a considerable amount of additional equipment in Hamburg, Germany. The German Atlantic Expedition aboard the *Meteor* had returned only a year before and the *Carnegie*'s sponsors were consciously imitating parts of that expedition's successful program. Most of the scientific staff of the *Carnegie* went from Hamburg to Berlin to visit the Institute of Oceanography and to talk there with the staff of the *Meteor*. In consequence, more Richter and Wiese reversing thermometers were ordered, the *Carnegie*'s winch was modified, and the *Meteor*'s gifts of a huge reel of aluminum-bronze wire and a glass-lined coring device (a type of bottom sampler) were gratefully accepted.

Harald Sverdrup of the Bergen Geophysical Institute, who had spent seven years aboard the *Maud* in the Arctic on a drift similar to the *Fram*'s and who now, among many other positions, was the "consulting oceanographer" for the *Carnegie*, also met the ship at Hamburg and brought with him a new plankton pump. With all this help, advice, and gear the *Carnegie* was soon fully equipped for her duties, and in July she sailed from Germany to Iceland and thence south toward the Panama Canal.

In flat, blue calms and in gray, heavy weather the *Carnegie* hove to at eight o'clock every other morning to occupy an oceanographic station. Captain Ault made shallow- and deep-water casts with a string of Nansen bottles while the ship's doctor manned the plankton pump, which was used to draw water, with its inhabitants, from three specific depths. A bottom sampler given by the Scripps Institution (or occasionally the *Meteor* model) was let over the stern while at the other end of the ship a sailor lowered a plankton net from the forecastle. Another net drifted amidships, and with four lines and a pump hose dangling over the ship's sides it was inevitable that, upon occasion, the lines got fouled. Sometimes, as when a string of Nansen bottles tangled on the oscillator mounting near the ship's keel, Captain Ault saved the gear by the dangerous procedure of swimming under the rolling ship in a crude diving helmet to disentangle the line by hand. This was not often possible, for tangles usually caused the lines to break suddenly and unexpectedly, and the *Carnegie*'s work was sometimes delayed for months until the lost gear could be replaced.

After the *Carnegie* had passed through the Panama Canal, she began prescribing a number of large loops which took her through the least-known portions of the Pacific, where no oceanographic measurements of any kind had ever been made. She circled into the southeastern Pacific, sailed north again to touch at Callao, Peru, then headed west across the equatorial Pacific. Next she turned north, to circumnavigate the North Pacific, narrowly missed several typhoons near Japan, and continued north, then east, past the Aleutian Islands and the tip of Alaska, until, after a year and three months at sea, she sailed into San Francisco Bay. At this point the projected three-year cruise was a little more than a third over. The *Carnegie* was next expected to string a line of loops and curves diagonally across the Pacific between the coasts of California and Australia, but the ship never sailed beyond Samoa.

On Thanksgiving morning the *Carnegie* arrived in the harbor of Apia in Western Samoa and the usual scramble ensued as Captain Ault made his official calls and ordered fuel and supplies. The scientists went ashore to have their nets repaired, their clothes washed, and their letters and packages mailed home. But on the following day, most of the crew had free time, and several expeditions were arranged both for the collection of bugs and plants and for a visit to the tomb of Robert Louis Stevenson at Vaea. Only Captain Ault, Soule, and a few others stayed behind to load gasoline. By lunchtime only 150 gallons of fuel remained to be transferred from barrels into the ship's sloshing tanks, and the engine room was so full of fumes that it took no urging to get the men to leave for lunch. An hour or so later, Captain Ault returned to his customary chair on the quarterdeck while the engineer and mechanic went below to complete winching the barrels aboard. In the

The *Carnegie* burning at Apia, Samoa. (*Courtesy Department of Terrestrial Magnetism, Carnegie Institution of Washington*)

engine room the switch was thrown to start the electric winch, and with a low, shuddering rumble a ball of fire erupted through the *Carnegie*'s deck. The captain was blown from his chair into the water, and as Soule and the steward dove overboard to help him, the engineer and mechanic frantically scrambled from the burning engine room through the hole blown in the deck. Small boats pulled away from every ship in the harbor. Several took the *Carnegie* in tow and beached her before she could drift into other ships; other cutters and skiffs picked up the men in the water. Captain Ault, apparently with only minor injuries, was rowed ashore to be driven to a hospital. On the way he died.

So ended the last cruise of the *Carnegie*. She had sailed nearly 45,000 nautical miles over the loneliest, least familiar stretches of the Pacific and had sent data back to Washington, D.C., from almost every port of call. Only a few of her records were lost in the explosion at Apia, yet in spite of this, the slim four-volume *Scientific Results of Cruise VII of the "Carnegie" during 1928–1929* did not appear for 15 years— *after* other ships had begun to visit many parts of the Pacific, *after* new methods had replaced some of the ones her staff had used, and *after* American oceanographers had become deeply involved in the problems presented by World War II.

It is difficult to reconstruct the circumstances that combined to strand the *Carnegie*'s data on the office shelves of the Department of Terrestrial Magnetism. Perhaps it lay there because the department had no oceanographic section and, quite naturally, placed a higher priority on publishing the magnetic data. Harald Sverdrup was asked to write the report on physical oceanography, and this he did in the early 1930's. Still, the problems of the Depression and the confusion of another war delayed its publication. Luckily, a great deal of the *Carnegie*'s findings made its way into Sverdrup, Johnson, and Fleming's classic volume, *The Oceans, Their Physics, Chemistry and General Biology*, which was published in 1942. In that one-volume compendium of oceanographic knowledge, the distribution of chemical compounds in the Pacific, the currents and water masses in the same area, and even a brief survey of the marine sediments are all credited to the seventh cruise of the *Carnegie*.*

The loss of the *Carnegie* posed a particular problem for the Scripps Institution for Biological Research, which by 1925 had expanded its program and accordingly changed its name to the Scripps Institution of Oceanography. It had been intended that Scripps would use the *Carnegie* for two years as soon as the vessel had completed her seventh cruise, for since 1917, when the old *Alexander Agassiz* had been sold, the institution had not had a seagoing ship and had had to make do with a small purse seiner and the occasional use of chartered vessels.

The director of Scripps at this time was T. Wayland Vaughan, a dapper Texan who had gone straight from Harvard to the U.S. Geological Survey, and who had accepted the directorship of Scripps in 1924, on the condition that he be allowed to change the laboratory from a biological station to an oceanographic institution. His plans were agreeable to the University of California, and within a few years programs were started for the study of marine sediments, chemical oceanography, and marine bacteriology. Vaughan also encouraged a growing number of cooperative studies whereby Scripps' scientists went on cruises planned by the California State Game and Fish Commission, the Coast and Geodetic Survey, and the Naval Hydrographic Office. These cooperative programs made up in part for the missing *Carnegie*, but did not, of course, help Vaughan raise the

---

* The only other oceanographic ship working in the Pacific at that time was the Danish vessel *Dana*. On board was Johannes Schmidt, who, after solving the riddle of the eels' spawning grounds in the Atlantic, had turned his attention to the eels in the Indian and Pacific oceans. The Carlsberg Foundation sponsored the *Dana*'s productive two-year cruise. Much of the work was done in the western and southwestern sections of the Pacific.

money he needed to house all the new activities he was importing. His most pressing need was for a new laboratory, which he estimated would cost between $100,000 and $120,000. The obvious person to approach was Ellen Scripps (her brother had died), and after Vaughan had received her promise for approximately a third of the necessary funds and had convinced the state legislature to add another third, he turned his thoughts to Washington, D.C., a city whose workings he knew well, where plans were afoot for channeling several million dollars of Rockefeller money into America's expanding but still rather lethargic program of marine research. Vaughan got the money he needed for a new laboratory, but a new and in some ways rival institution got much more.

The possibility that such massive aid might be available for oceanography seems first to have occurred to Frank R. Lillie, the director of the Marine Biological Laboratory (MBL) in Woods Hole. In 1925 Lillie began talking about the benefits that could be had from the establishment of a private oceanographic institution on the East Coast. Such a center, he thought, should be modeled after MBL, whose facilities were available each summer to a great number of university students and professors.

In the late 1920's the time appeared right for the introduction of such a plan. Bigelow's Gulf of Maine studies were being completed, the Naval Research Laboratory had been built in Maryland, Scripps had officially broadened its scope, and the Ice Patrol had begun making important physical investigations. It was the time, wrote Bigelow, "when, in a word, oceanography appeared to be stirring in the U.S."☥

Lillie knew that the General Education Board, a part of the Rockefeller Foundation, was interested in his ideas for an oceanographic institution, but he knew too that the plan would have a better chance of success if it were thought through and presented by the prestigious National Academy of Sciences rather than by any group of individuals. Therefore, he suggested that the NAS appoint a committee on oceanography to look into the possibilities. In 1927 such a committee was duly appointed, with Lillie as its chairman. Members were asked "to consider the share of the United States of America in a world-wide program of Oceanographic Research . . ."☥ which was to say—are American oceanographers keeping up with the Petterssons, the Murrays, the Hjorts, and the Heinckes? Lillie predicted that his committee would find "that the interests of oceanography have not advanced as they should in the United States of America in comparison with several other countries."♆ It was not surprising, therefore, that the committee's *Report on the Scope, Problems, and Economic Importance of Oceanography* . . ., written by committee secretary Henry Bigelow, showed that the United States did, in fact, lag far behind. According to the report, there were only three universities in the country where courses in oceanography were offered, but none of these institutions granted degrees in the subject. (In 1930 Scripps began giving a degree in biological oceanography.) Reflecting in part this lack of instruction, there were only 124 oceanographers in the whole of North America, but, continued the report, there weren't enough jobs available in the field to attract young scientists from other disciplines into marine science. The committee's list of ills continued with a description of the short-term, practical nature of all government-supported studies and the biological bias of most private ones. The report closed with a lengthy list of recommendations, which, in addition to philosophical advice, suggested the expansion of three existing institutions and the creation, as Lillie had hoped, of an oceanographic laboratory on the Atlantic Coast.

The General Education Board responded favorably to many of the requests made by the committee, and in 1929 and 1930 provided $265,000 for a laboratory and boat needed by the University of Washington's Puget Sound oceanographic laboratories, $245,000 for the expansion of a biological station on Bermuda, $40,000 for the proposed Ritter Hall of physical and chemical oceanography at Scripps, $24,000 for two marine laboratories in France, and the breathtaking sum of $3 million for the establishment of an oceanographic institution at Woods Hole.♆

Parenthetically, this was not the first time that the Rockefellers had given money for the study of oceanography. In 1926 John D. Rockefeller had given a substantial grant to the American Petroleum Institute for a five-year study of oil-bearing sediments. Parker D. Trask of the U.S. Geological Survey undertook the study and had 2,000 samples of marine sediments from 1,600 locations analyzed. The sediments came from the *Challenger* collection, the International Ice Patrol, the Bureau of Fisheries, the Finnish Hydrographic Service, the Hudson's Bay Company, and several dozen other sources. The object of the study was to find out how petroleum deposits are formed, which in turn would help geologists find new oil fields. "The science of oceanography is being called upon to aid the oil industry," wrote Trask.⸸ It was the beginning of a long and close association.

In January, 1930, the Woods Hole Oceanographic Institution (WHOI) was incorporated "to prosecute the study of oceanography in all its branches, to maintain a laboratory or laboratories, together with boats and equipment and a school for instruction in oceanography and allied subjects. . . ."⸸ Frank Lillie was elected president of the board of trustees, and Henry Bigelow accepted the position of director in spite of his genuine distaste for administrative duties. (The new institution was to be operated only in the summers, which would leave Bigelow free to work on his fish and other interests for most of each year.)

By early summer, work had already begun on a large, red-brick laboratory.* At the same time, plans were being made for acquiring a research vessel, and Bigelow was so convinced of the importance of having a fine, seaworthy vessel capable of making long voyages in the stormy North Atlantic that he tactfully dodged all offers of old pleasure yachts, sidestepped the economical compromise of converting tired fishing vessels, and held out for the best ship that Rockefeller money could buy.

"No other American marine laboratory, independent of the government, is able to do this," he noted as the trustees (their personal fortunes still quivering from the stock-market crash of 1929) agreed to spend $175,000 on the largest steel-hulled ketch in the world.⸸

A sailing ship with a powerful auxiliary engine was chosen over a steamship because the ketch would be less dependent on fueling stations and would be more stable. This latter attribute was important both in handling cumbersome equipment and in executing delicate chemical tests and microscopical studies. The contract for a slim, two-masted vessel, with two laboratories, two winches, and quarters for six scientists and 17 crew, was awarded to a shipbuilding company in Denmark. In anticipation of the ship's delivery (promised for the summer of 1931), Bigelow hired his former student Columbus O'Don-

---

* At Lillie's suggestion, the Marine Biological Laboratory had tried to give WHOI the three-fourths-acre lot needed for its first building, but again the laboratoy's infamous legal intricacies made such a donation impossible. The Carnegie Institution again came to their aid, this time with close to $27,000, the purchase price of the land.

The ketch *Atlantis*, research vessel for the Woods Hole Oceanographic Institution, fills away on a port tack. (*Courtesy Woods Hole Oceanographic Institution*)

nell Iselin as master of the research vessel *Atlantis* and as research associate in physical oceanography.

An unexpected problem was created by the impossibility of working during the summer of 1930, before the boat or the main building was ready. Under the terms of the Rockefeller endowment, a fund (not to exceed $50,000) was already being given to WHOI to cover its annual operating expenses, but there was nothing to operate. Discussion ensued as to how to incur some first-year expenses, and it was finally decided to give some of the embarrassingly superfluous money to the *Nautilus* expedition that was being equipped for a projected cruise beneath the Arctic ice. In return, some of the data collected on the voyage would be sent to WHOI, and the expedition's reports would be considered the first contributions of the new institution.

Earlier in 1930, when the Australian explorer Sir George Hubert Wilkins had announced his intentions of voyaging beneath the North Polar ice cap in a submarine, and hopefully of making a submerged run for the pole, the U.S. Navy had offered him the use of an overage submarine. Renamed the *Nautilus*, the vessel was modified and duly equipped with a Fathometer, an ice drill, some instruments for measuring the earth's magnetic field (not to be used from the metal sub), and a very special forward pressure hold. This forward compartment could be pressurized until the force of the air pushing out exactly balanced the pressure of the water pushing in, at which time a hatch in the floor of the compartment could be opened and instruments lowered into the sea. In this way water bottles, reversing thermometers, bottom samplers, plankton nets—all the usual gear—could be used to gather data from what was normally an inaccessible region, the sea beneath the ice cap.

The money from the new oceanographic institution was used to buy additional equipment, and in June, 1931, the *Nautilus* finally crossed the Atlantic to pick up her small scientific staff in Bergen, Norway. The crossing was so hard on the ship that she required extensive repairs, and it was not until the middle of August that she was ready to travel north. This delay of two months forced Wilkins to cancel his plans for a trip to the pole, but he gamely insisted on undertaking as much scientific work as would be possible in the remaining three or four weeks of the summer season. Accompanied by Harald Sverdrup, Floyd Soule from the Carnegie Institution, and a German physician, Wilkins and the *Nautilus* set off for the island of Spitsbergen and after a brief call there headed north and northwest for the edge of the drifting polar ice cap. The edge of the pack ice was much farther north than usual that summer, and the *Nautilus* churned through the waves for five days before she came upon it. At this point the *Nautilus* was ready to dive beneath the pack ice to investigate the deep waters of the polar sea, which had not been measured since Nansen had drifted over the area in the *Fram* in the 1890's. There was another reason for heading the *Nautilus* beneath the ice: the vessel had to submerge in order to use her forward compartment, her ice drill, and in fact all of her gear except the Fathometer and magnetic instruments.

The ice floes at the edge of the pack rose and fell, and the *Nautilus*, like a plane racing its engines prior to takeoff, whirred and buzzed as batteries were tested, the ballast system checked, and her diving rudders waggled to and fro. Only then was it discovered that the submarine had lost a diving rudder and would not be able to cruise beneath the ice.

Wilkins refused to return to Spitsbergen, although this seemed the only sensible thing to do, but insisted that the submarine follow the edge of the pack ice still farther north. For several days the *Nautilus* proceeded with only her Fathometer working—Soule rediscovered the Spitsbergen-Greenland ridge that Nansen had crossed some 30 years before—until she had gone as far north as the ice permitted and was, surprisingly, over the deep water of the polar sea. Magnetic instruments were landed and used on ice floes, and the submarine scraped and pushed her way beneath several small floes, and there, free from the action of surface waves, lowered her gear through the forward compartment. After almost a week and a half of fairly successful work in the ice, the *Nautilus* returned to Spitsbergen and the cruise was over.*

The *Scientific Results of the "Nautilus" Expedition, 1931, Under the Command of Captain Sir Hubert Wilkins* was published in 1933 and comprised the first five contributions to the Woods Hole Oceanographic Institution's series of collected reprints. Harald Sverdrup wrote on the general oceanography of the area, Soule compiled the echo-sounding data, the physician reported on gravity measurements, and Henry Stetson, a new research associate in submarine geology, described the bottom deposits.

The use of submarines for studying the ocean did not become popular as a result of the hectic cruise of the *Nautilus*. The expedition's technical innovations were still too new and unreliable to be followed by others, and although submarines had been used for marine research prior to the *Nautilus* expedition and were used occasionally for similar purposes afterward, they did not become a common tool for oceanographic work until the mid-1960's. The *Nautilus* expedition did, however, set a fine example of the way to wring the most information from the least number of observations, and the expedition's report was an admirable model for the new oceanographers arriving at Woods Hole.

The summer of 1931 was the first regular season for the "Oceanographic," as WHOI was commonly called, and a dozen scientists, their students, and assistants arrived to occupy the new laboratory. The *Atlantis* was still in Denmark, but the institution's new 40-foot launch, *Asterias*, was moored at the pier and served as a collecting vessel for the summer's work.

One of Bigelow's major problems was to find oceanographers who

---

* In 1958 the first nuclear-powered submarine, *Nautilus*, namesake of the one Wilkins used, cruised beneath the ice all the way to the North Pole.

would be willing to work at Woods Hole. His report for the National Academy of Sciences had indicated that there weren't many more than 100 or so in the United States, and these men were already employed. In addition, most of the professors among them had already made arrangements with Scripps or some government agency for summer field work. If oceanographers were not immediately available, at least not in the United States, there remained two possibilities. As Bigelow saw it, "You must either import him or bring him up."▼

Bigelow chose the latter course, and his staff in the early 1930's consisted of a physiologist, a paleontologist, several chemists, a meteorologist, and various biologists—all, however, with an interest in oceanography and a willingness to go to sea. Within several years those who returned each summer to Woods Hole could certainly claim to be oceanographers, and many of the graduate students they brought with them underwent the same metamorphosis.

In the summer of 1932 the *Atlantis* was at last in Woods Hole, and the routine that would become characteristic of the institution's prewar years was established. Throughout July and August the *Atlantis* went out on weekly cruises, and every student and professor in the laboratory, regardless of the delicacy of his or her digestion, was expected to go out at least once during the summer. Bigelow wandered through the laboratory helping Stetson to organize the *Nautilus*' bottom samples, stringing a plankton net on a frame for a surprised lady graduate student, or squinting through a microscope at the plankton that Alfred Redfield had collected.

Bigelow encouraged a wide variety of studies; and projects concerned with "The Role of Bacteria in the Cycle of Life in the Sea," "Methods for the Determination of Dissolved Organic Carbon and Nitrogen in Sea Water," and "Marine Erosion of Glacial Deposits in Massachusetts Bay" were the kinds of things that summer workers at the institution worked on.

But when Labor Day came each September the institution was deserted. A procession of overloaded cars took scientists and their families back to inland universities, Bigelow returned to his office and his fish in the Museum of Comparative Zoology, and even the business manager (who was also an ichthyologist) left Woods Hole for the more stimulating atmosphere of greater Boston.

In 1932 the Ice Patrol moved its headquarters to Woods Hole, and their oceanographers were given space at the institution which they occasionally used during the long winter months. From the laboratory windows facing Vineyard Sound the dim outline of the idle *Atlantis* could sometimes be seen through the winter fogs.

The *Atlantis* didn't spend every winter in Woods Hole, however. When funds were available and scientists willing, she sailed south to the Caribbean Sea or the Equatorial Atlantic and there collected deep-sea fish around Cuba, took salinity measurements off the mouth of the Amazon River, or collected eels from the Sargasso Sea for the Carlsberg Laboratory in Denmark.*

Also in the 1930's the *Atlantis* made a series of short cruises for the Bureau of Fisheries to help their scientists continue a program of mackerel investigations. The effects of the country's economic depression had hit government science as well as private projects, and the fisheries' budget was so badly cut that their vessel, *Albatross II*, had been laid up. Other laboratories and bureaus were facing similar problems, all caused by rapidly diminishing funds. By 1933 the situation had become so bad, and there were so many scientists unemployed, that President Franklin Roosevelt created a Science Advisory Board to recommend a solution. This organization promptly drew up a Recovery Program of Science Progress—a New Deal for scientists—but this was never put into effect. Oceanographic work, which among scientific pursuits was particularly expensive, slowed down. By the mid-1930's, however, most of the radically pruned budgets, whether of a university department or a naval laboratory, had returned to pre-Depression levels, and only a year or two later, preparations for World War II turned the Depression's famine into a hectic sort of feast.

This time American oceanographers were better prepared. As the government hastily resumed the research it had dropped after World War I and initiated a host of new projects, the scientists at Scripps, Woods Hole, and elsewhere were ready to turn their oceanographic knowledge into instruments of war and defense. This they did, but the effect that oceanography made on the war, though significant, was dwarfed by the tremendous changes that the war made on oceanography. After 1945, rambles and reveries by the seaside and long collecting trips aboard the *Albatross* were not even a memory.

---

* Iselin, who often commanded the winter cruises and who himself was most interested in ocean currents, was happy to spend time collecting eel larvae, for the animals' trip from the Sargasso Sea back to Europe was basically a passive drift, and "eels," maintained Iselin, "make fine drift bottles."

CHAPTER VIII

Oceanography and World War II

The product of the intellectual effort appears to be conditioned by
the time of planting and the character of the soil as well as by
quantity of seed.

—Herbert Gregory, "A Century of Geology"

During World War II a new and intimate relationship evolved between government and science which profoundly affected the growth,
form, and even the content of oceanography. The foundations for this
new alliance were laid in haste in 1940.

In the spring of that year, at a time when there were great differences of opinion as to what the nation's role in the war should be, there
was accord on one significant point—were the country to enter the
war, her scientists would play an extremely important part. It was
further agreed that the application of scientific knowledge to the
contingencies of war would take time and would require effective
administration, and from the desire to supply the latter came the
National Defense Research Committee (NDRC) and, a year later, the
more comprehensive Office of Scientific Research and Development.

The purpose of NDRC was to focus scientific attention on problems
of weapons research, and the committee's method, known as the
contract system, was designed to distribute a great variety of military
problems among an equally varied assortment of private institutes,
university departments, and industrial laboratories. The selection of

problems to be solved came largely from the armed services, but the way in which each task was to be handled was a matter for the scientists to decide. This flexible system was essentially a new way of relating the government's needs to the nation's scientists and was one which blurred the traditional distinction between government science and private science.

By the time the United States entered the war in December of 1941, the NDRC had let about 450 contracts which involved an estimated 2,000 scientists. In addition, many men and women who did not work under these contracts were brought directly into the scientific sections, units, and laboratories that were proliferating at a bewildering rate within the armed services. For oceanographers, this rush of activity meant that virtually all their numbers (with the exception of some marine biologists) became involved in war research.

The conditions under which these men found themselves working were markedly different from those which had characterized their prewar occupations, and the changes were bound to affect their work. The most obvious change was the sudden expansion of oceanography due to a vast infusion of men and money. Considering the latter first, a lack of funds had traditionally been the limiting factor for both public and private marine research. For a century the Baches, Maurys, and Bairds had tried to balance the funds they needed for research with the appropriations they could wring from Congress, while the administrators of private institutions had tried to live within the gifts and endowments which for them were financial boundaries. For both groups the problem each year was to tailor their research to fit their budgets. Then, suddenly, that problem was gone. As the war radically reordered the nation's priorities, the importance of science grew, and research appropriations soared. The funds that the government spent on war research rose from approximately $100 million in 1940 to $1.6 billion in 1945; during the same period, expenditures at the Naval Research Laboratory grew from $1.7 million to $13.7 million; and the budget of the Woods Hole Oceanographic Institution rose so spectacularly that as of 1942 publication of the annual treasurer's report was suspended for fear of revealing "the magnitude, if not the nature, of the investigations being undertaken at the request of the government."⋇

The rise in the number of scientists working on war research was equally impressive. At the Naval Research Laboratory the number of employees rose from 400 to well over 2,000; at WHOI a summertime staff of some 60 persons expanded into a year-round staff of 335; and on the West Coast the number of persons working for the University of California's Division of War Research and the Scripps Institution of Oceanography increased by several hundred.

In addition to this influx of funds and people, war research introduced some changes of a less productive nature. For one thing, in spite of the apparent growth in the number of persons involved in research, many of the newcomers had little or no training. Consequently, when an institution doubled in size it did not double its productivity. Secondly, there was a sense of urgency about war work that impelled, or at least encouraged, scientists to rush through each investigation in as short a time as possible. During the war there was rarely time to expand upon a study or to relate it to anything save the war's immediate needs. Finally, the inherent incompatibility between scientific freedom and military security proved to be a source of irritation and delay. Security clearance was a particular problem, and several knowledgeable but foreign-born oceanographers were never cleared and, therefore, could not work on projects for which they were eminently suited. Other security procedures were designed to keep large amounts of information compartmentalized, and this made it extremely difficult for scientists to exchange ideas and build on each other's work.

Perhaps the most important effect that the war had on the development of oceanography was the redistribution of interests within the field itself, and this was largely the result of the military's selection of research problems. For a period of three or four years it was the nature of the war, not the interests of the scientists, that determined which studies within oceanography would be pursued and which laid aside. Consequently, new interests arose within marine science and old ones sometimes changed or became relatively less important. The requirements of the war affected the development of each of the disciplines within oceanography, and, generally speaking, affected them in one of three different ways.

First, the problems generated by the war sometimes pointed out important scientific questions or areas of investigation, such as wave research or submarine acoustics, which had been neglected. When scientists found themselves unable to answer the military's questions because a basic understanding of the subject had not yet been worked out, hasty work was usually done on the more fundamental aspects of such problems, at least until enough information had been gathered to meet the military need. After the war, when scientists had time to examine these areas of neglect thoroughly, their studies often opened a new field of research.

The progress of ocean science was affected in a very different way when war work required only the compilation or reorganization of existing data, for such projects essentially asked scientists to go over what they already knew. During the war hundreds of master lists had to be prepared from thousands of partial lists, and special charts had to be made from the scattered notations that appeared on other charts. At best this type of data transcription served to corroborate the understanding that scientists had already gained from their prewar work, but even then it did little to advance the field in terms of new data or new ideas. After the war the scientists who had been marking time resumed their prewar work at about the same point where they had dropped it three or four years before.

A third way in which the war could affect oceanographic investigations was to suspend their progress entirely. If a project had no bearing on the war effort, and if in addition it required the use of a research vessel for the study of deep offshore waters, then its chances for continuation were poor indeed. After 1942, when even the coastal waters around the United States were threatened by enemy mines and submarines, research vessels were either taken over by the Navy and armed or were retired for safekeeping in some inland waterway. Investigations dependent upon their use were suspended for the duration of the war. This was not, however, as drastic an interruption as might appear, for even if no work could be done on a project, that same study was almost bound to profit later on from one or another of the hundreds of new instruments and techniques that were developed at a fantastic rate during the war.

With all these wartime changes—more people, more money, more instruments, and more military selection of topics—it is not surprising that oceanography changed both in size and in character. It emerged from the war a bigger, richer field and one of changed proportions, aims, and allegiances. Physical studies assumed a new importance, and marine biology gradually lost its former preeminence. Economically, oceanography no longer had the qualities of a small private enterprise but had become irrevocably tied to government support and government policy.

Among the most pressing problems of the war were those related to submarine and antisubmarine warfare, and it was with these that the greatest number of oceanographers were involved. That this would be so had been anticipated to some degree years before the war began, and a few persons in the Navy, remembering the scramble that had accompanied the development of submarine detectors and listening gear during World War I, had suggested that some war research be continued into the 1920's and 1930's. Their advice had largely gone unheeded, and work on submarine devices had continued only at the Naval Research Laboratory (NRL) in Maryland and at the Submarine Signal Company in Boston. At NRL the underwater sound group, still inspired by Harvey Hayes, worked to develop a reliable sonar system or, as it was called then, an echo-ranging system. After World War I, a great improvement in this type of gear had resulted from the development of high frequency transmitters which produced a sharp beam of sound analogous to the cone of light from a searchlight, rather than a hazy glow of sound. One advantage to having a beam rather than a glow was that the former could be aimed and swept in all directions. When such a well-directed beam of sound struck an object and echoes were returned, it was possible for the first time to ascertain the location of the target or obstruction. In other words, the new sonar systems were directional.

In the 1930's directional sonar systems had been designed and tested, redesigned and retested, with the understanding that, once properly put together, the systems would function reliably both as submarine detectors, for use aboard destroyers and patrol boats, and as surface craft detectors, for use aboard submarines. Yet, in the late 1930's there was still doubt as to how effective sonar might be in case of war, for the systems suffered from some curious problems. At times they worked and at other times they didn't, and when they didn't, no amount of knob twiddling or readjusting would correct their mysterious malfunctions.

These intermittent problems were encountered aboard the old four-stack destroyer *Semmes* in 1936. The *Semmes*, temporarily stationed at the Guantánamo Bay Naval Base in Cuba for a routine training session, had been equipped with a sonar system and given a tame submarine to chase around the bay. After a week or so of practice, her sound men reported that although the sonar worked well every morning, it would not always function properly later in the day. Especially on those

The four-stack destroyer U.S.S. *Semmes* worked with the R/V *Atlantis* to solve the riddle of the "afternoon effect." The *Semmes* carried no guns, so as to conform to the agreements made on naval limitations after World War I. (*Courtesy U.S. Navy Department*)

afternoons when the sun was bright and hot and the air still, the sonar's sound beam could not detect the submarine, even when the destroyer steamed right over her. One of the *Semmes*' sound men, Lieutenant William Pryor, thought he knew why this was so.

When the *Semmes* returned to her home port in New London, Connecticut, Pryor drove to the Woods Hole Oceanographic Institution to seek corroboration for his hypothesis, which argued that the sonar was being jammed by a screen of bubbles let off by plant plankton. He thought that on a sunny afternoon these plants would photosynthesize so rapidly and produce so much oxygen that millions of tiny bubbles would be released. In even a slightly effervescent sea, the sound from the sonar system would be scattered and distorted.

The Oceanographic's director, Henry Bigelow, and his assistant, Columbus Iselin, were intrigued by Lieutenant Pryor's problem, and although they did not believe that effervescing plankton were responsible for the sonar's misbehavior, they did agree with Pryor that the sea and not the sonar was probably causing the Navy's troubles.

Iselin was anxious to examine the sonar difficulties, and while Bigelow felt strongly that the Institution should not become too in-

volved in deep-water plumbing, as he sometimes called applied research, he agreed to offer the Navy the combined services of Mr. Iselin
and the *Atlantis* for a short period of time. In October the Navy
accepted his offer, and, as predicted, began to unwind the yards of
letters, orders, and directions that were necessary to permit the *Atlantis* to operate from Guantánamo Bay "in connection with underwater
sound investigations. . . ."

Shortly after Christmas the *Atlantis* sailed south, and having completed her previously scheduled investigations along the way, arrived
at Guantánamo Bay on January 30. The *Semmes* had already arrived
for her annual maneuvers, and scientists from NRL and the Submarine
Signal Company had also come south. The plan was for the *Semmes*
and the *Atlantis* to "ping on each other" with similar sonar sets, and
while tracking each other, the scientists aboard the *Atlantis* would
study the character of the surrounding sea water. A "sound stack" was
duly attached to the port side of the *Atlantis,* and for nearly two
weeks the ketch and the destroyer moved back and forth together in
and out of the bay. From the *Atlantis,* string after string of closely
spaced water bottles and thermometers were let into the sea.

The results were as Iselin had suspected. The water samples, when
analyzed for oxygen content and salinity, gave fairly consistent values,
which did not suggest that plankton were producing an excess quantity
of gas. The water temperatures, on the other hand, varied considerably
from hour to hour within a single day, and sound transmission was
most difficult when the topmost layer of the sea was warm and
unruffled.

"I do feel that we are on the track of something,"[*] wrote an officer
aboard the *Semmes* when he proposed to Bigelow that the same kinds
of investigations be repeated off Long Island later in the year. With
Iselin favoring the plan, Bigelow agreed to another joint investigation,
and in August the *Atlantis* and the *Semmes* set off again for a more
sophisticated series of trials. After the data from this cruise had been
studied, Iselin was ready to suggest to the Navy the reasons for its
sonar problems. Under certain conditions, he told them, there existed
well-defined layers of cooler and warmer water near the surface of the
sea, and due to the way sound traveled through water, these layers
were responsible for bending and distorting the sonar's beam of sound.
Iselin called this phenomenon the "afternoon effect."

It was already well known that the speed of sound in the sea varied according to the character of the water, and especially in accordance with its temperature. The early users of echo sounders had been aware of these relationships, and although the variable temperatures in the sea did not incapacitate their equipment, they did affect the speed of the sound pulses, and tables relating the speed of sound to the temperature and salinity of the water had been made to ensure more accurate soundings. When sonar was introduced and sound began to be sent horizontally instead of vertically through the sea, then differences in water temperatures became a serious problem. Even slight temperature changes created distinct layers of warmer and cooler water, and in each of these, sound traveled at a different speed. Because of this, a sonar's beam of sound traveling obliquely through these layers was deflected from a straight path. The beam was bent either upward or downward each time it passed from one layer into another (as light is bent in passing through a prism), and consequently it could fail to detect a submarine which lay directly in its path. The implications both for submarines and submarine hunters were not difficult to grasp.

With Iselin's explanation of the "afternoon effect" circulating slowly through naval channels, a growing number of persons in that service began to realize the importance of learning everything possible about the variable conditions of the topmost four or five hundred feet of ocean water. (The "safe depth" for World War II submarines was 412 feet.)

Were there worldwide charts, they asked Iselin, that showed how the temperature of sea water varied with depth and with the different seasons of the year? Only the most general ones, he had to reply, for although oceanographers had been interested in the structure of the ocean for a long time, they had not concerned themselves with detailed and time-consuming studies of a thin, superficial layer. Their general studies had shown, however, that in temperate and tropical seas there generally existed a layer of relatively warm, well-mixed water beneath which was a transitional layer called the thermocline, in which the temperature of the water dropped very quickly. Beneath the thermocline was the deep ocean water, where the temperature changed hardly at all. The thermocline, which affected the transmission of sound most radically and therefore was of the greatest interest to the Navy, varied both in depth and in definition, depending on the weather, the season, and a number of other factors. Often the phenomenon disappeared entirely, especially during the winter, when the sun's warmth was not great and when frequent storms stirred the waters. Interesting as this information was, such generalities were of little use to a submariner who might wish to hide from an enemy's probing sonar or to a destroyer's sound man whose job it was to track a submarine. What the Navy needed was some method of measuring or predicting the character of the thermocline in all the oceans of the world at all times of the year. This seemed an impossibly ambitious task, but in Iselin's opinion there was a new and incompletely refined instrument known as the bathythermograph that might be able to furnish the information that the Navy needed.

In 1934 Carl-Gustav Rossby, a meteorologist at the Massachusetts Institute of Technology and a summer resident of Woods Hole, designed and built an unwieldy device called an "oceanograph," which was supposed to continuously register the temperature of a body of water from its surface to a depth of several hundred feet. After struggling all one summer with the bulky and cantankerous instrument, and deciding that the idea was right but the design all wrong, Rossby gave the "oceanograph" to his friend Athelstan Spilhaus to redesign. Spilhaus was a young meteorologist from South Africa who was studying at MIT, and upon receiving the "oceanograph" he went to work and designed an instrument that was almost as inconvenient as Rossby's original model, but that could be used from a stationary ship with fair results, provided it was lowered and raised very slowly. The device

Columbus O'Donnell Iselin, second director of the Woods Hole Oceanographic Institution, favored close cooperation with the Navy during World War II. With Maurice Ewing, he wrote *Sound Transmission in Sea Water*, a report that was supported by one of the first contracts let by the National Defense Research Council. Iselin is shown here after his retirement near the WHOI small-boat dock. (*Courtesy Woods Hole Oceanographic Department*)

was used aboard the *Atlantis* in August, 1937, for the second series of sonar studies with the *Semmes*, and the Navy was extremely interested in the results. Unfortunately, the process of obtaining records of the water's temperature was so messy, time-consuming, and complicated that the device was not seriously considered as an instrument that would be of real value in case of war.

Spilhaus was not put off by the instrument's liabilities. He made further improvements and in 1938 applied for a patent for what he called a "bathythermograph," or BT for short. The instrument consisted of an open, rectangular frame within which was mounted a compressible bellows with a pen arm and a stylus at one end. The stylus, which moved in response to temperature changes, rested upon a smoked-glass slide, and across this it scratched its record of the sea's temperatures. The stylus not only moved horizontally in response to changes in temperature, but also vertically in response to changes in depth, and consequently a wobbling profile that related temperature to depth was scratched on the slide.

The BT still had some serious problems, however, and in 1940 two scientists from Lehigh University, Maurice Ewing and Allyn Vine, who were working at the Oceanographic for the summer, began redesigning the instrument so that it might respond much more quickly to temperature changes. They also wanted to streamline the instrument so that it could be lowered and raised from a fast-moving ship. With these attributes, the BT would be an invaluable aid to the Navy's sonar operators, who had to know the temperature structure of the water if they were to judge how their sonar systems would function.

One Sunday during that same summer of 1940, Columbus Iselin, who had just succeeded Bigelow as the director of WHOI, sat with his neighbor Frank Jewett on the latter's porch overlooking Vineyard Sound and discussed the prospects of the newly organized National Defense Research Committee. Jewett, who was president of both the National Academy of Sciences and the Bell Telephone Laboratories, was a founding member of NDRC and director of one of its several divisions. Since his responsibilities included the development of submarine and antisubmarine weapons, he was interested in hearing about the work that Iselin and the Navy had started at Guantánamo Bay and about the improved BT. Both men agreed that the Navy needed a way of predicting a sonar's behavior, and both believed that oceanographers were best suited to work on the problem.

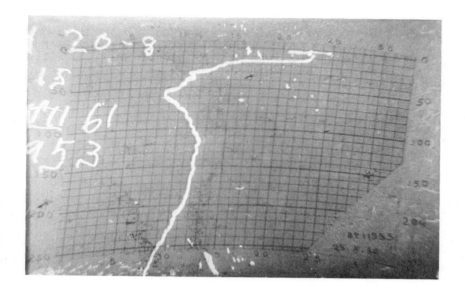

The bathythermograph record, top, shows that beneath a narrow layer of warm water lies a region of rapid transition some forty meters thick in which the temperature drops from 23° C. (73° F.) to 8½° C. (47° F.). Below this region—called a thermocline—the temperature changes only gradually. The instrument directly below the record is an early BT designed by Athelstan Spilhaus. It had all of the principles but none of the refinements of the Ewing and Vine BT pictured below it. (*Courtesy Woods Hole Oceanographic Institution*)

As a result of these conversations, Iselin asked the trustees of the Oceanographic to formally offer the Institution's facilities to the government for military research. This they did, and NDRC awarded one of its first contracts to WHOI for an investigation of the transmission of sound through the sea. In October the Institution's first year-round staff was assembled to work on the project.

Among the dozen or so persons to arrive at Woods Hole that autumn, or more precisely to return to the village, were Maurice Ewing and Allyn Vine. The former, an able and energetic scientist, had unhesitatingly left Lehigh, where he taught physics, to join Iselin on the sound project. The two were well suited to work together, for Iselin's knowledge of the internal arrangement of ocean waters was complemented by Ewing's familiarity with the intricacies of sound propagation. Together they wrote a report, *Sound Transmission in Sea Water*, which was delivered to the Navy in February, 1941. This was not a manual on how sonar would behave under certain circumstances, but a treatise on a new and unexplored subject—submarine acoustics. Not only did it set down what was then known about the transmission of sound underwater (and this was later incorporated into manuals for sonar operators), but it also pointed out what remained to be learned. The second category was almost as important as the first, and experiments were begun to supply some of the missing information. How, for instance, was sound reflected off different types of sea-floor sediments? Other investigations suggested by the report were undertaken after the war, and the study of submarine acoustics became an increasingly important part of oceanography.*

Although *Sound Transmission in Sea Water* was promptly published just a few months after the United States had entered the war, the information it contained was not of much use to sonar operators, for they had neither BTs nor temperature charts and without these could not tell how the surrounding waters were arranged or how their gear would work. Efforts to supply both BTs and charts proceeded simul-

---

* One of the postwar studies motivated in large part by Ewing himself was the investigation of the Sound Fixing and Ranging Channel, the SOFAR Channel, which is a naturally occurring stratum in the sea through which sounds are able to travel for thousands of miles. During the war it was proposed that a network of coastal listening posts be established to monitor this channel in order to locate underwater distress signals that could be sent by downed airmen or shipwrecked sailors. Unfortunately, the problems of positioning a receiver some half-mile or more below the surface were so great that the first permanent listening stations were not ready for use until 1947.

taneously, and BTs were soon modified for use aboard destroyers and submarines.

From a fast-moving destroyer it was decided to try lowering BTs on a wire which could be run off a short boom amidships. During the first trials the BTs went down into the water with no trouble, but on being reeled in, the brass torpedo-shaped instruments swerved and skipped over the waves and seemed bent on wrapping themselves around the ship's propellers. The problem was rectified by adding fins and non-twisting towing shackles which made the BTs' path through the water considerably less erratic.

The next step was to modify the BT for use aboard submarines, and in the spring of 1942 work was begun on this project at Woods Hole. The temperature-sensitive part of the instrument was redesigned to ride on the outside of the submarine, while the smoked slide and stylus were to be installed in the control room. By June the first dozen submarine BTs were ready, and the scientists who had designed and built them took the instruments to naval yards for installation and testing.

New submarines and vessels in for repairs and refitting were equipped with the BTs, and their crews were shown how to operate them and how to take advantage of the information they recorded. The Woods Hole oceanographers suggested that a submarine on patrol try to dive at least once each day to see how the water layers around it were arranged. This served several purposes. First, a BT profile could indicate how effective an enemy destroyer's sonar system was likely to be. If the smoked slide showed a strong thermocline, it was often possible to compute how a destroyer's sound beam would be bent and where the "shadow zone" or area of minimum sound penetration would lie. This shadow zone was the best place for a submarine to hide once it had shot off its torpedoes and called down upon itself the wrath and the depth charges of the destroyer.

Since the temperature of the water had a great effect on the buoyancy or trim of a submarine, a BT profile could also help a submariner estimate the amount of ballast he would need to take on in order to slide from periscope depth into the shadow zone. In winter, for example, when the water was well mixed, a submarine might have to let in 3,000 to 6,000 pounds of water to descend to a depth of 400 feet, while in the summer, in water with a well-developed thermocline, it was often necessary to take on between 6,000 and 18,000 pounds of ballast to execute the same maneuver.

"Found a 'best escape depth' at or below 200 feet with a temperature 'knee' [thermocline] so pronounced that I do not believe echo gear exists capable of driving down through it," wrote the commander of the submarine *Scorpion* in May, 1943. "The bathythermograph inspires feeling of confidence when in contact with enemy escorts. . . ."ᵠ

"The bathythermograph was quite helpful in . . . maintaining depth control," wrote the *Pomfret*'s commander. "By carefully watching for sudden temperature changes, the Diving Officer could make the necessary ballast changes before depth control was lost."ᵠ

By the spring of 1943 BTs were in use aboard many submarines, and at the end of each mission the smoked slides were sent either to Scripps, if the submarine had been operating in the Pacific, or to Woods Hole, if the profile had been made in the Atlantic. Thousands of BT slides were received as the war continued, and from the information scratched upon them, charts were constructed to show the average temperature structure of the oceans both in winter and in summer. These were sent back to the submariners and could be used when only general information was needed or when there was no time to make a BT slide.

By the end of the war, when some 60,000 BT slides had been col-

The U.S.S. *Pomfret* was one of many submarines to be equipped with a special BT designed at the Woods Hole Oceanographic Institution. (*Courtesy U.S. Navy Department*)

lected from the North Atlantic alone, and when scientists finally had time to examine them, some new and interesting things were learned about the uppermost layer of the ocean. Even more was learned when scientists began gathering their own BT data, not from random points scattered across the oceans, but from carefully arranged sections that cut through the axis of a current or followed the outline of an eddy. Since the pattern of a current could be deduced from the temperature structure of the water, bathythermograph records made in the Gulf Stream, for example, helped physical oceanographers see how the current changed its shape and position, not in the course of a month or a season, but from day to day. The movements of the Gulf Stream were exceedingly complex, and the whole current was found to wobble or meander from side to side as it flowed through the western Atlantic. There were eddies and countercurrents too which swirled and backed along the sides of the Stream. These peripheral movements had been too transient to study until the BT made rapid sampling possible. Like Iselin's and Ewing's report on sound, the huge collection of BT slides and the new instrument itself encouraged oceanographers to turn their thoughts and investigations in a new direction.*

To return to the first years of the war, before BTs had been per-fected, there was a feeling in the Navy that some way ought to be found to give submariners at least a rough idea of the temperature structure of the water they were operating in, for, with or without BTs, these men still had to deal with the environmentally inspired

---

* Theoretical work in physical oceanography advanced as did observational studies, and after the war Harald Sverdrup and Walter Munk at Scripps and Henry Stommel at WHOI began studying the dynamics of wind-driven cur-rents in hopes of explaining such phenomena as counter-currents, eddies, and the curious westward intensification of wind-driven currents. Concerning the last, even before the war Iselin had made the surprising discovery that north of the Straits of Florida the Gulf Stream crowds against the American coast, becomes narrower and more intense, and yet transports *more* water than at its apparent source. This intense crowding toward its western boundary, a characteristic shared by the Kuroshio and the Agulhas Current, had never been satisfactorily explained until 1948 when Stommel showed how in theory at least a wind-driven current would vary according to the investigator's concept of the Coriolis effect. If the Coriolis effect was thought of as operating uniformly over the entire ocean then oceanographers must envision circulation in the form of a simple, symmetrical gyre, but if the Coriolis effect was considered to vary with latitude and was in-troduced into the dynamic equations as a variable force, then the oceanographic model showed a gyre pushed and flattened against the ocean's western boundary. This approximation seemed to Stommel to be much closer to the observed circula-ation in real oceans. Other studies of this sort advanced rapidly after the war.

idiosyncrasies of sonar systems. The quickest way to supply the desired information was to construct temperature charts from lists of data that had already been collected; and this task was given to the Scripps Institution of Oceanography.

In 1936, the directorship of Scripps had passed from T. Wayland Vaughan to the Norwegian oceanographer Harald Sverdrup, and with Sverdrup had come a new emphasis on physical oceanography and on open-ocean work. Within a year of his arrival the institution had purchased a movie star's pleasure yacht and converted her into a research vessel. Rechristened the *E. W. Scripps*, the ship had been used for an average of six long cruises each year, and on many of these, numerous temperature and salinity measurements had been made. Using these data and, to an even greater extent, using the many observations made by the Japanese prior to 1939, the physical oceanographers at Scripps began working on temperature and salinity charts. (Since the temperature and salinity of water were usually studied together, both properties were included on the charts, even though the first affected sonar operation much more than the second.)

Unintentionally, the Japanese had furnished American oceanographers with an ample supply of information on the waters of the Pacific Ocean. In the decade preceding the war, oceanography had received much greater recognition and support in Japan than it had in the United States, and when the war began, the island nation had approximately 95 vessels capable of doing some marine research, 17 public marine laboratories, seven private ones, and more than a dozen fisheries stations. The temperature and salinity measurements made by all the research vessels for each of the organizations were published routinely in fishery station reports and in the bulletins and reports of the Japanese Hydrographic Department. These documents had been conscientiously distributed to marine scientists all over the world. Most of this data had been collected from the rich commercial fishing grounds near the islands of Japan—just the areas in which U.S. submariners wished to operate. To make the needed charts, then, it was only necessary to sort and recombine the wealth of information pertinent to each strategic region and to present it in chart rather than tabular form. Hundreds of such charts were duly produced at Scripps, but "on the whole," wrote Sverdrup, "the work was of such a nature that it did not lead to any new contributions to our understanding of the regions."*

Other of the wartime projects assigned to Scripps, or to the University of California's Division of War Research, were more closely related to academic interests, and one such project was the construction of sea-floor sediment charts for the eastern and southeastern continental shelves of Asia.

The nature of the sea floor, as Iselin and Ewing had pointed out, influenced the behavior of sound, and this was especially true in shallow waters of less than several hundred feet. Consequently, investigations had been started off San Diego to find out what happened to sound when it was bounced off various sediments. A smooth, firm sand bottom, it was found, reflected sound without much distortion, and was, therefore, a poor place for a submarine to try to hide. A rough, rocky bottom, on the other hand, sent a sonar's pulses ricocheting from boulder to boulder in a confusing pattern and was a fine place for a submarine to conceal itself. A soft mud bottom absorbed a great deal of sound and presented conditions somewhat analogous to the deep sea, where little or no sound was reflected from the bottom.

With this understanding of how sea-floor sediments affected sonar pulses, it became imperative for the Navy to supply its submariners (and antisubmarine personnel) with charts showing what kinds of sediments lay in the areas where they were to operate. Work on such charts began early in 1943; it was an enormous undertaking.

First, hundreds of ordinary navigation charts and field sheets which had been made over the past 100 or so years by the Japanese, British, Dutch, French, and Americans were assembled, and on each, bottom notations of sand, gravel, mud, etc., were circled; each kind of sediment was marked with a particular color. These bottom notations were especially numerous on Japanese charts, for the Japanese had continued the practice of measuring depths with a lead line, a method which brought in a sample of the bottom, well into the 1930's, years after other nations had begun to rely almost exclusively on sonic sounding devices. Next, all of the 400,000 bottom notations which had been circled were plotted on master sheets, and finally boundaries were drawn between areas of sand, sand-and-mud, rock, and so forth. The accuracy of these maps depended in large part on the density of bottom notations available for each area, and this ranged from fewer than one to more than 100 per square mile.*

---

* In a similar manner, detailed sediment maps of the eastern and western Atlantic were made by the Germans, largely from British and American navigation charts.

The 43 sediment maps that were finally constructed showed the distribution of shallow-water sediments all the way from the Kamchatka Peninsula off Siberia, south to Japan and China, west through the Indonesian archipelago, and north again to the head of the Bay of Bengal. Ambitious as this project was, it was not the first to show what sediments lay on large portions of the world's continental shelves. In the 1930's Japanese, Dutch, and American geologists had made a few general sediment maps of some shelf areas, and in doing so had begun to find that the traditional concept of continental shelves as being well-sorted piles of continental debris was not supported by the data they assembled. The idea had been—and still was during World War II— that the tailings of continents had washed into the sea for millions of years and that the coarsest of these (the sands and gravels) had settled near shore, while the finer ones, such as muds and silts, had been carried farther seaward. As this process continued, great shoulders or shelves of well-ordered sediments had supposedly built out around each continent, and on each of these should be found concentric bands of graded sediments. But the maps of the 1930's hadn't shown these bands of gravels, sands, and muds. Instead, there were extensive mud flats near shore, where the coarsest sediments were supposed to lie, rocky, current-swept zones where nothing was being deposited, and a disturbingly common aggregation of coarse sediments at the outer edge of the shelf where only the finest silt was supposed to settle. These anomalous discoveries were confirmed by the Asiatic war maps, which showed similar patterns of sediment distribution. Although there were few opportunities during the war to study the new sediment charts from anything but a military point of view, in later years some of the geologists who had helped construct them obtained a private grant in order to extract different sorts of information from the same charts. Tables were made showing the relationships between types of sediments and depth, and between sediments and their distance from shore. Reasons for the observed distributions were sought too, and it was found that some sediment patterns were inherited from the past. During the Ice Age, for example, sea level was much lower, and rivers flowed across the exposed continental shelves, depositing

---

The maps were published in atlases which accompanied the "Uboothandbüch." The latter contained information on water temperatures, currents, tides, and shore features, and even gave the locations of lighthouses, navigation signals, church steeples, and war installations.

coarse debris along the way. Some of these old river deposits can still be found beneath coastal waters, as can the beaches which formed at the edge of the continental shelf, or wherever the sea met the land, many thousands of years ago.

The pattern of sediments was also found to be influenced by currents and wave action, the position of coral and algal communities, and by the arrangement of submarine hills and canyons. From the study of all these factors came a much clearer understanding both of the distribution of sediments on continental shelves and of the processes which effect this distribution.

The outrageously diverse trials and tribulations of sonar operators gave oceanographers some biological as well as physical and geological problems to solve during the war.

"The Japs may have some newfangled gadget that they drop [into shallow water],"⍦ reported a submarine operator early in the war when he heard a strange crackling noise all around his vessel which shut out the thrum of an enemy vessel's engine.

"Noises . . . as if the deck grating over the boat storage had been lifted and dropped . . ." reported an officer on the U.S.S. *Tarpon*, who then distinctly heard the echo-ranging pings from two ships but listened in vain for the sound of their propellers and noted that "nothing was in sight through the periscope."⍦

Within the first six months of the war, American submariners anxiously reported hearing staccato taps, clanks, rumblings, hammerings, croakings, whistlings, and, most eerie of all, the rhythmic beat of sonarlike beeps and pings. As hundreds of unidentified sounds were reported, each having emanated from some invisible source, the National Defense Research Committee let contracts for studies which together were intended to constitute a "survey of underwater sound." There were three kinds of sounds to investigate: naturally occurring, physical ones such as the sound of rain falling on the sea or of whitecaps curling and splashing down upon themselves, manmade sounds from ships and harbor installations, and animal sounds. Of all these, the prodigious variety of noises produced by fish, crustaceans, and mammals were the most diverse and confusing.*

---

* The Japanese National Research Council was also concerned with underwater sounds, especially with a "noisy sound" that was reported each evening at their

Of course marine animal noises had been heard long before the war, and when NDRC asked the U.S. National Museum for a list of all known marine noise producers, that well-stocked repository that had been cared for by Spencer Baird came up with an inventory which included 14 families of fish and 17 families of crustaceans. (Mammals were not mentioned, in part because nineteenth-century anatomists, having dissected a few and having found no vocal cords, had incorrectly proclaimed the animals mute.) Still, a partial list of noise producers was hardly a first step toward solving the submariners' problems, for it provided no reliable way of matching sounds to animals or vice versa. There were plenty of opinions as to what sounds were produced by which animals, but little information. For example, there was a remarkably widespread crackling noise that had disturbed the owners of wooden-hulled vessels for well over a century and that had been variously ascribed to shipworms chewing on the timbers and to communities of clacking mussels living on the vessel's bottom. Both guesses were wrong, as was the Coast and Geodetic Survey's more modern hypothesis that related a different noise—like coal rumbling down a metal chute—to road traffic on a coastal highway. These puzzling sounds and contradictory explanations had not, however, inspired marine zoologists to investigate the sounds in the sea, had not, that is, until the Navy's sound men and NDRC called for help.

In the summer of 1942 Martin Johnson, a zoologist at Scripps and a biological consultant at the University of California's Division of War Research, was asked to study underwater animal noises. One of the first things he did was dangle a hydrophone over the end of Scripps' 1,000-foot-long pier, and in doing so he discovered a "chorus of fish noises. . . ."

"The chorus," he reported, "begins about sunset and increases gradually to a steady uproar of harsh froggy croaks, with a background of soft drumming."⸙

He also heard the mysterious crackling noise.

To identify a particular animal as the producer of a certain sound, Johnson collected many kinds of fish, shrimp, clams, and squid, and each of these he placed in a large concrete tank in which was sus-

---

coastal listening posts and that was encountered with annoying frequency by their submariners. For a long while it was assumed that microturbulence in the water was responsible for this noise, but later, a war research project related the "noisy sound" to fish and shrimp.

pended a hydrophone. Any noise produced in a tank was picked up by the hydrophone, amplified, and recorded. These sounds could then be analyzed electronically and one or more characteristic acoustic signatures ascribed to each kind of animal. Without this sophisticated gear, which certainly had not been used by marine zoologists before, many animal sounds had been inaudible, and others had been indistinguishable—a croak was a croak.

Among the most interesting underwater noise producers to be studied in this manner were several shrimp known as pistol prawn or snapping shrimp. These animals emitted a very occasional but a very sharp "snap" or "crack," and after listening for a year to their infrequent pops, Johnson finally identified them as the producers of the mysterious crackle.* Extensive communities of snapping shrimp were found to live in rocky areas all along the coasts, and from each of them emanated a continuous crackle, day and night, summer and winter. In a similar manner Johnson traced the chorus of froggy croaks to three kinds of croakers but was not able to determine what produced the soft drumming until after the war.

The information thus gained by Johnson and other marine zoologists was incorporated into Navy manuals, which were used to identify underwater sounds and also to predict in a general way the noises that would be encountered in various localities. Snapping shrimp were common in rocky areas, toadfish in shallow, muddy bays, whales and dolphins in deep water, and so forth.

After the war, marine zoologists had time to study the noises they had discovered from a biological point of view, and instead of asking how the snap of a shrimp would sound over a destroyer's hydrophones, they wanted to know how the shrimp made such a loud snap and why it snapped and when. In answering these questions zoologists made use of the instruments that had been introduced to them during the war, and soon a whole new field of marine biology was being explored. Animals thought to belong to a single species were separated into two groups because of the different sounds they made; the distribution of snapping shrimp was mapped by sonic means; and the annual

---

* Johnson discovered that these snapping shrimp were responsible for a multitude of previously inexplicable sounds, including the timber chewing and the coal rolling. The "shrimp crackle . . .," he said, was so "conspicuous in the area where these observations were made . . . [that it] is loud enough in some spots to be audible to the unaided ear above the surface during calm water."⸸

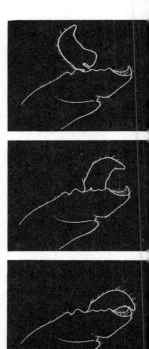

One species of the ubiquitous snapping shrimp whose cracklings and poppings so disturbed sonar operators on submarines. The animal pictured, *Alpheus longimanus*, has the snapping shrimp's characteristically enlarged claw. Silhouettes show snapping of claw. (*From the* Challenger *Report*, Zoology, *Vol. 24, and courtesy Moody Bible Institute*)

migration of croakers was found to begin first with males—their distinctive croaks diminished first—and later by females, whereas it had previously been assumed that the two sexes departed together.

More was learned too about when fish vocalize, and postwar studies at the Bureau of Fisheries Laboratory in Beaufort, North Carolina, found that members of the incredibly noisy toadfish family only grunt for five or six weeks each spring and that during this time they remain in shallow water. During the war, when it had been discovered that the noise from a single large toadfish could detonate a certain type of acoustic mine that was being sent to the Pacific, the Naval Ordnance Laboratory had delayed the shipment in order to change the actuating mechanism on the mines. In view of what was learned later about the habits of toadfish, the changes were unnecessary.

Different sorts of marine biological research were stimulated by other of the Navy's problems, but usually these troubles were not acute enough to warrant intensive studies. Investigations of bioluminescent organisms which could silhouette a surfacing submarine in a halo of greenish light or studies of a mobile layer of animals which appeared on echo-sounding records as a false sea floor were only barely begun during the war but were followed up with enthusiasm in later years.*

On occasion, the effects of war research on marine biology were very different, and while intensive wartime studies were initiated on antifouling compounds, for example, this work neither added much information to the biologist's store nor opened new areas for him to study. (The studies were more productive in the field of paint chemistry.) In general, marine biological investigations profited less from war research and hence emerged in a relatively less important position within oceanography after the war.

---

* A notable example of the delayed effect that military problems often had on marine studies is the investigation of the deep scattering layer (dsl). The dsl is a concentration of animals which moves toward the surface of the sea at night, then returns to depths of approximately 1,000 to 1,500 feet during the day. These organisms scatter the sound pulses emitted by echo sounders and thus appear on echograms as hazy bands at intermediate depths. These bands, or false sea floors, were only infrequently encountered before the war because the echo sounders of the 1920's and 1930's were not designed to run continuously, and their widely spaced soundings did not give the kind of picture of the water column or of the bottom that a continuous recording does. Then, when continuous, self-recording echo sounders came into use during the war, the dsl was encountered with surprising frequency in almost all parts of the Atlantic and Pacific. Observations of the layer were kept confidential by the Japanese and Americans in case it proved useful as a hiding place for submarines. After the war the dsl was declassified and a considerable number of fascinating studies begun.

In 1945 scientists aboard the *E. W. Scripps* spent a 24-hour period following the rise and fall of the false bottom on the ship's echo-sounding gear. Guided by the echograms, 14 hauls were made with closing nets above, in, and under the layer in an attempt to capture the animals responsible for the phenomenon. The results were disappointingly inconclusive. Few animals were caught, and the investigators were led to suspect that the echo sounder's record gave a misleading impression of the animals' numbers and concentration or that the layer was really an aggregation of fish which were following and feeding upon migrating plankton but which were too fast to be caught in the slow-moving nets.

Scientists at Woods Hole became intrigued with the problem a few years later and began using fast, mid-water trawls and deep-sea cameras to study the deep scattering layer. These investigations, and further studies on the West Coast, have shown that the dsl is composed primarily of small mid-water fishes with gas-filled swim bladders (such as lantern fish) and occasionally of certain types of jellyfish.

If, during the war, the greatest number of oceanographers worked to solve problems related to submarine warfare, then certainly the next greatest number were concerned with amphibious warfare. In this instance, the need for oceanographic knowledge and services was not anticipated before the war; but by the fall of 1942, when plans were being made for the invasion of North Africa, it became apparent that some way was needed to forecast the wave and surf conditions that would so considerably influence the success of scheduled landings. Consequently, the Oceanographic Section of the Army Air Force Weather Directorate asked that the possibility of wave forecasting be studied, and the project was given to Harald Sverdrup, director of Scripps, and to Walter Munk, a young Austrian geophysicist at the same institution. The two men were asked to work out a practical way of forecasting wave conditions, using the information given on standard weather maps. Considering the almost nonexistent state of wave research at the time, the request was an immensely difficult one.

The study of water waves had not always been neglected, and at various times in the nineteenth century men had studied waves in two, essentially different ways. Early in the century, deep-water waves had been considered as problems in theoretical hydrodynamics, and theories had been formulated which expressed how such waves ought to be formed, how they should move, and how their dimensions should be interrelated. Toward the end of the century, other men had chosen to study waves from an empirical point of view, and they had spent their time at sea and on shore measuring not the ideal but the real waves that came rolling across the ocean. The next step was to relate the two approaches, but this was difficult to do, and early attempts to tie theory to experience met with little success.

One such attempt was made by a Britisher, Vaughn Cornish, who was particularly noted for years of conscientious wave watching. The results of his efforts were published in his book *Ocean Waves and Kindred Geophysical Phenomena* in 1934. Some 40 years before, Cornish had begun watching the waves that broke on the beach below his house on the south coast of England, and from there he had carried his investigations to foreign shores and far out to sea. On one occasion he found himself at sea on a P. and O. Orient liner during a fierce North Atlantic storm. The gale sank several ships with all hands on board, and although it put his own liner "for a time in peril . . ." he considered that "the opportunity for which I had longed for years had arrived, and at last I had a stationary post of observation amidst the great waves of the ocean. Moreover, by a fortunate and rare coincidence, the tremendous wind had come on to blow exactly in the direction of the heavy swell already running, and so there was no confused turmoil, but one magnificent procession of storm-waves sweeping across the sea from horizon to horizon."[*]

Cornish, huddled on the promenade deck, estimated the height of these awesome waves to be 40 feet from trough to boiling crest. Next he timed the successive arrivals of the waves, estimated the distance between them, and from these two measurements calculated their speed to be 43 miles per hour.

He made all these observations and calculations, he maintained, to provide mathematicians with a vast collection of actual measurements so that they might compare the properties of real waves to those of their theoretical models. By integrating the two, Cornish hoped that the theory of water waves could be greatly refined.

A "magnificent procession of storm-waves" became an important phenomenon to understand if naval operations were not to be disrupted and delayed during wartime. (*Courtesy Woods Hole Oceanographic Institution*)

"But there are fashions in mathematics as in all human pursuits, and I [have] had to wait no less than thirty years before a younger generation of mathematicians . . . began to develop the material which I had collected."[Ψ]

Even when theoreticians began to notice his work, there was no single opinion on how observation and theory should be related, and the results of similar studies were often contradictory. Some scholars insisted that the speed of waves could never exceed the speed of the wind that generated them, while others maintained that waves could travel half again as fast. But such questions troubled few minds until the war began. Then these issues became important to resolve.

Sverdrup and Munk, working at Scripps, had little time to reexamine nineteenth-century wave theories or to make many additional observations of the long Pacific swells that swept in upon the California coast. In spite of their hurry, they did get the theories and observations put together, and in doing so generated such enthusiasm for wave research that scientists began studying waves after the war and have done so in greater detail and with greater sophistication ever since.

Very generally, Sverdrup and Munk found that the size of waves did not depend solely on wind speed but was also affected by the size and shape of the body of water over which the wind blew (the fetch) and the length of time that the wind had been blowing (the duration). They studied too how the wind's energy was transferred to the water, and again found that more factors were involved than had been suspected. They arranged all this information in a useful form so that the character of waves and swell could be computed from wind speed, fetch, and duration—information that was commonly available. As soon as their report *Wind, Waves, and Swell; A Basic Theory for Forecasting* was completed in 1943, they began to study the action of waves in shallow water so that surf conditions might be forecast from wave predictions. Again they tried to pull theory and observation together. They described how waves slowed down when they entered shallow water, became steeper, higher, and closer together, and finally broke. The depth at which they broke (in other words the location of the surf zone) could be calculated from the original dimensions of the waves. This information appeared in a report on *Breakers and Surf* . . . which was followed by a score of short wave reports in which Sverdrup and Munk described how waves would behave in specific situations. Field manuals were prepared from all these reports, and in

Harald Sverdrup, director of Scripps from 1936 to 1948, adjusts a current meter that is being tested at the end of Scripps' pier. During World War II Sverdrup worked on many projects and wrote a series of reports on waves, swell, and surf. (*Courtesy Scripps Institution of Oceanography, University of California, San Diego*)

addition, a course in military oceanography was begun at Scripps to train the wave and surf forecasters needed for amphibious operations.

The forecasting service which thus evolved was made available (although not fully used) for the invasion of Normandy in June, 1944. Months before the operation began, a Swell Forecast Section was established, and the men in this group set up a network of 51 wave-reporting stations which extended along the south and southeast coasts of England. At each station the speed and direction of the wind, the wave height, and the wave period (the interval between successive waves) were measured three times a day, and from these observations the relationship between weather and waves in the English Channel became familiar. Forecasts were then begun for the invasion beaches on the Normandy coast, and when possible the accuracy of these early predictions were checked with aerial photographs. To make the forecasts more accurate, secondary factors such as the effect of tidal currents on the shape and height of waves and the influence of coastal irregularities on surf conditions were also considered. By the spring of 1944, with D Day approaching, wave and surf conditions on the four assault beaches could be predicted two and three days in advance with an accuracy of about 80 percent.

Normandy was to be invaded on the morning of June 5. The tides, both of war and the sea, were considered favorable, but as the day approached a cold front swept through the Channel, stirring up winds of 12 to 18 knots and rough, unfriendly seas. The Swell Forecast Section confirmed what anyone walking outside could see and feel, and D Day was reluctantly postponed. Later in the day the weather began to clear, and there was hope that the assault could be launched early on the sixth. The forecasters worked that day and night, and as observations came in from their wave-reporting stations, they coordinated this information with predicted tidal conditions, with the different exposures of the assault beaches, and with every bit of information that might have some bearing upon their most important forecast. Shortly after midnight on the sixth, they were asked to describe the waves and surf that would be racing and breaking onto the invasion beaches as dawn broke later that day. They predicted that waves three to four feet high would move through the embarking or unloading zone that lay some ten miles offshore, and that on the beaches themselves the waves would peak to heights of four and even six feet before breaking. They also predicted a longshore current of 1.4 knots and a rising tide that would quickly cover the beaches, with their tangles of barbed wire and piles of concrete obstacles.

The forecast was admirably correct, but the consequences of rough seas and an unseen current on landing procedures were underestimated, and the decision was made to go ahead with the invasion. On the morning of the sixth, British and American landing forces encountered almost exactly the conditions that the Swell Section had predicted. Choppy seas hampered the transfer of men from transports to landing craft, and when crowds of small boats finally converged upon the narrow, littered beaches, the inshore current caused collisions and confusion. Only in the following weeks and months, as thousands of men and tons of equipment were ferried across the Channel, did studies reveal that a three- to five-foot sea commonly meant a reduction of up to 80 percent in the tonnage that could be transferred and landed in a given period of time.

With a better understanding of how the sea state affected the efficiency of landings, the wave forecasts became more meaningful, and late in 1944 the Swell Forecast Section was moved to Ceylon in preparation for the invasions of Burma and Indonesia. In this case ships and landing craft were not to be taken across a narrow channel as before, but were to sail across the wide Bay of Bengal through waves that were sometimes generated by storms thousands of miles away. To complicate the job of forecasting still further, the invasions were scheduled to take place during the southwest monsoon season, a weather pattern unfamiliar to Western scientists, and the location of the invasion beaches was kept secret until two weeks before the landings. In spite of these handicaps, the Swell Forecast Section set up a ring of 20 observation stations, initiated weather flights over the Andaman Sea, and for the first time employed a few subsurface wave recorders that had been designed and built at Woods Hole.*

For the Burmese invasion, however, the greatest forecasting aid was a wave report that had been written at Scripps, for the landings were to be made in a most unusual area—the delta of the Irrawaddy River. Allied forces were to come ashore near the capital city of Rangoon, which lay near one of the river's many mouths, and to reach their destination they had to cross tens of miles of shallow water that lay just off the delta. Back and forth across this extensive submerged plain ran treacherous seven-knot tidal currents. These sometimes damped the swells that came in from the sea, but at other times the swift streams aggravated the waves' orderly procession and caused ferocious tide rips and a choppy, unsettled sea. Luckily, Scripps Wave Report Number 11, "On wave heights in straits and sounds where incoming waves meet a strong tidal current," had been completed earlier that year and could be used, with modifications, to help predict the wave and surf conditions that were encountered on the delta at the time of the invasion. A successful landing was made in the spring of 1945, and British forces, swarming into Burma both by sea and by land, drove out the Japanese.

As more and more amphibious invasions were launched, it became

---

* A hydrodynamic theory worked out prior to the war had formulated the relationship between the fluctuations in pressure at the sea surface that were caused by differences in wave heights and the much smaller variations in pressure on the sea floor that resulted from the same cause. In hopes that this theory was adequate to account for what really happened in the sea, subsurface wave recorders were built to record underwater pressures at depths of about 75 to 125 feet. From these records the height and period of surface waves were calculated. The accuracy of the system was not really known until after the war, when subsurface recorders were installed and tested off Woods Hole and Bermuda.

Through low waves U.S. marines make an easy mock landing on Green Beach on Taiwan. During World War II swell-forecast groups tried to predict the landing conditions that marines would encounter as real invasions were planned in lands bordering the Atlantic, Pacific, and Indian oceans. (*Courtesy U.S. Marine Corps*)

evident that wave forecasting was only one of the oceanographic services needed for successful landings. The invasion of Tarawa in November of 1943 provided the most appalling example of what a disregard for environmental considerations could mean.

The assault on this small, heavily defended atoll in the Gilbert Islands was a major amphibious operation and involved a great number of planes, ships, and troops. In planning the landings, the only chart of the atoll available was one made by the U.S. Exploring Expedition in the 1840's. Understandably, this chart did not show the lines of submerged reefs that lay off shore, nor did it indicate the patches of soft sand and the piles of jagged coral heads that the landing craft would have to cross. With almost no knowledge of these hazardous conditions, marines were ferried toward Tarawa on an ebbing tide. One after another, the landing craft ran aground on unsuspected reefs, and heavily laden marines were dumped into the surf under fire to wade and swim across a long series of intervening channels and bars. The losses at Tarawa were awesome—1,000 men killed, more than 2,000 wounded. Such a blind approach to enemy shores could not be repeated.

Earlier in 1943 the task of furnishing relevant oceanographic information to military services in all parts of the world had been given to the Navy's Hydrographic Office, and to gather the requested information the Office had established an Oceanographic Unit. This group, directed by a Radcliffe zoologist, Mary Sears, began preparing comprehensive reports on strategic enemy territories. Each report was part of the confidential *Joint Army–Navy Intelligence Studies* (JANIS), and each represented an attempt to bring together between two covers all the oceanographic information that would be useful in planning amphibious invasions or submarine operations. Like the temperature and salinity charts and sediment maps that were being prepared at Scripps and Woods Hole, the JANIS reports incorporated quantities of information that had been gathered before the war. Each of the studies contained a section on waves and swells, diving conditions, the character of the sea water (its temperature, salinity, color, transparency, etc.), bottom sediments, marine life, tides, and currents.*

Toward the end of 1944, just as many JANIS reports, sediment maps, and other charts were nearing completion, the theaters of war began to shrink, and the need for environmental information declined. Of the 43 sediment maps prepared for Asia, only 25 were published, for by the time the others were finished, the enemy had withdrawn from those areas. It was the same for other projects: as the need passed, the work stopped. The end of the war was approaching, and with it came new problems for the scientific community.

Probably the first group to take a serious look at these problems was the Office of Scientific Research and Development (OSRD), the organization that had encompassed NDRC. Fully a year before the war ended, members of OSRD began to consider how best to dismantle the huge war-research complex that their organization had built. Unlike

---

* Toward the end of the war, Japanese oceanographers came up with an ingenious plan for using ocean currents to transport goods. Their country was in desperate need of gasoline and soybeans, yet lacked ships in which to carry these commodities. The oceanographers suggested that gasoline from Formosa and soybeans from Korea be packed in drums that were weighted to float 10 or 15 feet below the surface of the sea and set adrift. The circulation of the China Sea and the Straits of Korea was fairly well known, and with a few additional investigations it was determined that 30 to 50 percent of the goods pushed off the islands of Formosa and approximately 90 percent of those rolled into the sea off the east coast of Korea would drift to Japan and strand upon the beaches of Honshu and Kyushu. In spite of this encouraging report, there was a reluctance among the military to try so bizarre a scheme, and only a single, unmanned vessel, loaded with soybeans, was cast off the Korean coast. As anticipated, it drifted to Japan.

their counterparts from World War I, these men realized that the
military would continue to need scientific research in times of peace as
well as war, and with this understanding, they divided the country's
war-research projects into two classifications, those which ought to be
continued and those which should be terminated. But here their responsibility ended. The list of projects to be continued was handed
over to the War and Navy departments in hopes that these services
would keep the research alive. The military did not have a clear idea of
how this should be done, nor did scientists, and there were many in
both camps who lacked the foresight of the OSRD planners and
assumed that oceanography and most other sciences would return to
prewar practices and procedures.

When the war finally ended in the summer of 1945, it did seem for a
while that a return to prewar conditions had begun. Scores of professors left Scripps, WHOI, the Hydrographic Office, government
laboratories, or wherever they had worked during the war and returned to their former positions in colleges and universities. Moving in
the opposite direction, toward the oceanographic institutions, were the
research vessels that had been laid up during the war and the marine
geologists and biologists who had left to work directly for the Army
or Navy. Apparently the old boundaries between private science and
government science were being reestablished.

Then Columbus Iselin, to pick just one of oceanography's several
postwar administrators, realized with some sense of shock that his
institution at Woods Hole could *not* return to its former ways. In part,
inflation had seen to that. The Oceanographic simply did not have the
money to operate as it had before.

"It has become evident," Iselin said in 1946, "that post-war costs of
oceanographic field work have more than doubled. . . . The costs of
operating our own vessels have become so high that without Government subsidy we could not hope to undertake offshore observations."∗

What a quandary. The institution that Bigelow had so jealously
protected from too deep an involvement in government research now
found itself unable to launch an open-ocean expedition without government funds.

There were two directions in which WHOI and similar institutions
could move. Either they could attempt to live within their endowments, and this would mean restricting their investigations to coastal,

summertime studies, or, if they wished to keep their expanded wartime programs more or less intact, they could accept, indeed encourage, government support. Predictably, the institutions chose to continue their involvement with the government, and the conjunction between private oceanography and public need that had developed during the war was on its way toward becoming the standard relationship during times of peace as well. This new situation, in which the use of public funds so considerably influenced the size and the character of private oceanography, was very different from the one of mutual independence that had existed before the war.

But in the years immediately following the war it was not clear just how the government planned to support oceanographic research. By the end of 1945 a partial adjustment to peace had been made—many military projects involving oceanographers had been terminated, and temporary oceanographers had returned to their former jobs—and private institutions found themselves employing a relatively large number of marine scientists whose services they could no longer afford. If these men were to continue investigating the sea—if, in other words, the government were to continue assuming responsibility for the health of oceanography—then new projects and dollars would have to be supplied without delay.

The Navy was first to respond to this need, and the war was hardly over before plans were being made to involve oceanographers in a program of atomic weapons tests that was being planned jointly by several government agencies. Operation Crossroads, as these tests were called, was the first of several large operations which utilized the talents of oceanographers and in doing so helped them to make the transition from war research to a more peaceful routine.

In 1945, when the decision was made to test atomic weapons on Bikini Atoll in the Pacific, almost nothing was known about the atoll and its surrounding waters, for the lonely scrub-covered island had not been an area of military activity during the war. Consequently, the Navy's Bureau of Ships (which during the war had initiated many oceanographic studies) took on the job of organizing a thorough survey of the area. According to the director of the Bureau's Underwater Studies Group, Roger Revelle, this was to be "an integrated investigation of all aspects of the natural environment within and around the atoll: the currents and other properties of the ocean and lagoon waters, the surface geology, the identity, distribution and abundance of living creatures, and the equilibrium relationships among all these."⸸

Such a detailed study would allow the Atomic Energy Commission to assess precisely the effects of atomic weapons and would give marine scientists a chance to learn a great deal about the structure and ecology of a coral atoll. In March, 1946, scientists from Scripps, WHOI, the U.S. Geological Survey, the Fish and Wildlife Service, the Smithsonian, and the Navy began to arrive, and most of these men embarked upon investigations that were closely related to the research they had done during the war. Walter Munk, who with Sverdrup had worked on waves and breakers, studied the relationship between the waves and the physical contours of Bikini; Martin Johnson, who had helped solve the mystery of snapping shrimp, studied the plankton of the surrounding Marshall Islands; and K. O. Emery, who had prepared many sea-floor sediment maps, investigated the atoll's submarine geology.

For several months dredges bounced and thumped along the flanks of Bikini Atoll, tow nets strained plankton from the blue lagoon, and BTs sank through the island's clear, warm waters. Then, early in summer, the field work completed, the scientists returned to their institutions to spend a much longer period of time studying and analyzing the muds or measurements that they had gathered. At Woods Hole, Iselin estimated that more than 40 members of his 300-plus staff were involved in some way in the Bikini study, and the resurveys of the atoll requested by the Navy continued to support the work of these and many other scientists.*

Another of these large, consciously scientific projects was Operation Highjump, a naval expedition to the Antarctic organized in 1947. As in the Bikini studies, oceanographers were involved in a long series of investigations.

In spite of the Navy's large operations which helped support oceanographers after the war, it was evident that marine science needed a more stable, dependable source of income than could be provided by an occasional project or operation. Again the Navy moved to meet this need, and in 1946 the well-known Office of Naval Research (ONR) was established to encourage, or one might say reactivate, the kind of

---

* The first of the resurveys, made in 1947, included a program of deep drilling. Much was learned of the atoll's geologic history, and many arguments concerning the theories of atoll formation were laid to rest. (See Chapter IV, p. 169.)

long-term studies that had been carried on before the war. Once further modifications had been made in the already flexible contract system, ONR was ready to make funds available for projects that would continue for an indeterminate number of years. In ONR's first year of operation, the Office initiated one of these open-ended contracts for the study of Pacific Ocean biology. This was to be a series of investigations which would continue some of the work that had been started during the war on bioluminescent organisms and marine sound producers. At the same time, a contract was signed with the Woods Hole Oceanographic Institution for a study of the hydrography of the western Atlantic, and funds were provided for monthly surveys of two offshore areas and for less frequent surveys of other areas.

There was still a feeling among scientists, however, that their work needed a broader base of support and that oceanography should not be too completely financed by the Navy. Some objected on philosophical grounds—"We cannot tie science to the military and hope to see it used for peace, no matter how ingeniously we write the contracts"[¥]— while others were concerned for the health of the science itself. With only military support, oceanography could easily grow in a lopsided manner, for there were certain studies within the field with so little relevance to even the most imaginative of military needs that they were not likely to be supported. Also, with the newly formed Department of Defense paying the bills, the kind of research that could be done in those areas marked "important to the government" was not entirely left to the scientists' discretion. Some funds, these scientists began to argue, should be made available for the support of investigations without regard to military value.

With the protection of basic research and the encouragement of science education in mind, a plan was put forward for the creation of a National Science Foundation (NSF), and from 1946 onward, some form of NSF Act duly appeared each year in the Senate or the House or even on the President's desk. A bill was finally passed and signed in 1950, a director appointed the following spring, and "a sum barely sufficient for the simple needs of . . . business operation" appropriated.[¥] In spite of its initial poverty, NSF grew rapidly and within a few years was using its funds to distribute grants, fellowships, laboratories, equipment, and for oceanography, even large research vessels.

With the establishment of NSF in 1950, the significant sources of oceanographic support which developed during and after the war and which exist today are accounted for, and although no two reports agree as to who spends what on marine research, the general picture is that the Department of Defense controls about 46 percent of each dollar spent on oceanography, the NSF 7 percent, other government agencies 28 percent, and private sources some 19 percent.*⸸ These figures vary from year to year, but in all instances they reflect the government's decision to become and to remain oceanography's chief supporter.

And since public funds continued to flow into private oceanographic institutions after the war, marine scientists were finally able to take advantage of the new instruments and techniques that had been devised during the war years. The postwar period became the time of real gain for oceanographers, for although the necessities of war had increased their numbers and the variety of their equipment, it was only in the years that followed that these new strengths could be used for a basic investigation of the sea. It was then that those scientific gains were realized that the war had in a sense prepared for but not allowed.

---

* The contributions to marine research made by oil companies and other private industries have not been included, for although the work financed by such companies is becoming extensive, the results are rarely made public and are therefore set apart from the rest of the country's oceanographic programs.

CHAPTER IX

Geophysical Studies and the New Theory of Sea-Floor Spreading

If earth scientists have been trying to fit the history of the earth
. . . into the framework of a rigid and fixed pattern of continents,
then it is not surprising that it has been impossible to answer the
major questions. It is not our methods nor our observations that
have been wrong, but our whole attitude.

—J. Tuzo Wilson, "The Movement of the Continents"

In the years following World War II there occurred a revolution in
the earth sciences. Earlier in the century, and for a long time before
that, geologists and geophysicists had believed the earth to be a fairly
stable planet, especially insofar as the position of its continents and
oceans was concerned. Then, within a few bustling decades, this concept gave way to a radically new view, and now scientists believe that
the surface of the earth is being constantly rearranged on a grand scale.
Continents drift, the sea floor spreads, the poles wobble and wander,
and the magnetic North and South poles may even switch places
several times in a million years.

This conceptual revolution has had two stages. The first, concerning
the mobility of continents, was introduced most energetically by the
German geophysicist Alfred Wegener about 1915, and the evidence he
used to support his controversial theory of continental displacement,
or continental drift, as it is called today, was gathered almost entirely
from the land. The second stage in the revolution has been the suggestion that it is the sea floor, moving like a giant conveyor belt, that is

carrying the continents across the globe, and this idea, first formally expressed in 1960, is based on evidence gathered largely at sea. The two parts of the new thinking, like the movements of the continents and the sea floor themselves, are inextricably bound together, and so, although oceanographers had little to add to the early discussions of continental drift, the story of sea-floor spreading must go back at least as far as Alfred Wegener, and perhaps a little beyond.

The idea that continents move, or might have moved in the past, did not originate with Wegener. The speculation had been around at least since the seventeenth century and had been suggested most often to explain the interlocking shapes of Africa and South America. The apparent separation of these two continents had first been ascribed to the Biblical flood, which had supposedly washed the two lands apart, then later related to George Darwin's theory of the creation of the moon. This theory, proposed in 1879 and accepted by many scientists, stated that the moon had been created from a piece of earth torn from what is now the Pacific Ocean basin. Other men were quick to point out that such a catastrophic event would have subjected the earth to stresses and strains sufficient to pull Africa and South America apart, and they suggested how this might have occurred. Osmond Fisher, for example, author of *Physics of the Earth's Crust*, wrote an article in 1882, "On the Physical Cause of the Ocean Basins," in which he argued that the separation of the moon must have created a hole 41 miles deep. In order to partially fill in this gaping scar, and thus make it compatible with the present size of the Pacific, Fisher imagined that molten magma from the earth's interior seeped into the hole and that, at the same time, the sides of the hole drew in as if to heal the wound. The slow but irresistible motion that he thought must have accompanied this drawing in would have broken up the earth's original landmass "into fragmentary areas, now represented by continents. This," continued Fisher, "would make the Atlantic a great rent, and explain the rude parallelism which exists between the contours of America and the Old World."* As appealing as this idea of a single original landmass was—and it was appealing to paleontologists and geologists who wished to explain the occurrence of similar fossils and rocks on either side of the wide oceans—nineteenth-century physicists would have none of it, and their arguments seemed unassailable. The earth, they insisted, was far too young to have undergone such drastic and time-consuming changes. As they understood it, the planet had formed as a molten mass of material and had been cooling ever since. Once the crust had consolidated, 100 million years ago, or less, further cooling had resulted in shrinking, and this had wrinkled the earth's surface like a shriveling apple. On the earth, these wrinkles were ranges of mountains.

Geologists were not in a position to argue with physicists over the age of the earth, and given such a relatively short time in which to imagine the crust of the earth evolving, most did not even try to explain the present distribution of land and sea in terms of vast drifts and rearrangements.

Then, at the turn of the century, the withered-apple concept of the earth began to run into trouble. First, new information on the structure of mountain ranges indicated that if the earth's cooling and shrinking were solely responsible for the formation of mountains, then a primordial earth of gigantic proportions must have been shrinking at an incredible rate. Moreover, why would the earth wrinkle into mountains only at certain places and not at others? More important, the discovery of radioactivity gave the earth its own supply of heat and put an end to the idea that the planet's age could be determined by the amount it had cooled. In the same years, still other kinds of evidence were gathered that added to the suspicion that the earth was neither young nor shrinking, and that its continents might not be so fixed as formerly believed. Geologists were being asked to make a lot of changes in their picture of the earth's history, but where, some wondered, was a new theory of the earth's evolution that could make sense of the growing number of apparent incongruities?

Alfred Wegener first introduced such a theory—his idea of continental displacement—to members of the Geological Association of Frankfurt am Main in a lecture he delivered on January 6, 1912. His interest in the new hypothesis, he said, went back another two years, "when, on studying the map of the world, I was impressed by the congruency of both sides of the Atlantic coasts, but I disregarded it at the time because I did not consider it probable."*

Later, however, he became acquainted with the paleontological evidence used to show that Brazil and western Africa must once have been connected, and this inspired him to review his own idea and present it first as a lecture and then as an article in scientific journals. But Wegener's thoughts on continental drift had hardly been pub-

Alfred Wegener, the German geophysicist and explorer whose forceful presentation of the theory of continental displacement in 1915 triggered a long and emotional debate between believers in continental permanence and "drifters." (*Courtesy Johannes Georgi*)

lished when the young man left his study to join an expedition that intended to traverse the Greenland ice cap, and as soon as he had returned from that successful venture, he became involved for a short time in World War I. Finally, in 1915, on an extensive sick leave, he was able to set down in detail his ideas on continental drift in a book he called *The Origin of Continents and Oceans.** The book, which was translated into many foreign languages, enjoyed a wide circulation and touched off a controversy within the scientific community greater than any since Darwin's *Origin of Species*.

Wegener's hypothesis reached far back to the early history of the earth, when the planet, he said, might have been completely covered by a crust of continental material. This crust had become warped and compressed into high mountains and had slowly grown thicker, but less extensive, until, in the early Mesozoic, some 100 to 200 million

---

* In his book Wegener acknowledges that several other writers, notably F. B. Taylor and H. B. Baker, had had somewhat similar ideas and had gotten them into print before he had. "I became acquainted with all these works only when the displacement theory . . . had already been worked out [in my mind]."Ψ

years ago, there existed a single, large continent, which he called Pangaea. (See Geologic Timetable, p. 91.) For reasons which Wegener did not pretend to know, Pangaea began to break up, and its pieces— our present continents—plowed slowly into a single world ocean, Panthalassa. As the continents moved apart, rifts and rents opened between them, and these, breached by the ancient sea, grew into the Atlantic and Indian oceans. (Wegener's followers usually substituted two supercontinents, Laurasia in the northern hemisphere and Gond-wanaland in the southern, for Wegener's Pangaea.)

Wegener used evidence drawn primarily from land to argue for his hypothesis, but he did learn enough about the ocean basins to realize that the fit he proposed among continents would be much closer if he used the edges of continental shelves rather than coastlines as the true perimeter of land areas. Although this refinement did not appear until the book was in its fourth edition, all editions included a chapter on "The Floor of the Ocean," in which Wegener sought to disprove the common idea that the oceans had once been continental areas which had subsequently sunk. In an effort to show that the ocean floor could not have been a former continent, Wegener overemphasized some of the differences between the two areas and insisted that ocean basins were basically flat and featureless. "We do not know . . . a single feature . . . in the deep sea which we could claim with any certainty as a chain of mountains," he wrote.[*] This left him with the Mid-Atlantic Ridge to explain, and he claimed that that feature was a sort of refuse pile composed of continental debris which had formed as the Americas were separating from Africa and Europe. Ocean trenches, which he knew commonly lay off inland chains, were easier to explain, and these, said Wegener, were either very dense areas of the sea floor which had sunk or were the scars left by island arcs as they pushed through the sea floor.

Wegener's hypothesis contradicted much of what men thought they knew about both the sea and the land, but because so much more work had been done on land, it was the geologists, paleontologists, botanists, biologists, and physicists—not oceanographers—who reacted so violently to Wegener's new scheme. Generations of scientists had built a large and complex superstructure upon the assumption that continents and ocean basins were permanent features, and the pulling apart and reassembling of all the conclusions which rested on this assumption did not appeal to many.

322

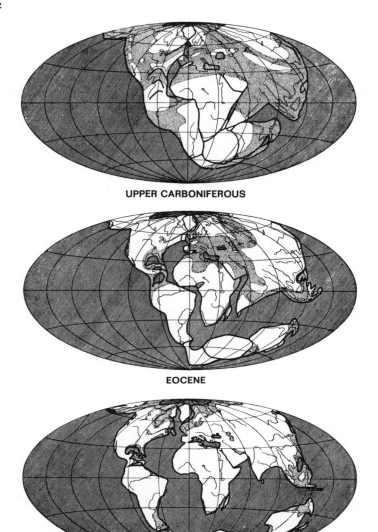

UPPER CARBONIFEROUS

EOCENE

LOWER QUATERNARY

Alfred Wegener's reconstruction of the world as he believed it to exist approximately 275 million years ago, top, 60 million years ago, and one million years ago. The dotted areas are shallow seas, and the outlines of modern continents and rivers are included only for purposes of identification. (*From* The Origin of Continents and Oceans, *Dover Publications, Inc., New York, 1966. Reprinted by permission of the publishers.*)

"If we are to believe Wegener's hypothesis," complained a scientist at one of the numerous symposia held on continental drift, "we must forget everything that has been learned in the last seventy years and start all over again."[Ψ]

Even in 1950, 35 years after the hypothesis had been introduced and 20 years after Wegener had died, continental drift could still elicit an emotional debate. The British Association for the Advancement of Science held a conference on drift in that year which "drew a crowd which packed the biology lecture room." Six papers were read and, as usual, biologists in favor of drift lined up against physical scientists who denied its very possibility. After a heated discussion, in which all the well-known arguments were brought forth on either side, the chairman of the meeting "invited those present to vote for or against continental drift and the result, roughly assessed by a show of hands, was a dead heat."[Ψ]

What seemed to be stalling the argument was the problem of finding a plausible mechanism for the continents' drift. If they moved, what caused them to do so?

"The question as to what forces have caused these displacements . . . cannot yet be answered conclusively,"[Ψ] Wegener had written in 1915, and the same could be said in 1950 as well. Yet actually some exciting new mechanisms for continental drift had been suggested in the 1920's and 1930's. One, which Wegener himself had favored in the last editions of his book, and which became increasingly important later on, was known as the "theory of thermal cycles." This suggested that radioactive heat within the earth built up until some parts of the earth's mantle (the thick layer between the crust and the core) became molten. Roughly circular patterns of convection currents, called convection cells, then developed, and within these the hot, molten material moved upward toward the earth's crust, then horizontally for hundreds or even thousands of miles, then finally down again toward the center of the earth. (The depth of these cells is still a matter of conjecture.)

An Irish physicist and geologist, John Joly, endorsed this idea in the 1920's (it was not original with him), and scientists in favor of continental drift welcomed his new presentation of the theory, for they took it to mean that continents could slide through the temporarily molten mantle wherever a convection cell happened to form. It was much easier to imagine such a slippage than it was to envision a continent grating and grinding its way through a solid mantle as Wegener had originally supposed.

A more comfortable and refined relationship between the theory of thermal cycles and continental drift was worked out in the early 1930's by the British geologist Arthur Holmes. Holmes, described as a quiet man of extraordinary charm, believed that continents did not actively plow their way around the globe but were passively drawn along by the moving mantle. In an article, "Radioactivity and Earth Movements," published in the *Transactions of the Geological Society of Glasgow* in 1931, Holmes suggested that continental blocks might be "carried apart on the backs of the [convection] currents [and] the intervening geosyncline [the large subsiding trough that formed between them] would develop into a new oceanic region."♈

According to Holmes, the continents moved farther and farther apart, perhaps leaving a few remnants behind as a mid-ocean rise, and the ocean between them became progressively wider. Without putting a name to it, he had proposed a theory of sea-floor stretching which had many points in common with the sea-floor-spreading theory held by many scientists today. But Holmes' ideas, which he referred to only infrequently after 1931, did not stimulate others to consider the possibility of sea-floor spreading. For one thing, the testing of the hypothesis seemed awesomely difficult, and for another, Holmes did not push his ideas forward, for he doubted the validity of continental drift, the very process that his hypothesis was meant to explain.

In 1953, as a man in his sixties, Holmes confessed that, "despite appearances to the contrary, I have never succeeded in freeing myself from a nagging prejudice against continental drift; in my geological bones, so to speak, I feel the hypothesis to be a fantastic one."♈

Holmes was not the only scientist to be discouraged by the dilemmas posed by continental drift. The arguments both for and against it had become so well known as to be stale and uninspiring, and the questions raised with wearisome repetition seemed unresolvable. After 1950, most scientists were content to let the matter lie. The topic was worn out.

Then, just a few years later, the issue of continental drift bounced back into the center of scientific discussion, for two new kinds of investigations had begun to uncover exciting information which some scientists believed could break the deadlock. One of these investigations was the study of paleomagnetism, or fossil magnetism, in which infor-

Maurice Ewing, formerly director of the Lamont-Doherty Geological Observatory, organized some of the earliest seismic studies of the continental shelf. (*Courtesy Lamont-Doherty Geological Observatory*)

mation about the earth's past is gathered by studying the magnetic properties of old rocks. The second, geophysical investigations of the sea floor, helped scientists examine the structure of that portion of the earth's crust that lay beneath the oceans. Augmenting the older techniques of sounding, dredging, and coring—methods that had been used to chart the sea floor and examine the topmost layer of marine sediments—the new geophysical methods measured the seismic, magnetic, and gravitational properties of a much thicker portion of the suboceanic crust of the earth.

Certainly one of the first and most enthusiastic advocates of geophysical studies was Maurice Ewing, a tall, energetic Texan who, as previously mentioned, had worked at Woods Hole during World War II. Ewing's interest in marine geophysics predated the war by nearly a decade, however, and he had first become involved in seismic work as a graduate student at the Rice Institute in Texas. There, one summer, Ewing was hired by an oil company to help look for oil beneath the shallow lakes and bayous of southern Louisiana.

The theory on which these oil men based their seismic techniques—and with which Ewing, as a physics major, was familiar—states that sound travels at different speeds through materials of different densities. The denser the material, the faster goes the sound. Although this theory had been formulated in the seventeenth century, it had not been regularly applied until the 1920's, when oil companies realized that they could send sound through the ground (i.e., use seismic methods) to obtain a rough picture of subterranean structures. This, of course, helped them find oil.

By 1930, at about the time that Ewing got his job, the seismic techniques used on land began to be extended into shallow water. Ewing worked on one of these early marine projects, and he learned how to set homemade bombs into the lake mud, detonate them, and record on a small, portable seismograph the sounds that traveled through the water, the mud, and quickest of all through the dense, underlying rock. From the seismograph records, or "wiggle traces," it was possible to determine the thickness of the lake sediments and to learn something about the geologic structures buried beneath them. The records could indicate a large mass of less dense material, for example, which might be a salt dome where oil was trapped. Ewing was fascinated, and several years later, when it was suggested he undertake his own seismic experiments, not in lakes, but in the sea, he jumped at the chance.

It was in the summer of 1935 that Ewing, who then taught at Lehigh University, finally boarded the Coast and Geodetic Survey's vessel *Oceanographer* and put out to sea off Cape Henry, Virginia, to complete a line of seismic stations that he had begun on land. To his intense disappointment, he was given almost no time at all to experiment with his bombs and listening devices. He returned from the short cruise more convinced than ever that his contraptions would work, but with almost no data to show that they really would. His friend and adviser, Professor Richard Field, of Princeton, suggested he try to work from the *Atlantis* out of Woods Hole, and consequently Ewing drove north to talk with Henry Bigelow and Columbus Iselin. After discussing his plans, Ewing was promised some time aboard one of the Institution's cruises, provided he could convince all involved that his methods were not unduly risky. Anxious to cooperate, Ewing set up an experiment in a nearby sand pit, and his demonstrations with bombs and blasting caps persuaded a still wary Bigelow to let him aboard the *Atlantis*.

The following October, Ewing, two co-workers, and several other scientists sailed on the *Atlantis* and successfully made seismic measurements at four stations in relatively shallow water off Woods Hole, and at another four off Cape Henry. At each of these, the procedure was to load a whale boat with packages of blasting gelatin and then fire these at intervals as the small boat pulled a mile or more away from the *Atlantis*. The sounds from these explosions were picked up by geophones which rested on the sea floor beneath the ketch.

These seismic refraction studies showed that a layer of relatively loose or unconsolidated sediments some one to two thousand feet thick lay above a second, thicker layer of denser material that Ewing called semiconsolidated sediments. Beneath both these layers was a still denser material, and this Ewing assumed to be bedrock.

Simpler seismic reflection measurements were also made, and for these, the sound was bounced vertically off the sea floor as if by an extremely powerful echo sounder.* Using this technique, Ewing could see many discontinuities or horizons in the semiconsolidated layer which indicated minor changes in density within each of the sedimentary layers.

Ewing returned to the Oceanographic Institution in subsequent summers to improve his seismic techniques, and one of the first things that he realized he had to do was devise a safer, more efficient way of detonating bombs. Setting them off from a whale boat was not only risky and time-consuming, but totally impracticable in all but the calmest weather. By September, 1940, shortly before the war all but stopped seismic investigations, Ewing had perfected what he called the flotation technique. Each homemade bomb, with a timer and a firing device, was attached to its own float and ballast rig, and several of these bomb packages were dropped from the *Atlantis* onto the sea floor. The ship was then positioned some distance away, and Ewing and his col-

---

* A seismic reflection technique is actually used to measure the depth of very deep water when the sound sent out by a conventional echo sounder is not strong enough. In 1949, for example, the British research vessel *Challenger II* used the reflection technique to sound a deep trench off Guam.

"As we approached the southern end of the Mariana Trench our Hughes echo sounder went off soundings—that is, the water was too deep to return an echo. Undismayed, we hove to and tossed a 1¼ pound charge of TNT off the fantail. Using a hydrophone for the echo detector, we recorded the echo from the jarring explosion with our seismic apparatus. To our amazement the travel time for the echo was over 14¼ seconds and corresponded to a depth of 5,900 fathoms [almost seven miles], nearly 1,000 feet deeper than the previously known greatest depth in the Philippine Trench."ᵠ

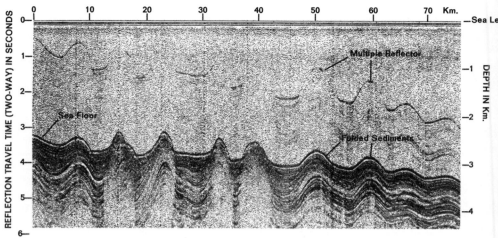

A seismic reflection profile made in the Gulf of Mexico with an electric sparker shows layers of sediments below the sea floor that have been compressed (folded) until they have buckled into a corrugated pattern. Multiple reflectors, which are actually duplicate images of the bottom, appear on the record when sound waves bounced off the sea floor are unintentionally reflected from the sea surface and bounced a second time against the bottom. (*Courtesy U.S. Geological Survey and Naval Oceanographic Office*)

leagues recorded the sounds that the bombs sent through the sediments, and finally retrieved the bomb-shooters, which had detached themselves from the ballast and floated to the surface. The flotation method worked better than the whale boat, but still it had its drawbacks. The floats, for example, were 30-gallon tanks filled with high-octane gasoline, and they tended to leak. For most of each voyage they were lashed to the rail of the *Atlantis,* and since the ship had no explosives locker, the TNT for the bombs rode in the chief scientist's cabin. The night before tests were to be made, Ewing and his students carried the TNT into the deck lab and packed the yellow flourlike substance into meteorological balloons. (They worked within sight of whomever was lying in the "Jesus crib," a half-covered bunk tucked into a niche near the deck lab, and it is said that the crib's occupant often experienced strange feelings as he watched the bombs being made on the rolling ship at night.) Once the balloons were filled with TNT, they were sent overboard on the hydrographic wire for the night so that the water pressure might squeeze them into solid balls. The next morning the hardened bombs were drilled and fitted with blasting caps.

One of the first cruises on which Ewing's flotation method was used was one which left Woods Hole in mid-September, 1940.

"Friday, 20 Sept. Hove to at 0400," reads the log of the ship's first officer, Tom Kelley. "Gang up and began work at 0545. Gear over 0830. 1 bomb off prematurely. 1 on time and lost the other. Rest of gear back and aboard 1730. No luck. . . .

"Got three sharks—blew one to pieces with T.N.T.—never knew what hit him. Very efficient method."*Ψ

On Saturday, the crew was again up before dawn, ready to follow the same schedule.

"Lost first bomb by bad splice. . . . Got one to go off at 1000," but the results were poor. "Something haywire with the apparatus," adds Kelley.Ψ

On Sunday, it was "blowing fresh" so the program was canceled temporarily, but instead of improving, the weather became worse. The wind hauled quickly to the northwest, and a heavy swell was soon rolling the *Atlantis* from side to side and making the handling of gas tanks and explosives impossible.

By Monday, when the "wind dropped considerable . . . began dropping over bombs at 0745. Usual story hunting for them all day. Only one went off. We got a record from it."Ψ

After two more days and nights of work, and two more good records, the *Atlantis* started back to Woods Hole. "By evening wind so strong that sails struck and ½ speed," wrote Kelley, and, on the following day, she was "still plugging along under bare poles. . . ."Ψ

On such a cruise, which was in no way out of the ordinary in terms of foul weather and lost or malfunctioning equipment, Ewing measured layers of sediments 1,200 to 1,500 feet thick. He could already see that the continental shelf off the eastern United States was formed in part of a thick wedge of sediments. (Later work upheld his belief.)

Before Ewing had made seismic measurements, estimates of the thickness of shelf sediments had varied wildly and had ranged from a

---

* In another shark-hunting incident, the sailors aboard a Coast and Geodetic Survey vessel took an extra charge of TNT armed with a time fuse, wrapped it in meat scraps, and chucked it overboard. As planned, a gluttonous shark swallowed it whole, and then, with two flicks of his tail, swam right up to the ship in hopes of finding more. For a few frantic seconds there was a great waving and shouting by the rail, but the shark would not be scared off. His violent demise a moment later did not seriously damage the ship but shook the vessel hard enough to decommission all her more delicate instruments.

few hundred feet to a few miles. The difficulty in predicting the thickness was partly due to the fact that the shelf was known to be an area of transition between continents and ocean basins, and no one knew with any certainty what processes had been rearranging the sediments there. It would be easier, many scientists thought, to predict the amounts of sediments that had accumulated in the deep ocean basins, for there the clays and oozes had presumably lain for several billion years with little to disturb them. Seismic techniques, however, were not ready to be moved into the deep sea, and when World War II began, the experimental investigations were dropped, with only rare exception, in favor of more immediately useful studies.

Seismic shooting from the *Atlantis* involved setting off bombs, then timing the arrivals of their reverberations, which had traveled through layers of mud and rock beneath the sea floor. (*Courtesy Woods Hole Oceanographic Institution*)

After the September, 1940, cruise, Ewing returned to Lehigh, only to be recalled to Woods Hole the following month to start working on projects related to the impending war. Ewing's interests and duties at the Oceanographic soon began to proliferate at a remarkable rate, his reputation ascended accordingly, and in 1943 Columbia University in New York City offered him a chair in geophysics. Ewing agreed to accept it as soon as his war work was finished, and in 1944 he "reported to a building with no floor and with the worst collection of broken furniture you've ever seen."*

The facilities at Columbia left something to be desired, but Ewing struggled somewhat fretfully through each academic year and escaped with relief in the summers to his laboratory at WHOI. His heavy teaching load did not give him as much time as he wanted for research, and in 1948 he was about to leave Columbia when the university accepted the Thomas W. Lamont estate in Palisades, New York, and gave it to Ewing to use for a geophysics program. Ewing was delighted, and the following year he and a half-dozen graduate students carried their books and instruments up a shaded, winding drive, through a formal rose garden, and into the old stone mansion. The estate, perched high on the western shore of the Hudson River, became the Lamont Geological Observatory. (It was renamed the Lamont-Doherty Geological Observatory in 1969 in honor of support received from the Henry L. and Grace Doherty Charitable Foundation, Inc.)

Within the next year or two, Ewing, his students, and his brother John, a capable geophysicist, organized a broad program of geological and geophysical investigations, and soon the estate's greenhouse had become an instrument shop; the seven-car garage, a storage building for deep-sea cores; an indoor swimming pool, a luxurious test tank; and the Lamonts' root cellar, a dark, quiet home for Columbia's sensitive seismographs.

Seismic studies of two sorts were, in fact, central to the observatory's program. The seismographs in the root cellar were used to measure earth tremors caused either by earthquakes or by large explosions, and such studies had already given scientists some information on the gross features of the earth's structure. The records had shown, for instance, that the crust beneath some portions of the deep ocean was considerably thinner than the 22-mile crustal layer typical of continental areas.

To examine this difference further, the second type of seismic study

The *Vema*, a 202-foot iron-hulled schooner, was built in Copenhagen in 1923. Thirty years later she was purchased by the Lamont Geological Observatory and has gradually been converted to an entirely diesel-powered research vessel. The *Vema* continues to work in all the world's oceans for the observatory. (*Courtesy Lamont-Doherty Geological Observatory*)

begun at Lamont was marine. At first Ewing continued to use the *Atlantis*, or any other available research vessel, until, in 1953, a three-masted auxiliary schooner, the *Vema*, was purchased for the observatory. The 202-foot iron-hulled *Vema* had been built as a pleasure yacht in the 1920's. (The *Atlantis* was later built in the same Danish shipyard.) During World War II the *Vema* had been converted into an American training ship, and after the war the ship was bought by a Canadian who hoped to use her as a charter vessel. His first customer was Doc Ewing, and in spite of all the gear that broke and all the things that went wrong on the *Vema*'s first cruise, Ewing saw she was a good ship and persuaded Columbia to buy her.

With the *Vema*, and with funds and surplus explosives from the Office of Naval Research and the Bureau of Ships, scientists at Lamont extended their seismic measurements further and further into the deep sea, and by 1956 several hundred deep-water refraction measurements had been made. These shots, made with powerful bombs and depth charges instead of homemade bombs, penetrated all the way through the oceanic crust to the earth's mantle.*

The picture that emerged from these studies was one of remarkable regularity. The total thickness of the dense, oceanic crust was about four miles, or less than a fifth the thickness of the much lighter, less dense continental crust. In one way, the marked differences in thickness and composition seemed to reaffirm the traditional view of continental permanence, for there was no known geological process that could turn one type of crust into the other, nor could the continents have plowed through the oceanic crust without leaving some mark. Scientists favoring drift had hoped that seismic investigations would discover great mounds of sediments pushed up along the leading edges of continents and thinly covered scars and scrapes along the trailing edges. Such deformations, however, had not been found.

Yet, in another way, the seismic evidence from the deep sea did not fit comfortably with the conventional view. If continents and ocean basins had existed as presently arranged throughout the long history of the earth, then the ocean basins should be full of sediments. But they were not. Seismic measurements showed that only one or two thousand feet of unconsolidated sediments lay in the deep sea, and this was not nearly enough to account for continuous sedimentation at even the slowest rate. This was certainly a puzzle, and scientists, realizing that seismic work could not give all the answers, had already begun to study the sea floor's gravitational and magnetic properties.

In the early 1900's, some unsophisticated but fairly accurate measurements of the earth's gravitational pull over ocean basins had been made, and these had shown that the force of gravity was approximately the same at sea as it was on land. This meant that the two areas, land and sea, were in balance, and since the continental crust obviously stood higher than the oceanic crust, it seemed reasonable to assume

---

* At the base of the earth's crust there is a boundary called the Mohorovičić Discontinuity (the Moho), where the speed of sound abruptly increases. The nature of this boundary is not known, but it is considered to be the dividing line between the crust and mantle. The Moho usually lies about 22 miles beneath continents and only some four miles beneath ocean basins.

that the former was composed of relatively light material and the latter of a much denser material. The seismic investigations begun by Ewing and others bore out this assumption.

But although large areas might be in balance, scientists found that by studying a region in considerable detail they could find gravity values that were higher or lower than what they should be. These deviations they called gravity anomalies, and the anomalies were interpreted as being caused by underlying geologic structures. A mass of abnormally dense material, for example, would appear as a positive anomaly, a wedge of light material as a negative one. Further, gravity anomalies could also be interpreted as meaning that a certain area was not in balance (in isostatic equilibrium) with its surroundings and that forces in the earth were working to push it up or hold it down. Like seismic studies, gravity measurements could not indicate precisely what kind of materials lay beneath the sea floor or just how they were arranged, but the data could be used to limit the number of possibilities.

Until the late 1950's, gravity measurements were difficult to make at sea, for the up-and-down motions of ships masked the very force that the delicate gravimeters were trying to measure. To get around this problem a Dutch geophysicist, Felix Andries Vening-Meinesz, designed a multiple pendulum gravimeter, early in the 1920's, which could compensate for a little of a ship's motion. To make sure that there would only be a little to compensate for, he installed his pendulums aboard submarines and made each gravity measurement at the quiet depth of about 250 feet. In 1923 he began a series of oceanic measurements for the Netherlands Geodetic Commission, and many of his first cruises were made in the Dutch East Indies. Consequently, he discovered, and became interested in, long strips of negative anomalies that seemed to parallel the deep ocean trenches that lay most notably off Java and Sumatra.

In the 1930's Vening-Meinesz briefly turned his attention to a trench in another part of the world—one that lay off Puerto Rico—and to study this one he accepted an invitation to join the Navy-Princeton Gravity Expedition of 1932. For this investigation the Navy provided U.S. submarine *S-48*, duly fitted out with special equipment, and Princeton supplied several scientists. Among them was a graduate student, Harry Hammond Hess, whose job it was to learn all he could about the theory and practice of making gravity measurements, help

Vening-Meinesz make these measurements, then interpret the expedition's data once Vening-Meinesz had returned to Holland.

The expedition started from Guantánamo Bay, Cuba, and from there the *S-48* made three long loops which took her into the Caribbean and up among the Bahamas. Her dives were spaced about 60 miles apart and thus could be made every four or five hours, night and day. Fifty-three stations were occupied during the expedition.

Of particular interest to Vening-Meinesz was a thin strip of negative anomalies which were found north of Haiti and Puerto Rico near what is now known as the Puerto Rican Trench. As was the case in the East Indies, the strip of anomalies paralleled the islands and lay just shoreward of the axis of the deep trench. Vening-Meinesz, and with him Hess, sought to explain this particular configuration in terms of convection cells which they believed existed in the earth's mantle.* Over the downward plunging limb of a convection cell, they said, the oceanic crust is pulled down, and this forms a trench like the ones in the East and West Indies. Beneath each trench is an excessive amount of oceanic crust, and although this material is dense, it is lighter than the underlying mantle material which it displaces. According to Vening-Meinesz, the mass of relatively lighter material caused the strong negative anomalies found near each trench.

There were other ways, however, of interpreting the same anomalies. At the Lamont Geological Observatory, a former student of Ewing's, J. Lamar Worzel, had begun amassing quantities of gravity data with the help of the Office of Naval Research and Navy submarines. In 1954 he published a paper with Ewing in which he argued that trenches were tensional features, caused by a pulling apart of the ocean's crust, and not, as Vening-Meinesz had suggested, by a pulling down and squeezing together of the crust. (It was not until about 1958 that gravimeters were developed for continuous use aboard surface ships, and even when the data from these more efficient instruments became available, opinion remained divided on the origin of ocean trenches.)

Still another kind of geophysical investigation which could supply

---

* If a liquid or semiliquid is heated unevenly, a roughly circular pattern of gyres —convection cells—will be generated as the hotter, less dense material rises on one side and the cooler, denser material sinks on the other. The motion itself is known as a convection current. (See illustration, p. 351.)

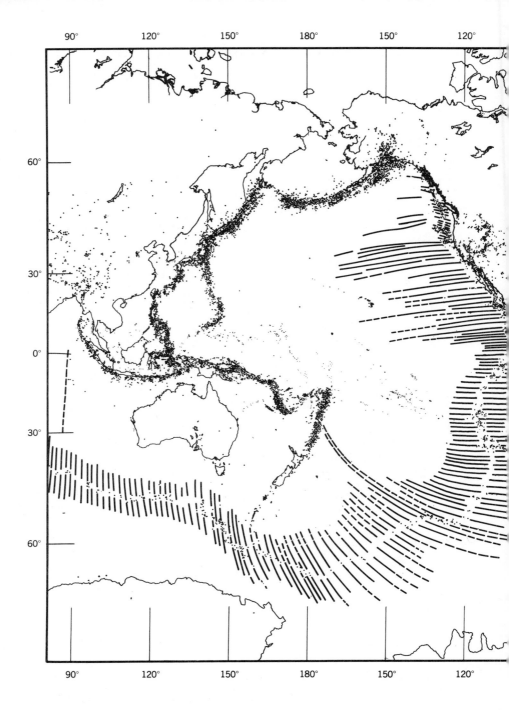

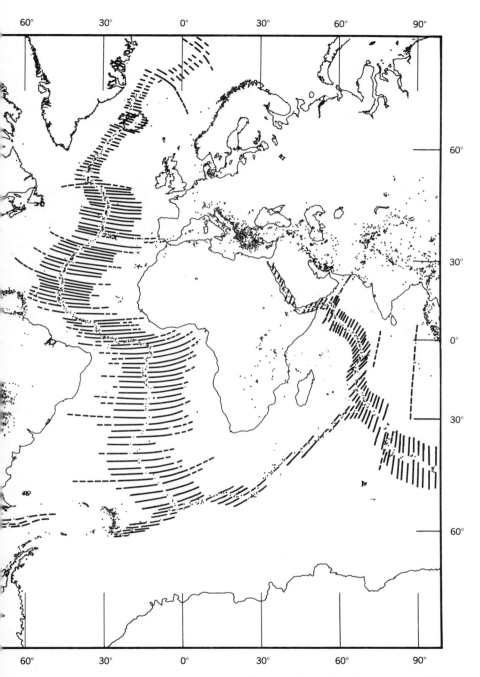

The worldwide system of undersea ridges and earthquake centers. The close relationship between the phenomena was pointed out in 1956 by Maurice Ewing and Bruce Heezen (*From* The Face of the Deep *by Bruce C. Heezen and Charles D. Hollister. Copyright © 1971 by Oxford University Press, Inc. Reprinted by permission*)

some, but not all of the information needed to determine the composition and arrangement of the material beneath the surface of the sea floor was the study of geomagnetism. Before World War II, special nonmagnetic ships like the *Carnegie* were used for these studies, but during the war a Magnetic Airborne Detector was developed which facilitated these investigations considerably. The magnetic detector had originally been designed for use on the small planes which flew back and forth across the Strait of Gibraltar hunting for submerged submarines. Later it was modified so that it could be towed behind a ship.* Using such a device, Ewing initiated a program of magnetic investigations at Lamont in the late 1940's, but the most spectacular results came a few years later from surveys made in the East Pacific by scientists from the Scripps Institution of Oceanography.

Early in the 1950's, sounding surveys had revealed several parallel bands of rough, irregular topography that extended westward for thousands of miles off the coast of California. These were the Mendocino and Murray fracture zones (several more were discovered later), and scientists at Scripps wanted to make a detailed magnetic survey over at least one of them in hopes of learning more about their structure. The chance to do this came in 1955, when the Coast and Geodetic Survey vessel *Pioneer* was scheduled to do some work in the area and it was agreed that Scripps scientists and their magnetometer could go along. With the magnetometer trailing behind her, the *Pioneer* ran a tight grid of lines about five miles apart over an area which included part of the Murray fracture zone. When plotted, the measurements that had been made showed that the pattern of magnetic variations was not at all similar to the random arrangement found on land. Instead, the anomalies seemed laid out in long, narrow strips some 13 or 14 miles wide and hundreds of miles long. These mostly ran in a north-south direction, but at the Murray fracture zone the pattern was interrupted and offset by about 84 miles in an east-west direction. When the survey was extended, and data were gathered from the larger Mendocino fault zone, a much greater displacement—one of

---

* Another wartime invention of great use to oceanographers making surveys of any kind was the Long Range Navigation System (LORAN), which was developed both in Britain and in the United States for determining the precise location of ships and planes that were hunting enemy submarines. The LORAN system, a very accurate one, depends on electronic signals sent from land stations, and so must be used within range of these stations.

more than 700 miles—was discovered. These astonishing displacements indicated that certain sections of the sea floor had moved hundreds of miles in an east-west direction. This was the first indication of such large-scale movements, but scientists were not immediately able to suggest a plausible explanation for the motion. Later, magnetic records became important evidence for sea-floor spreading.

Another intriguing question arose in conjunction with the shape of the anomalies. What caused the long striplike pattern? The strips seemed to be related neither to the topography of the area nor to the underlying rocks, which seismic studies had suggested were fairly uniform slabs of volcanic lavas, all of which would be expected to have approximately the same magnetic characteristics. Perhaps, thought several scientists at Scripps, the pattern was at least partly due to former variations in the earth's magnetic field which had caused a single type of rock—in this case volcanic lava—to have a variety of magnetic characteristics. That this could happen was due to the fact that as the effusions from periodic eruptions cooled, they acquired a part of their magnetic signature from the prevailing conditions of the earth's magnetic field, and these conditions changed.

Scientists had known for many years that a potentially magnetic material, such as lava, or even pottery and brick, was not affected by the earth's magnetic field when heated above the Curie point (575°C.). But as the material cooled below that point its atoms lined up in accordance with the earth's magnetic field and stayed that way unless reheated. The magnetism thus acquired was called remanent magnetization. With knowledge of this phenomenon, geologists were able roughly to estimate the age of a lava flow by matching its remanent magnetization to the period in the earth's history when the geomagnetic field was aligned in similar directions. (Anthropologists used remanent magnetization in much the same way to date pot shards and bricks.)

When studies of remanent magnetization were just beginning, early in the twentieth century, a French investigator found a rock that was reversely magnetized; that is, the north was where the south should have been, and vice versa. Similar rocks were soon found all over the world, but they were not all of the same age, and scientists were led to suspect that the earth's north and south magnetic poles had traded places several times. Using radioactive methods to date a large collection of normal and reversely magnetized rocks, it became apparent that at least three reversals had occurred in the past three or four million years. (By 1968, reversals had been plotted in much greater detail, and more than 100 were found to have occurred in the past 30 million years.)

In addition to helping scientists work out a timetable for magnetic reversals, remanent magnetization could also be used to estimate a rock's original orientation and position on the earth. Scientists working in Britain in the 1950's, for example, found that in some very old rocks, such as those of the Triassic Period formed about 200 million years ago, the magnetic orientation was not north and south but rather northeast and southwest. This meant that either the earth's poles had moved since the Triassic or that Britain had rotated some 30 degrees clockwise. Both alternatives found favor within the scientific community.

For those who assumed that the earth's poles did not move, but that islands and continents did, the study of remanent magnetization could be used to indicate a land's north-south motion in addition to its rotation. It was found that the British Isles had traveled hundreds of miles northward since the Triassic, but the real show was India. There, a study of fossil magnetism suggested that the subcontinent had traveled well over 4,000 miles from a position south of the equator to its present position in the northern hemisphere.

The other alternative was to believe that the poles had wandered, while lands remained fixed, and this point of view was adopted by a second group of British scientists led by S. Keith Runcorn at the University of Newcastle-on-Tyne.* These men studied the remanent magnetization of rocks in Britain and Western Europe, then plotted the path that the poles must have traveled over the past two billion years. Runcorn and his colleagues published this polar wander curve in 1955, and shortly afterward turned their attention toward the rocks in North America in order to corroborate their initial findings. The polar wander curve drawn from these data, however, was not the same as the first but lay roughly parallel to it some 30 degrees to the west. The

---

* The concept of polar wandering was not new; in fact, it was quite well known in the 1870's when young George Darwin chose it as the topic of his first scientific paper. He discussed "the fixity or mobility of the earth's axis of rotation . . ." and concluded that the poles could change their positions if the interior of the earth were plastic, and he, for one, thought that it was. Physicists disagreed with the young man, and their sharp criticisms prompted Charles Darwin to write to his son, "Hurrah for the bowels of the earth and their viscosity and for the moon and for the Heavenly bodies . . . and for my son George. . . ."Ψ

surprised scientists realized that the two curves would coincide only if the continents on either side of the Atlantic had drifted apart.

The rock magnetists in both groups, then, were suggesting that continents had rotated and drifted and, with added support from oceanographers whose studies of magnetic patterns indicated that the sea floor had moved too, the issue of continental drift was reopened with a vengeance. As before, a great deal of work remained to be done on the investigation of mechanisms that might cause the poles to wander, the sea floor to migrate, and the continents to drift. The idea that convection cells in the earth's mantle effected these crustal rearrangements was still among the more plausible explanations, but this was a questionable hypothesis, and more evidence for the existence of convection cells was badly needed. Some of the desired information was obtained by studying the heat that constantly radiates from all parts of the earth.

In the 1930's, E. C. Bullard, a British geophysicist, started measuring the heat that flowed through continental materials and found that the amount was consistent wherever he measured. This heat came from the breakdown of radioactive elements within the earth's crust, and since it was known that these elements are more abundant in granitic rock—continental material—than in oceanic basalts, it was logical to assume that the flow of heat through the ocean floor would be much less than through land. It was a difficult matter to substantiate this assumption, but heat flow measurements in the ocean were finally attempted in 1947 and 1948 by members of the Swedish Deep Sea Expedition. A "geothermometer" 36 feet long was dropped into the ocean bottom on several occasions but the device did not work very well, and only two reliable measurements were made. "Both gave unexpectedly high values. . . ."[*]

A few years later Bullard helped design a shorter, more manageable heat probe while working at Scripps with Roger Revelle and Arthur Maxwell. With this 10-foot probe, measurements were made in the Pacific in 1950 and in the Atlantic in 1952. All the readings obtained were much higher than expected; in fact, they were close to the average continental heat flow. Where was the extra heat coming from? The oceanic sediments were carefully examined for radioactive materials, but there was nothing in them that could account for the high readings. Perhaps, suggested Revelle and Maxwell, "heat is being transported outward by convection in the mantle."[*]

This possible indication of convection cells became even more exciting when it was found that the heat flow measurements made at sea varied much more than those made on land and that the highest values came from mid-ocean ridges. By 1956 Bullard had found that five to six times the normal amount of heat flowed through the ridge areas, and this suggested to him that ridges might sit over the rising limbs of convection cells. (An abnormally low heat flow was sometimes, but not always, found in ocean trenches.)

It was possible, of course, that the heat measured in the ridge areas did not come from the deep interior of the earth, but was merely the result of local volcanic activity. To try to answer this and other questions, scientists steered their research vessels over mid-ocean ridges, for more and more it appeared that the most active part of the ocean basins, and the place where the key to the ocean's evolution might be found, were these immensely grand and imposing submarine mountains.

The discovery of mid-ocean ridges had usually followed a predictable pattern. In the nineteenth century, when soundings in the open ocean were commonly made a hundred or more miles apart, only the broadest of rises or depressions in the ocean's floor could be discerned. The first of such features to be shown on a map was the Dolphin Rise, discovered in 1853 by Maury's surveyors. Later, as more detailed surveys were made, still with mechanical sounding machines, the boundaries of the original rise were often extended. In the case of the Dolphin Rise, the *Challenger* expedition found that it was really part of a "long sinuous ridge, with a depth of only 1,000 fathoms over it, [which] separates the two deep troughs on either side of the Atlantic. . . ."[*] It remained for ships of the twentieth century, equipped with sonic sounders, to discover that these rises and ridges were chains of rugged mountains. For the Mid-Atlantic Ridge, the thousands of soundings made in the 1920's by the *Meteor* gave the first detailed picture of the area's mountainous topography.

In a similar manner, other ridges in other oceans were discovered. Alexander Agassiz found a plateau in the eastern Pacific which the *Carnegie*, with her echo sounder, later showed to be a broad range of mountains, and Johannes Schmidt, on the *Dana*, found a rise in the Indian Ocean which was later shown to be the Carlsberg Ridge. One curious aspect of this ridge's structure was not discovered until the 1930's.

In 1933 and 1934, scientists from the British Museum of Natural History and a surveyor from the Royal Navy's Hydrographic Office set forth on the John Murray Expedition to map "some 22,000 miles of the floor of the Arabian Sea and the neighboring areas of the Indian Ocean . . . [with an] Echo-sounding apparatus."⸸ This was the Carlsberg Ridge area which, with its southern extension, forms what the scientists understood to be "a Mid-Indian Ridge, running from north to south down the Indian Ocean, which reminds one of the similar Mid-Atlantic Ridge in the western ocean."⸸

As the expedition crossed and recrossed the ridge, the British scientists found to their surprise that "throughout the greater part of its length . . . this ridge appears to be double," and that between the parallel crests a "deep gully . . . runs along the length of the northern part of the Carlsberg Ridge."⸸ This was the first well-documented report of a rift valley found within a submarine ridge. (A rift valley is a long, narrow depression formed when a section of the earth's crust drops down along approximately parallel lines of weakness.)

The leaders of the expedition compared the long, winding gully they had found to the Great Rift Valley of Eastern Africa, and the latter, it seemed to them, formed a mirror image of the arcuate undersea ridge. They further noticed that a belt of earthquake centers, their positions determined by seismograph centers on land, extended "down the middle of the Arabian Sea where we now know that the Carlsberg Ridge runs its course. A similar belt can be traced down the length of the Mid-Atlantic Ridge."⸸

These earthquake belts indicated that the mid-ocean ridges were active, unstable regions, and the dredging done on the John Murray Expedition brought up samples of basaltic lava which further confirmed this assumption.

Since the Mid-Indian and Mid-Atlantic ridges seemed in so many ways similar, it was natural for the museum scientists to plan an expedition to the Mid-Atlantic to look for a rift in that mountainous ridge. The expedition was scheduled for 1940, and the choice of a ship had been made and approved, when the war halted the plans. Later, a rift was discovered in the Mid-Atlantic Ridge, as the British scientists had suspected, but this did not happen until 1953, when Bruce Heezen and Marie Tharp at Lamont, making a map of the ocean floor, noticed the consistent appearance of a notch in the echo-sounding profiles made across the ridge area. The notch proved to be the Mid-Atlantic Rift Valley.

A physiographic diagram of the North Atlantic constructed by Bruce C. Heezen and Marie Tharp clearly shows the sinuous Mid-Atlantic Ridge and the rift valley that runs for great distances along its crest. The map was first drawn in the 1950's, then revised, as shown above, in 1968. (*Courtesy Bruce C. Heezen*)

PHYSIOGRAPHIC DIAGRAM OF THE

NORTH ATLANTIC OCEAN

BY BRUCE C. HEEZEN AND MARIE THARP

Revised 1968

Also at Lamont, and at about the same time, Ewing and his colleague Bruce Heezen noticed the belt of earthquake centers that lay along the Mid-Atlantic and Mid-Indian ridges. Convinced that the combination of rifted ridge and earthquake centers was more than a coincidence, and that, in fact, the two phenomena were related, they predicted that a rifted ridge would be found wherever a belt of mid-ocean earthquake centers was shown to lie. This prediction was made in 1956, and at that time seismologists had already outlined a fairly continuous belt of earthquake zones that ran for more than 34,000 miles through the world's oceans. It seemed incredible that a ridge of that length could exist beneath the sea, but on the numerous expeditions launched in the late 1950's (many as a part of the International Geophysical Year*) soundings were made along the earthquake belt, and in almost all cases an undersea ridge or rise was discovered.

In 1958 a marine geologist at Scripps, Henry Menard, became curious to see just how close to the middle of the oceans these mid-ocean ridges really were. On a globe he drew a median line through the oceans, then superimposed the position of all known segments of the world-circling ridge. He found that "the median line follows the sinuous crest of the oceanic rises and ridges through the Atlantic, Indian, and Antarctic oceans with remarkable consistency."* This strongly suggested that the formation of the ridges was causally related to the formation, or at least to the positioning, of continents.

Menard went a step further and suggested that submarine ridges might not be permanent structures, and he based this assumption on the varied topography of mid-ocean areas. He knew that there were broad, gentle rises with no rift valleys, as in the eastern Pacific; narrow, steep-sided ridges like those in the North and South Atlantic; and finally, as in the western Pacific, irregular, spread-out areas dotted with islands and submarine mountains (seamounts) that could hardly be called rises at all. It occurred to Menard that all of these might represent different stages in a continuous process of ridge development. The formation might begin as a gentle rise, then, becoming increasingly

---

* The IGY, as it is commonly called, was a 30-month period from July, 1957, through December, 1959, during which a worldwide program of geophysical research was conducted. The largest and most comprehensive science project ever organized, IGY included 11 fields of study, one of which was oceanography. Seventy ships from 35 nations participated in the marine investigations, which were primarily planned to examine tides, deep-water circulation, and other phenomena related to the earth's climate and weather.

active, progress to its most impressive stage as a chain of rough and precipitous volcanoes. Finally, the mountains might quiet down, slowly subside, and leave behind only a scattering of old volcanic peaks, the higher ones forming islands, the others seamounts.

Menard's two suggestions, taken together, pushed him toward a belief in constantly shifting continents, for if the land's position was truly related to mid-ocean ridges, and if the ridges themselves were developing in different places and at different times, then the position of the continents must be equally variable. This was a difficult line of reasoning to follow, however, for in 1958, when these ideas were expressed, the traditional belief in the permanence of continents and ocean basins was so strong that most scientists, including Menard, moved only cautiously toward a new understanding.

But move they did. So much new information had been gathered from ocean basins—the discovery of rifted ridges with high heat flows and belts of earthquakes, the remarkably thin veneer of young sediments, the deep trenches, and the apparent movement as indicated by displaced magnetic patterns—and so much of this information could not be fit easily into the traditional concept of oceanic permanence, that some scientists felt that it was time to formulate a new history of the continents and ocean basins. What was needed, these men argued, was a broad and basic explanation that would provide a framework within which to arrange, further examine, and, if possible, test the increasing number of hypotheses that purported to explain some aspect of the oceans' evolution. Only a few scientists worked toward this synthesis. Each used a different array of data for each had had a different experience with the sea.

Bruce Heezen put his pieces of the puzzle together and suggested that the earth was expanding, and that this expansion was taking place along the mid-ocean ridges. New crustal material, he said, was slowly, but constantly being formed in the rift valleys of the submarine ridges, and this caused the oceans to grow wider and wider. As they did so, the continents became farther and farther removed from each other.

Ewing did not entirely agree with his colleague Heezen, for he did not believe the earth to be expanding nor the continents to be moving relative to each other. In making his own synthesis, from the impressive amounts of data at his command, Ewing suggested that the sea floor was moving, but that the continents were not. Convection currents in the earth's mantle, he said, were rising under the mid-ocean

In 1960 Harry Hess, head of the geology department at Princeton University, offered one of the first thorough and well-ordered explanations of the currently popular theory of sea-floor spreading. (*Copyright © 1952 by Princeton University Press.*)

ridges, adding new material there, and thereby forcing the sea floor to move away from the ridge toward the continents on either side of the ocean basin. The slowly moving sea floor passed under the continents without moving them, yet the motion did exert enough compressive force to push mountains up along the coast of the land.

Harry Hess, who in the 28 years since his participation in the Navy-Princeton Gravity Expedition had become chairman of the geology department at Princeton, offered still another explanation, and his formed the nucleus of the exciting and now extremely popular theory of sea-floor spreading. Hess had gathered the ideas needed for his new picture of the ocean basins' evolution from his early work with Vening-Meinesz on gravity anomalies and trenches, and from a later study of the extensive Darwin Rise in the western Pacific. From the former Hess had acquired the belief that there are convection cells in the mantle capable of moving the oceanic and continental crust that lies above them. Vening-Meinesz had also introduced him to the idea that the position of a downward-flowing limb of a convection cell might be marked by a great ocean trench. These ideas gave Hess a mechanism that could explain the movement of the sea floor and the continents, and also provided him with a sink—the trenches—into which to dump some excess crustal material. (The latter allowed him to avoid the complexities involved in postulating an expanding earth.) It required much more than these two speculations, however, to convince Hess that the sea floor or the continents were really moving, for even in the 1950's he did not believe in continental drift. Hess needed more information about mid-ocean ridges, where new oceanic crust might be generated, and about the structure and arrangement of whole ocean basins. These kinds of data he was able to gather as an officer in the Naval Reserve.

Hess joined the Reserve early in the 1930's, reportedly to make his gravity work on Navy submarines proceed more smoothly, and when World War II began, he left Princeton to work in the service full time. He was made a navigation officer aboard a large transport vessel, the *Cape Johnson*, and in that capacity was able to keep the ship's echo-sounding equipment operating almost continuously as the vessel moved through the Pacific. When there was time, and no threat from Japanese submarines or aircraft, Hess was even able to use the ship to survey some interesting topographic features. As a result, he discovered more than 100 curious, flat-topped seamounts whose shapes, dimly outlined by the echo sounder, reminded him of volcanic cones with their tops cut off.

At first Hess considered these structures to be drowned atolls, then refuted his own arguments in favor of the idea that they were really ancient, Pre-Cambrian islands formed more than 500 million years ago. He named the structures "guyots," after the Swiss geologist and glaciologist Arnold Guyot, who had taught at Princeton in the 1850's. The guyots that Hess found were scattered in a wide band that extended diagonally across the Pacific from the vicinity of Guam and the Mariana Islands southeast toward, but not to, the west coast of South America. The guyots roughly paralleled the band of islands and atolls that marked the Darwin Rise.

After the war Hess stayed in the Reserve and for a time each year worked in the Naval Hydrographic Office, adding thousands of new echo-sounding profiles from all parts of the oceans to master charts. Poring over the new profiles, he expected to discover hundreds more of the truncated outlines so typical of guyots. But he did not. For no apparent reason, the structures seemed almost entirely confined to the band in the Pacific.

While Hess was discovering this curious pattern of distribution, scientists at Scripps were dredging rocks from the tops and sides of the guyots, dating the fossils found in the rocks, and announcing that nothing older than the Cretaceous Period (70 to 135 million years ago, see timetable, p. 91) was to be found on them or, for that matter, was to be found anywhere in the oceans. These findings not only forced Hess to give up his idea that guyots were ancient islands, but also troubled scientists who believed that the ocean basins were permanent and should, therefore, contain ancient sediments.

Hess had always considered that guyots, like atolls, were an indication of an area's subsidence, and postwar drilling on Eniwetok had shown that a volcanic island could sink some 4,000 feet or more.* This suggested that the band of guyots and atolls had once been a ridge of great height, but if this were true, there remained two stubborn questions to answer. First, why was the band so wide? In some places it was almost 2,000 miles across, fully twice as wide as either the Mid-Atlantic or Mid-Indian ridge. And second, why were there no ancient sediments to be found on what clearly seemed to be a very old structure? Hess considered these questions in the light of what he believed about convection cells and what he knew about the structure of the sea floor, and it seemed to him that only if convection in the mantle was causing the sea floor to spread and the continents to drift could he explain the curious array of data that he had examined from the sea.

Hess first presented his ideas on sea-floor spreading (although he did not use the term) in a paper on "The Evolution of Ocean Basins." He described the work as an "essay in geopoetry," and credited the ideas it contained partly to Arthur Holmes, Vening-Meinesz, and many contemporary workers, and only partly to his own imagination. The paper was distributed privately as a preprint in 1960, then published formally as the "History of Ocean Basins" in 1962.

"Long ago Holmes suggested convection currents in the mantle to account for deformation of the Earth's crust," wrote Hess, and "if it [mantle convection] were accepted, a rather reasonable story could be constructed to describe the evolution of ocean basins. . . ."ᵛ

According to Hess's arguments, at some time in the history of the earth, many, many millions of years ago, convection cells developed in the earth's mantle, and their rising limbs came up beneath both the land and the sea. Where the ascending currents developed beneath the land (as they must have done beneath Gondwanaland), the land split apart and its huge fragments were carried passively away from each other.

---

* Darwin had recognized that the area in the western Pacific, dotted with atolls, must once have stood much higher, and in deference to his insight, the area was named the Darwin Rise.

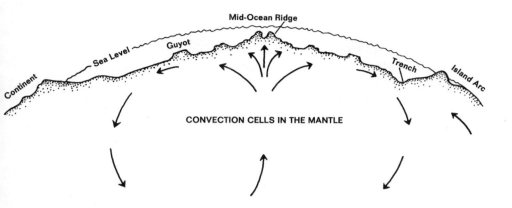

According to the theory of sea-floor spreading, convection currents in the earth's mantle are capable of moving great sections of oceanic and continental crust at a rate of one or two inches per year. Where these currents rise, a mid-ocean ridge is often created and toward the center of the ridge new oceanic crust is formed and is then moved to either side by the diverging currents. Eventually this slowly moving crust may be pushed beneath the edge of a continent or may be pulled back toward the mantle in a deep ocean trench. (*After David A. Ross,* Introduction to Oceanography)

Between these pieces—the earth's new continents—deep rifts opened, and these, filled by an ancient ocean, became narrow seas and eventually broadened into oceans. As this process went on, the ascending currents, which were bringing molten magma up from the interior of the earth, formed a ridge between the separating continents. For as long as the convection cell remained active—and Hess estimated that to be between 200 and 300 million years—the sea floor continued to spread, the ridge to split, and the magma to well up within the rift in a long and irregular series of volcanic eruptions. Hess felt that this process not only accounted for the mountainous topography of the mid-ocean ridge area, but also explained the high heat flow there. (It also explained the belt of earthquakes found along the ocean ridges, although Hess did not mention this.)

Some of the volcanoes formed at the ridge crest, said Hess, rose above the level of the sea, where the force of the ocean's waves eventually leveled their crests. As the sea floor spread, these truncated volcanoes moved down the flank of the ridge and became guyots or atolls rising from the deep-sea floor. In time, these volcanic peaks, as well as the sediments that had been accumulating on the sea floor, encountered the downward-flowing limb of the convection cell, and then, said Hess, they "ride down into the jaw crusher of the descending limb, are metamorphosed, and eventually probably are welded onto continents."[*] The destruction of the ocean floor in this manner explained why no ancient sediments had been found in the ocean basins and finally resolved the problem that had seemed to exist between the rate at which sediments were being deposited and the unaccountably thin cover that had accumulated.

Hess further maintained that the continents, like the sea floor, stopped their traveling when they encountered the descending limb of a cell. "Because of its much lower density it [the continent] cannot be forced down, so that its leading edge is strongly deformed and thickened when this occurs."[*] (The example that was later given was the western coast of South America where the Andes Mountains are being pushed up as the sea floor to the west rides down into the Peru-Chile Trench and perhaps pushes under the continent.)

"This long and devious route leads to the conclusion that the present shapes and floors of ocean basins are comparatively young features," said Hess.[*] In fact—and this would have surprised the nineteenth- and early-twentieth-century scientists who had argued so strenuously over the permanence or impermanence of continents and ocean basins—"the ocean basins are impermanent features, and the continents are permanent although they may be torn apart or welded together and their margins deformed."[*]

"In this chapter the writer has attempted to invent an evolution for ocean basins," finished Hess. "It is hardly likely that all of the numerous assumptions made are correct. Nevertheless it appears to be a useful framework for testing various and sundry groups of hypotheses relating to the oceans. It is hoped that the framework with necessary patching and repair may eventually form the basis for a new and sounder structure."[*]

At Lamont, Scripps, Woods Hole, and elsewhere, the authors of those various and sundry hypotheses relating to the oceans struck up a lively debate over Hess' imaginative proposals. A few, such as Robert

Dietz in California and J. Tuzo Wilson in Canada, piled right in and began to reinterpret their data in the light of the new hypothesis. Dietz, who coined the term "sea-floor spreading," pointed out that fracture zones, such as the ones he knew off California where the magnetic pattern was offset for hundreds of miles, might "mark shears between regions of slow and fast creep."[*] Wilson, who had been studying the geology of islands in the Atlantic, reasoned that if Hess' ideas were essentially correct, then islands such as Iceland, the Azores, and Bermuda should all have originated as part of the Mid-Atlantic Ridge and subsequently should have moved away from it. The youngest islands should be found near the ridge, and the oldest at the edges of the ocean basins. Using radioactive methods, Wilson had determined the ages of many island rocks, and putting these together, he found that the ages fit neatly into the pattern that sea-floor spreading had led him to expect. He found, for example, that Iceland, sitting right on the ridge, was composed of rocks all less than 10 million years old. The Azores, however, located hundreds of miles to the east, had rocks 20 million years old. Still farther to the east were the Canary Islands, composed of rocks extruded 32 million years ago, and in Bermuda, which lay about as far to the west as the Canaries lay to the east, the oldest rocks were 36 million years old. Using these ages, Wilson calculated that the islands had been carried away from the ridge by the spreading sea floor at a rate of one to two inches per year.

Scientists who objected to Hess' sea-floor spreading were not won over by Wilson's island arguments.

"Dear Professor Wilson," wrote the respected Russian geologist V. V. Beloussov, "What is the situation now with the new theory that you are defending? There is simply no foundation to it . . . nobody has shown that deep convection currents, which are necessary for your theory, really exist or even that they *can* exist. . . ."[*]

"In the theory that you are so ardently advocating, the geological development of continents is much more schematized. . . . The old, tested and rather precise geology of the continents is being sacrificed for the sake of as yet indefinite data on the structure of the oceans! You propose to use the process of repeated opening and closing of oceanic basins as the basis of historical geology. Would not it be better to wait until we get more direct evidence on the structure of the ocean floor?"[*]

In less formal terms other scientists dismissed sea-floor spreading as impossible or published long and thought-provoking lists of problems and objections.* Most, however, adopted a rather philosophic attitude, knowing that they would have to wait to see how well and how completely the concept of sea-floor spreading could make sense of the facts that were known about the earth.

"These hypotheses are interesting and may have true merit," wrote the marine geologist Francis Shepard cautiously, "but at present there is no adequate way to test them."⁺

Then, in 1963, just a year after Hess had formally published his ideas on sea-floor spreading, a twenty-five-year-old graduate student discovered a way to test the new hypothesis. When Frederick Vine, a student at Cambridge University in England, was casting about for a topic suitable for his doctoral dissertation, it was suggested that he investigate the sea floor's magnetic anomalies. Agreeably, he began to study magnetic data brought back by H.M.S. *Owen* from the Carlsberg Ridge in the Indian Ocean and similar profiles made over the Mid-Atlantic Ridge. Like the scientists working on these data before him, Vine found the anomalies laid out in long strips which apparently ran parallel to the ridges. It was difficult to imagine how this pattern was created, for it seemed either to require that bands of dissimilar rocks be arranged in long parallel strips or that rocks of the same type exist in a highly magnetized state in one strip and then in a weakly magnetized state in another. Both explanations seemed improbable.

While Vine pondered the ambiguous anomalies, two scientists with the United States Geological Survey, Allan Cox and Richard Doell, were studying the magnetic properties of lava cliffs at the edge of the sea. They had chosen a place where waves had cut into and exposed layers of sequentially extruded lavas, and in studying these they had found that the top two layers were magnetized normally, the third layer was reversely magnetized, and that lower, and therefore older, layers were again normally magnetized. Cox and Doell used radioactive methods to determine the ages of the lavas, and with these dates published a tentative time scale for the earth's most recent magnetic reversals.

---

* In 1969, A. A. Meyerhoff, a geologist, wrote an article, "Spreading Sea Floor: By Convection or Conviction," in which he listed 26 observations *not* explained by drift and spreading. "Data do suggest that sea-floor spreading and continental drift belong more properly in the realm of mythology than in the science of geology."⁺

Vine read the Cox and Doell paper, and it gave him the last bit of information he needed. Putting the dated reversals, the striplike magnetic pattern, and the hypothesis of sea-floor spreading together, Vine and his dissertation supervisor, D. H. Matthews, came up with an explanation of the magnetic pattern which at the same time provided a test for sea-floor spreading.

If spreading of the sea floor occurred, reasoned the two scientists, fresh lava would be extruded along the axis of a mid-ocean ridge, and as it cooled below the Curie point, it would be magnetized in the prevailing direction of the earth's magnetic field. This material would then be pulled apart and carried away from the center of the ridge. Newer material would continue to well up from the mantle, and if, in the meantime, the earth's magnetic poles had changed places, the newer material would be reversely magnetized. As the spreading and pole switching continued, a pattern of normal and reversely magnetized lavas would be formed on either side of each active mid-ocean ridge. Furthermore, if the spreading were at all regular, then the pattern on one side of the ridge would be a mirror image of the pattern on the other side.

It seemed unlikely that another explanation of so precise a pattern of magnetic anomalies would be found, and consequently a test of the sea-floor-spreading hypothesis became the ability to discern the predicted magnetic pattern.*

Vine's and Matthews' work gave the concept of sea-floor spreading a tremendous boost, and by the mid-1960's the process of filling out the new hypothesis had begun in earnest. New evidence, new refinements, and new problems came pouring in from around the world. At Lamont, for example, where Hess' ideas had been coolly received and

---

* Apparently the Vine and Matthews explanation of the magnetic anomaly pattern was preceded by a similar one made, unbeknownst to them, by the Canadian geologist L. W. Morely. Using data collected by Scripps' scientists off California, Morely hit upon the idea that "the upwelling rock under the ocean ridges . . . must become magnetized in the direction of the earth's field prevailing at that time. If this portion of rock moves upward and then horizontally to make room for new upwelling material, and if, in the meantime, the earth's field has reversed . . . it stands to reason that a linear magnetic anomaly pattern of the type observed would result."Ψ

Morely stated his beliefs in an open letter which he sent to several scientific journals and a society in the spring of 1963. His former teacher, J. Tuzo Wilson, maintains that Morely's letter was rejected as an example of unsound science by both British and American journals. The Vine and Matthews paper was published in *Nature*, a British scientific journal, in September, 1963.

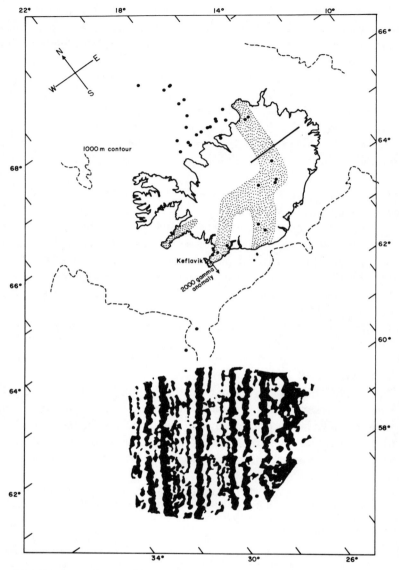

In this diagram of magnetic anomalies over the Reykjanes Ridge—a part of the Mid-Atlantic Ridge—the blackened strips represent positive anomalies and the white areas in between are negative anomalies. The strips are arranged with surprising symmetry on either side of the ridge crest. (*Reprinted with permission from J. R. Heirtzler, et al., "Magnetic Anomalies over the Reykjanes Ridge," 1966, Pergamon Press*)

only cautiously examined, scientists began reexamining the large stock of magnetic data that they had collected from all the oceans of the world, and soon they were able to recognize in the South Atlantic, the South Pacific, and Indian oceans the same pattern of magnetic anomalies that Vine had predicted should exist.

By matching the most recent portions of these anomalies to the pattern of reversals that the American geologists Cox and Doell had found and dated on land, dates could be assigned to the marine anomalies and a rate of spreading computed. Actually there was more than one rate to compute, for although the basic arrangement of the magnetic anomalies was the same across many ocean ridges, the width of the normally and reversely magnetized strips differed, which suggested that the sea floor was spreading at different rates in different parts of the oceans. In the Atlantic, for instance, it was found that the sea floor was carrying Africa and South America away from each other at a rate of more than four inches per year, while in the North Atlantic, near Iceland, the sea floor was spreading a bare half-inch annually.*

Once the rates of spreading had been calculated, James Heirtzler and several other scientists at Lamont who had been doing much of this work further assumed that the rates were essentially constant (although not everywhere the same). This assumption enabled them to date, by extrapolation, the older marine anomalies, and this they did, all the way back to the late Cretaceous, some 76 million years ago. The major anomalies were numbered for convenience.

As exciting as this line of inquiry was, and as persuasive as the magnetic evidence appeared to be, still more information remained to be gathered if the hypothesis of sea-floor spreading were to gain more than a tentative acceptance. Marine seismic records provided some of the necessary evidence, for they showed that in general the layers of sediment on the sea floor were thinnest over the young, ridge-crest area, thicker on the flanks of the ridge, and thicker still in the ocean basins to either side. This was not always the case, however, and scientists opposed to the idea of spreading pointed out that the more or less uniform thickness of sediments in the South Atlantic, in an area where the magnetic pattern was particularly clear and well defined, did not

---

* An extension of the central rift valley of the Mid-Atlantic Ridge cuts through Iceland, and British scientists have been using beams of light to measure the distance across the rift precisely. They have found it to be growing wider by some three-eighths of an inch each year.

seem compatible with the process of sea-floor spreading. How could the sediments be evenly distributed, they asked, if the crest of the ridge were 60 or 70 million years younger than the outer flanks, as spreaders and drifters claimed that it was. This was a difficult question to answer, and it appeared that it, and other questions raised by other discrepancies, could be resolved only by drilling through all the layers of marine sediments and actually sampling the oceanic crust below. If samples of the crust, or of the sediments immediately above it, could be collected at several points across a mid-ocean ridge, and if these samples could be dated, then it would be possible to determine if the sea floor did, in fact, become progressively older with increasing distance from the ridge. If it did, the process of sea-floor spreading would be hard to deny. To collect the samples, however, was no easy matter, for it would be necessary to drill through hundreds of feet of sediments that lay beneath several miles of water.

In 1964, scientists from Scripps, Lamont, Woods Hole, and the Institute of Marine Sciences (a part of the University of Miami) organized a Joint Oceanographic Institutions Deep Earth Sampling program—JOIDES—with the ambitious intent of collecting direct evidence for sea-floor spreading. This deep-sea-drilling project, financed by the National Science Foundation, had been preceded by two comparatively small projects, Mohole and LOCO, which had shown that deep-sea drilling was possible.

The first task was to launch a one-month test cruise, and this was sent off in the spring of 1965, on the drilling vessel *Caldrill*. Holes were successfully drilled and cores obtained at six sites off Jacksonville, Florida, in water as deep as 3,400 feet. On the basis of the experience gained, plans were then made to lease a much larger drilling vessel and proceed into deeper water.

In July, 1968, the 400-foot *Glomar Challenger*—named in part for the old British *Challenger* and in part for her owner and operator, the Global Marine Company—put out to sea on an open-ended expedition that scientists hoped would take her around the world. The first leg of her journey was a tour through the Gulf of Mexico and the Bahamas, where, as was expected, some extremely old rocks were sampled. Next she headed across the North Atlantic, where a line of drilling sites was laid across the Mid-Atlantic Ridge. Unfortunately, a combination of problems beset the drilling procedures, and only rarely was the oceanic

The research vessel *Vema* of the Lamont-Doherty Geological Observatory ran surveys of possible drilling sites for the JOIDES program. She is shown here rounding Cape Horn. (*Courtesy Lamont-Doherty Geological Observatory*)

crust—the basement rock beneath the sediments—reached. A second attempt to test the hypothesis of sea-floor spreading was planned for the third portion of the expedition (Leg 3), on which Arthur Maxwell and Richard Von Herzen, both from WHOI, were to be co-chief scientists. Seven drilling sites that lay in a line across the Mid-Atlantic Ridge in the South Atlantic were chosen.

"At 0800 hours on 1 December, 1968, the *Glomar Challenger* departed Dakar, Senegal, to commence Leg 3 of the Deep Sea Drilling Program."⁕ Ahead of her sailed Lamont's ship *Vema* with the intention of making quick surveys at each site in order to guide the *Glomar Challenger* to the most promising position.

Long after dark on the 20th of December, the *Glomar Challenger* reached a featureless spot in the South Atlantic that had been selected for the first of the ridge sites. The *Vema* had chosen a position over a small hill on the far western flank of the Mid-Atlantic Ridge near Magnetic Anomaly 13, which Heirtzler had suggested had an age of about 38 million years. The water at this spot was more than two and one-half miles deep, but the sediments, as measured by the *Vema*'s seismic profiler, were not much more than 300 feet thick. The *Glomar Challenger* steamed into position and dropped an acoustic beacon, which, in combination with a shipboard computer and auxiliary positioning motors fore and aft, would enable the vessel to maintain the fixed position required for drilling. Less than an hour later this system was in operation, and then, beneath the ship's bright deck lights, the drilling crew began the long and strenuous job of assembling and lowering the drill string. Ninety-foot sections of five-inch pipe were noisily slid from racks that covered the *Challenger*'s foredeck, pulled up into the tall derrick amidships, attached to the string, and lowered through a well in the center of the ship. Some 12 hours after the job was begun, the drilling head finally reached the sea floor. The drill string was spudded-in, and the bit began to grind down through layers of soupy, gravelly, or hard-packed oozes.

The drilling went surprisingly fast. At first the sediments were soft, and the bit went through nearly 60 feet every hour; then a hard bed of calcite was encountered at about 175 feet beneath the sea floor, and the drill string dropped only 15 or more feet in the same amount of time. Periodically, a core was taken and lifted to the surface through the drill pipe. Once on deck, it was taken to the ship's core laboratory, where it was split down the middle with a cheese cutter or with a machine saw if it were hard. The archive half of the core went into cold storage, and the working half was sent, after a battery of physical tests, to the paleontology laboratory. There, the three paleontologists on board made smear slides from the bands of sediments, and by identifying the microfossils in each, were able to estimate its age.

As the bit wore on toward the 300-foot mark, where the seismic record showed that the sediments stopped and the oceanic crust began, coring operations became continuous. The drill kept whirling lower and lower, until, 351 feet beneath the sea floor, there was a major break in the drilling rate. In the next hour only about five feet of material were drilled, and for two hours more the drill worked noisily on, but

The deep-sea-drilling ship *Glomar Challenger* began drilling and coring ocean sediments in 1968 for the Joint Oceanographic Institutions Deep Earth Sampling program. Much of the data obtained on the ship's numerous cruises has supported the theory of sea-floor spreading. (*Courtesy Scripps Institution of Oceanography, University of California, San Diego*)

the bit would sink no lower. The decision was made to abandon the hole, and the last core was brought up. In it, as hoped, were bits and pieces of basaltic basement.

The deep core was taken to the paleontology laboratory for examination. No microfossils could be found in the basement rock itself, but in the ooze just above it were certain Foraminifera which had become extinct some 39 million years ago, a date which agreed well with the 38 million years predicted by Heirtzler's study of magnetic anomalies.

The next two drilling sites occupied by the *Glomar Challenger* were closer to the center of the Mid-Atlantic Ridge and were centered over Magnetic Anomalies 6 and 5. At the first of these sites holes were successfully drilled in the basement, but at the second, only a small chip of basalt was recovered. At both sites, however, the sediments were examined and dated, and it was found that the ages suggested by the microfossils were within two or three million years of those predicted by magnetic studies.

The *Glomar Challenger* drilled at several more sites before going in to port at Rio de Janeiro, Brazil, and when Leg 3 was completed, close to 2,000 feet of core from seven points across the Mid-Atlantic Ridge had been collected, examined, and preserved for further study. The age of the sea floor, as determined by the fossils in these cores, ranged from 11 million years near the crest of the ridge to 67 million years at a point hundreds of miles to the west near the very edge of the ridge area. The rate of spreading computed from these ages was nearly identical to the rate which Heirtzler had calculated. To obtain essentially the same results from two entirely different lines of inquiry was considered such compelling evidence for sea-floor spreading that the concept finally attained the stature of theory in the minds of most of the world's earth scientists.*

As this book is completed, the *Glomar Challenger* is still drilling at sites all over the world and is still collecting evidence in support of sea-floor spreading. Even after only three legs of her journey, however, an impressive summary of JOIDES' scientific achievements was published. The first item in that long list is "the extremely strong substantiation of the concept of sea-floor spreading and continental drift," and the last is "proving that several academic institutions that are normally competitive, the industrial complex in America, and the federal government can work together to produce a project of large size and of profound results for many branches of science."⊻

And that too is where oceanography is today—at a point in its growth as a group of closely related sciences where it is recognized as being worth a great effort and productive of profound results.

---

* Most recently, the theories of sea-floor spreading and continental drift have been subsumed within the concept of "plate" or "global tectonics." This more inclusive theory postulates that the crust of the earth consists of six or more large, rigid plates that move relative to each other. These plates are continuously being created at one edge (along mid-ocean ridges) and destroyed at another (along trenches or elsewhere). The boundaries of these plates are most easily located by mapping earthquake centers. The boundary between the American and the Pacific plates, for example, is located in part along the San Andreas Fault in California.

References to sources of quoted passages and other supplementary or background information will be found in the notes that follow. Each note refers to a passage flagged by ⚓ on the text page indicated at the beginning of the note. In cases where more than one passage on a text page is flagged, the pertinent notes follow in the same order. An n following the page number indicates that the reference is to a footnote on that page.

Introduction

Page 13   Henry B. Bigelow, *Report on the Scope, Problems, and Economic Importance of Oceanography, on the Present Situation in America, and on the Handicaps to Development, with Suggested Remedies.* Report submitted to the National Academy of Sciences from the N.A.S. Committee on Oceanography, 1929, p. 1.

Page 14   Otto Müller, quoted in Torben Wolff, *Danish Expeditions on the Seven Seas.* Copenhagen: Rhodos, 1967, p. 43.

Chapter I

Page 24   American State Papers, Miscellaneous, Vol. 1, quoted in A. Hunter Dupree, *Science in the Federal Government.* Cambridge: The Belknap Press of Harvard University Press, 1957, p. 14.

Page 26   Lieutenant L. M. Goldsborough, Letter, 1830, in *Officers' Letters to the Secretary of the Navy,* Vol. 9, p. 53. (Files of the Navy Department).

Page 27   "A Brief Account of the Discoveries and Results of the United States Exploring Expedition," *American Journal of Science and Arts,* Vol. 44, No. 2 (1843), p. 393.

Page 29   Lieutenant Charles Wilkes, Memorandum to Commodore Isaac Chauncey, 1838, quoted in Nathan Reingold, editor, *Science in Nineteenth-Century America, A Documentary History.* New York: Hill and Wang, 1964, p. 119.

Page 29   Laurence Kirwan, *A History of Polar Exploration.* New York: W. W. Norton and Company, Inc., 1960, p. 131.

Page 31 "A Brief Account of the United States Exploring Expedition," p. 393.

Page 32 Senator Simon Cameron, quoted in D. C. Haskell, *The United States Exploring Expedition 1838–42 and Its Publications 1844–1874.* New York: New York Public Library, 1942, p. 23.

Page 32 Charles Pickering, Letter to Asa Gray, 1843, quoted in Haskell, *op. cit.,* p. 6.

Page 33 Asa Gray, quoted in Haskell, *op. cit.,* p. 20.

Page 34 Chaptain Charles Wilkes, "On the Depth and Saltness of the Ocean," *American Journal of Science and Arts,* 2nd series, Vol. 5, No. 13 (1848), p. 41.

Page 35 *Ibid.,* p. 42.

Page 35n Charles Bonnycastle, quoted in H. Drubba and H. H. Rust, "On the First Echo-Sounding Experiment," *Annals of Science,* Vol. 10, No. 1 (1954), p. 29.

Page 35n *Ibid.,* p. 32.

Page 36 Lieutenant Matthew Fontaine Maury, Letter to William Blackford, 1847, *Maury MSS., Letter Books,* Vol. 3 (Library of Congress, MS Division).

Page 38n William Redfield, "On the Prevailing Storms of the Atlantic Coast," *American Journal of Science and Arts,* Vol. 20, No. 1 (1831), p. 29.

Page 38 Matthew Fontaine Maury, *The Physical Geography of the Sea,* 6th edition. New York: Harper and Bros., 1858, p. xiii.

Page 39 Matthew Fontaine Maury, quoted in E. L. Towle, "Science, Commerce and the Navy on the Seafaring Frontier" (unpublished Ph.D. dissertation, University of Rochester, 1966), Diss. Abs. 27:173–A, p. 142.

Page 39 Maury, *The Physical Geography of the Sea,* p. x.

Page 41 Alexander Dallas Bache, "Lecture on the Gulf Stream . . . ," *American Journal of Science and Arts,* 2nd series, Vol. 30, No. 90 (1860), p. 313.

Page 43 *Ibid.,* p. 322.

Page 43 *Ibid.,* p. 323.

Page 45 Jacob W. Bailey, "Microscopical Examination of Soundings Made by the U.S. Coast Survey off the Atlantic Coast of the United States," *Smithsonian Contributions to Knowledge,* Vol. 2, Article 3, p. 8.

Page 47 G. M. Bache, Letter to Alexander Dallas Bache, 1846, Appendix 4, *Report of the Superintendent of the U.S. Coast Survey for 1846,* pp. 46–51.

Page 47 Alexander Dallas Bache, *Report of the Superintendent of the U.S. Coast Survey for 1846,* p. 23.

Page 48 Alexander Dallas Bache, "Lecture on the Gulf Stream," p. 329.

Page 48 Alexander Dallas Bache, *Report of the Superintendent of the U.S. Coast Survey for 1851,* p. 68.

Page 50 Quoted in Towle, *op. cit.,* p. 236.

Page 51 Lieutenant J. C. Walsh, Letter to M. F. Maury, 1850, quoted in Towle, *op. cit.,* p. 264.

Page 51 Log of the U.S.S. *Taney,* Naval Records, National Archives, quoted in Towle, *op. cit.,* p. 259.

Page 51 Walsh to Maury, quoted in Towle, *op. cit.,* p. 267.

Page 51n Log of the U.S.S. *Taney,* quoted in Towle, *op. cit.,* p. 267 .

Page 53 Matthew Fontaine Maury, *Annual Address Delivered Before the Maryland Institute . . . ,* October, 1825 (pamphlet), Baltimore, 1856, p. 23.

Page 53 Log of the U.S.S. *Dolphin,* Naval Records, National Archives, quoted in Towle, *op. cit.,* p. 285.

Page 54 Jacob W. Bailey, Letter to M. F. Maury, 1853, quoted in Towle, *op. cit.,* p. 287.

Page 58 Matthew Fontaine Maury, quoted in T. W. Corbin, *The Romance of Submarine Engineering.* London: Seeley, Service and Company, Ltd., 1913, p. 222.

Page 58 Maury, *The Physical Geography of the Sea,* p. 25.

Page 58 "The Physical Geography of the Sea," book review, *American Journal of Science and Arts,* 2nd series, Vol. 19, No. 57 (1855), p. 449.

Page 58 George Blunt, quoted in John Leighly, "M. F. Maury in His Time," *Bulletin de l'Institut Océanographique de Monaco,* numéro spécial 2, Vol. 1 (1968), p. 157.

Page 61 C. Wyville Thomson, *The Depths of the Sea,* 2nd edition. London: Macmillan and Company, 1874, p. 367.

Page 62 William Ferrel, "Autobiographical Sketch," *Biographical Memoirs,* National Academy of Sciences, Vol. 3 (1895), p. 295.

Page 63 Quoted in Frederick W. True, *A History of the First Half-Century of the National Academy of Sciences, 1863–1913.* Washington, D.C.: The Lord Baltimore Press, 1913, p. 224.

Page 65 F. M. Thorn's Report, quoted in Thomas G. Manning, *Government in Science, The U.S. Geological Survey 1867–1894.* p. 127.

Page 65 "The Present Condition of the Coast Survey," *Science,* Vol. 8 (1886), p. 360.

Page 66 Robert D. Gerard, "Oceanographic Measurements from Anchored and Dynamically Positioned Ships in Deep Water," preprint, p. 3. (For a volume in honor of Georg Wüst, in press. London: Gordon and Breach, Science Publications, Ltd.)

Page 66 John Elliot Pillsbury, "The Gulf Stream—A Description of the Methods Employed in the Investigation and the Results of the Research," Appendix 10, *Report of the Superintendent of the U.S. Coast and Geodetic Survey for 1890,* p. 473.

Page 66 *Ibid.,* p. 586.

Page 67 Joint Resolution for the Protection and Preservation of the Food-Fishes of the Coast of the United States, quoted in *Report on the Condition of the Sea Fisheries in 1871 and 1872,* p. xi. (This became the annual Report of the Commissioner in 1873.)

Page 72 Spencer Fullerton Baird, Letter to Captain Zera Tanner, 1883, quoted in Paul S. Galtsoff, *The Story of the Bureau of Commercial Fisheries Biological Laboratory . . . ,* Circular 145, Washington, D.C., 1962, p. 44.

Page 73 Edwin Linton, "The Man of Science and the Public" (an appreciation of Spencer Fullerton Baird), *Science,* Vol. 48 (1918), p. 33.

Page 74 "Professor Agassiz' School of Natural History," *Popular Science Monthly,* Vol. 3, No. 8 (1873), p. 123.

Page 75   E. Ray Lankester, "An American Sea-Side Laboratory," *Nature*, Vol. 21, No. 543 (1880), p. 498.

Page 75   Frank R. Lillie, *The Woods Hole Marine Biological Laboratory*. Chicago: University of Chicago Press, 1944, p. 116.

Page 79   Alexander Agassiz, "Exploration and Study of the Tropical Pacific Ocean," *Carnegie Institution of Washington Year Book No. 1*, 1902, p. 274.

Chapter II

Page 81   C. Wyville Thomson, *The Depths of the Sea*, 2nd edition. London: Macmillan and Company, 1874, p. 49.

Page 82   *The Ocean, a Description of the Wonders and Important Products of the Sea*. New York: G. Lane and P. P. Sanford, 1844, p. 17.

Page 83   Thomson, *op. cit.*, p. 1.

Page 84   *Report of the British Association for the Advancement of Science*, Vol. 9, 1839, p. xxvi.

Page 86   Sir James Clark Ross, *A Voyage of Discovery and Research in the Southern and Antarctic Regions*, Vol. 1. London: John Murray, 1847, p. 202.

Page 86n   Louis Dollo, *Poissons*, in *Résultats du Voyage du S. Y. Belgica*, 1904, p. 8.

Page 87   Ross, *op. cit.*, p. 26.

Page 87   Ross, *op. cit.*, p. 26.

Page 87   *Ibid.*

Page 92   Charles Darwin, excerpts from *On the Origin of Species by Means of Natural Selection*, in Marston Bates and Philip Humphrey, editors, *The Darwin Reader*. New York: Charles Scribner's Sons, 1956, p. 202.

Page 92   *Ibid.*, p. 195.

Page 94   Thomson, *op. cit.*, p. 434.

Page 95n   *Ibid.*, p. 423.

Page 95n   *Ibid.*, p. 426.

Page 96   John Macgillivray, quoted in Daniel Merriman, "A Posse ad Esse," *Journal of Marine Research*, Vol. 7, No. 3 (1948), p. 144.

Page 97   Thomson, *op. cit.*, p. 470.

Page 97   T. H. Huxley, "On Some Organisms Living at Great Depths in the North Atlantic Ocean," *Quarterly Journal of Microscopical Science*, New Series Vol. 8 (1868), p. 205.

Page 98   *Ibid.*

Page 98   Thomson, *op. cit.*, p. 409.

Page 99   *Ibid.*, p. 50.

Page 99   *Ibid.*

Page 99   *Ibid.*, p. 52.

Page 99   *Ibid.*, p. 56.

Page 100   *Ibid.*, p. 54.

Page 100   *Ibid.*, p. 57.

Page 101   *Ibid.*, p. 58.

Page 101   *Ibid.*, p. 64.

Page 101   *Ibid.*, p. 66.

Page 101n   *The Guardian*, London. June 25, 1873, quoted in Daniel Merriman, *op. cit.*, p. 145.

Page 102 Thomson, *op. cit.*, p. 70.
Page 102 *Ibid.*, p. 76.
Page 103 *Ibid.*, p. 79.
Page 103 *Ibid.*
Page 103 *Ibid.*
Page 103 *Ibid.*, p. 280.
Page 103 *Ibid.*
Page 104 Louis Agassiz, "Report upon Deep-Sea Dredging in the Gulf Stream . . . ," Appendix 10, *Report of the Superintendent of the U.S. Coast Survey for 1869*, p. 208.
Page 106 William B. Carpenter, in "Notes," *Nature*, Vol. 4 (1871), p. 107.

Chapter III

Page 110 Hugh Robert Mill, *An Autobiography*. London: Longmans, Green, and Company, 1951, p. 50.
Page 111 Henry Nottidge Moseley, *Notes by a Naturalist on the "Challenger."* London: Macmillan and Company, 1879, p. 578.
Page 112 C. Wyville Thomson, *The Atlantic, A Preliminary Account of the General Results of the Exploring Voyage of H.M.S. "Challenger,"* Vol. 2. London: Macmillan and Company, 1877, p. 328.
Page 113 *Report on the Scientific Results of the Voyage of H.M.S. "Challenger": Summary of Scientific Results*, Vol. 2. London: Her Majesty's Stationery Office, 1895, p. 1186.
Page 113 Thomson, *op. cit.*, p. 352.
Page 113 Moseley, *op. cit.*, p. 575.
Page 114 Alexander Agassiz, *Three Cruises of the U.S.C.G.S. Steamer "Blake,"* published in the *Bulletin of the Museum of Comparative Zoology*. Vol. 14, 1888, p. 37.
Page 117 Moseley, *op. cit.*, p. 594.
Page 118 *Ibid.*, p. 578.
Page 120 Thomson, *op. cit.*, p. 301.
Page 121 *Ibid.*
Page 122 Louis Agassiz, "A Letter Concerning Deep-Sea Dredgings, addressed to Professor Benjamin Peirce, Superintendent U.S. Coast Survey," *Bulletin of the Museum of Comparative Zoology*, Vol. 3, 1872, pp. 49–53.
Page 123 C. Wyville Thomson, *The Depths of the Sea*, 2nd edition. London: Macmillan and Company, 1874, p. 280.
Page 123 Moseley, *op. cit.*, p. 587.
Page 123 Sir John Murray, *The Ocean, A General Account of the Science of the Sea*. New York: Henry Holt and Company, [no. pub. date, but about 1913], p. 179 and p. 185.
Page 124 T. H. Huxley, "Notes from the 'Challenger,'" *Nature*, Vol. 12, No. 303 (1875), p. 316.
Page 124 Thomson, *The Atlantic*, p. 202.
Page 125 *Ibid.*, p. 270.
Page 125 *Ibid.*
Page 128 Alexander Agassiz, "On the 'Challenger' Collections," *Nature*, Vol. 15, No. 377 (1877), p. 256.
Page 129 Alexander Agassiz, *Three Cruises of the "Blake,"* p. 34.

Page 130   Ibid., p. 30.

Page 130   Alexander Agassiz, Letter to C. P. Patterson, Superintendent of the Coast and Geodetic Survey, in G. R. Agassiz, *Letters and Recollections of Alexander Agassiz*. Boston: Houghton Mifflin Company, 1913, p. 181.

Page 131   C. Wyville Thomson, quoted in G. S. Ritchie, *The Admiralty Chart, British Naval Hydrography in the 19th Century*. London: Hollis and Carter, 1967, p. 341.

Page 135   Albert 1er, Prince de Monaco, *Résultats des Campagnes Scientifiques du Prince de Monaco*, No. 17, *Céphalopodes* 1900, p. 17.

Chapter IV

Page 146   C. Wyville Thomson, *The Atlantic, A Preliminary Account of the General Results of the Exploring Voyage of H.M.S. "Challenger,"* Vol. 2. London: Macmillan and Company, 1877, p, 291.

Page 146   Ibid., p. 299.

Page 146n   Ibid.

Page 146   Ibid., Vol. 1, p. 197.

Page 147   John Murray and A. F. Renard, *Report on Deep-Sea Deposits . . .*, in *Report on the Scientific Results of the Voyage of H.M.S. "Challenger."* London: Her Majesty's Stationery Office, 1891, p. 30.

Page 147   Ibid., p. 191.

Page 148   Ibid.

Page 152   Stephen W. Mitchell, "The 'Lost Atlantis' and the 'Challenger' Soundings," *Nature*, Vol. 15, No. 391 (1877), p. 553.

Page 153   Charles Darwin, quoted in Laurence P. Kirwan, *A History of Polar Exploration*. New York: Norton and Company, Inc., 1960, p. 217.

Page 154   Murray and Renard, *op. cit.*, p. 323.

Page 154   Ibid., p. ix.

Page 155   Ibid., p. v.

Page 156   Sir John Murray and Dr. Johan Hjort, *The Depths of the Ocean*. London: Macmillan and Company, 1912, p. 144.

Page 157   Captain Cook, quoted in Henry O. Forbes, "The Great Barrier Reef of Australia," *The Geographical Journal*, Vol. 2, No. 6 (1893), p. 540.

Page 157   Robert Louis Stevenson, *In the South Seas . . .* (facsimile reproduction). Honolulu: University of Hawaii Press, 1971, p. 141.

Page 159   Charles Darwin, *The Autobiography of Charles Darwin 1809–1882*. Nora Barlow, editor. London: Collins, 1958, p. 98.

Page 159   John Murray, "On the Structure and Origin of Coral Islands," *Proceedings of the Royal Society of Edinburgh*, Vol. 10 (1880), p. 506.

Page 162   George Douglas Campbell, 8th duke of Argyll, "A Conspiracy of Silence," *Nature*, Vol. 37, No. 942 (1887), p. 53.

Page 162   T. H. Huxley, "The Duke of Argyll's Charges Against Men of Science," *Nature*, Vol. 37, No. 954 (1888), p. 342.

Page 163   Alexander Agassiz, *Three Cruises of the U.S.C.G.S. Steamer "Blake,"* published in the *Bulletin of the Museum of Comparative Zoology*, Vol. 14 (1888), p. 55.

Page 165n   Darwin, *op. cit.*, p. 139.

Page 166 Alexander Agassiz, quoted in Sir John Murray, "Alexander Agassiz—His Life and Scientific Work," *Science*, Vol. 33, No. 858 (1911), p. 882.

Page 166 *Ibid.*

Page 166 G. R. Agassiz, *Letters and Recollections of Alexander Agassiz with a Sketch of His Life and Work*. Boston: Houghton Mifflin, 1913, p. 395.

Page 166 Charles Darwin, Letter to Alexander Agassiz, May, 1881, in Francis Darwin, editor. *Charles Darwin . . . and a Selected Series of His Published Letters*. New York: D. Appleton and Company, 1893, p. 299.

Page 167 G. R. Agassiz, *op. cit.*, p. 323.

Page 167 *Ibid.*, p. 333.

Page 168 *Ibid.*, p. 261.

Page 169 Harry S. Ladd and Seymour O. Schlanger, *Drilling Operations on Eniwetok Atoll. U.S. Geological Survey Professional Paper 260–Y*. Washington, D.C.: U.S. Government Printing Office, 1954, p. 863.

Chapter V

Page 176 Otto Pettersson, "A Review of Swedish Hydrographic Research in the Baltic and North Seas," *The Scottish Geographical Magazine*, Vol. 10 (1894), p. 281.

Page 177n Fridtjof Nansen, *The Waters of the North-Eastern North Atlantic*. Leipzig, Germany: Dr. Werner Klinkhardt, 1913, p. 14.

Page 179 Otto Pettersson, *op. cit.*, p. 299.

Page 179 H. N. Dickson, "The Movements of the Surface Waters of the North Sea," *Geographical Journal*, No. 3 (March, 1896), p. 262.

Page 183 Fridtjof Nansen, *Farthest North, Being the Record of a Voyage of Exploration of the Ship Fram 1893–1896*, Vol. 1. London: Harrison and Sons, 1897, p. 206.

Page 183 *Ibid.*, p. 237.

Page 183 *Ibid.*, p. 205.

Page 184 *Ibid.*, p. 327.

Page 185 *Ibid.*, p. 214

Page 187n Vice-Admiral S. O. Makaroff, "The 'Yermak' Ice-Breaker," *Geographical Journal*, Vol. 15, No. 1 (1900), p. 33.

Page 187n *Ibid.*

Page 188 Fridtjof Nansen, *Scientific Results of the Norwegian North Polar Expedition, 1893–1896*, Vol. 3, No. 9 (1902), p. 406.

Page 189 V. W. Ekman, "On Dead-Water," in F. Nansen, editor, *Scientific Results of the Norwegian North Polar Expedition, 1893–1896*, Vol. 5, No. 15 (1906), p. 16.

Page 190 D'Arcy W. Thompson, "Dr. Otto Pettersson" (obituary), *Nature*, Vol. 147, No. 3736 (1941), p. 702.

Page 190 Fridtjof Nansen, *Scientific Results*, p. iii.

Page 191n William C. Redfield, "Summary Statements of Some of the Leading Facts in Meteorology," *American Journal of Science and Arts*, Vol. 25 (1834), p. 131.

Page 193 Fridtjof Nansen, "Some Oceanographical Results of the Expedition with *Michael Sars* in the Summer of 1900," *Nyt Magazin for Naturvidenskaberne*, Vol. 39, No. 2 (1901), p. 130.

Page 195 *Ibid.*, p. 136.

Page 195 *Ibid.*, p. 141.

Page 196 *Ibid.*, p. 140.

Page 196 *Ibid.*, p. 160.

Page 197 Bjørn Helland-Hansen, "Physical Oceanography," in Sir John Murray and Dr. Johan Hjort, *The Depths of the Ocean*. London: Macmillan and Company, 1912, p. 283.

Page 198 Bjørn Helland-Hansen, "Current-Measurements in Norwegian Fiords . . .," *Bergens Museums Aarbog*, No. 15 (1907), p. 3.

Page 203 Albert Defant, *Physical Oceanography*, Vol. 1. Oxford: Pergamon Press, 1961, p. 345.

Page 203 *Ibid.*, p. 346.

Page 205 Helland-Hansen, "Physical Oceanography," p. 285.

Page 205 D'Arcy W. Thompson, "The Voyages of the *Discovery*," *Nature*, Vol. 140, No. 3543 (1937), p. 530.

Chapter VI

Page 206 W. C. McIntosh, *The Resources of the Sea*. . . . London: C. J. Clay and Sons, 1899, p. x.

Page 209 D'Arcy Wentworth Thompson, quoted in Ruth D'Arcy Thompson, *D'Arcy Wentworth Thompson. The Scholar-Naturalist 1860–1948*. London: Oxford University Press, 1958, p. 137.

Page 213 P. T. Cleve, "Microscopic Marine Organisms in the Service of Hydrography," *Nature*, Vol. 55, no. 1413 (1896), p. 90.

Page 216 Captain Thor Iversen, quoted in Sir John Murray and Johan Hjort, *The Depths of the Ocean*. London: Macmillan and Company, 1912, p. 54.

Page 217 C. G. Joh. Petersen, "The Sea Bottom and Its Production of Fish-Food," *Report of the Danish Biological Station for 1918*, Vol. 25, p. 55.

Page 224 Johan Hjort, "Fluctuations in the Great Fisheries of Northern Europe," *Rapports et Proces-Verbaux*, Vol. 20 (1914), p. 38.

Page 225 *Ibid.*, p. 11.

Page 227n *Ibid.*, p. 138.

Page 228 *Ibid.*, p. 205

Page 230 Otto Müller, quoted in Torben Wolff, *Danish Expeditions on the Seven Seas*. Copenhagen: Rhodos, 1967, p. 10.

Page 231 Victor Hensen, *op cit.*, p. 772, quoted in Murray and Hjort, *The Depths of the Ocean*.

Page 232n James Johnstone, *Conditions of Life in the Sea* . . ., Cambridge, England: The University Press, 1908, pp. 160–63.

Page 232 *Ibid.*, p. 190.

Page 233 Ernst Haeckel, "Planktonic Studies . . .," U.S. Commission of Fish and Fisheries, *Report of the Commissioner for 1889–91*, p. 572.

Page 238 H. H. Gran, "Preservation of Samples and Quantitative Determination of the Plankton," *Publications de Circonstance*, No. 62 (1912), p. 3.

Page 238 Henry Bryant Bigelow, *Report on the Scope, Problems, and Economic Importance of Oceanography.* . . . Report submitted to the National Academy of Sciences from the N.A.S. Committee on Oceanography, 1929, p. 138.

Page 240n Petersen, *op. cit.*, p. 52.

Page 242n Otto Pettersson and C. F. Drechsel, "Mémoire d'une Expédition Internationale pour l'Exploration de la Mer," *Rapports et Proces-Verbaux*, Vol. 32 (1923), p. 64.

Page 243 Bigelow, *loc. cit.*

Page 243 The General Secretary, the International Council for the Exploration of the Sea (mimeographed report issued by ICES), 1970, p. 1.

Chapter VII

Page 245 First Lord of the Admiralty, Earl St. Vincent, quoted in C. P. Idyll, editor, *Exploring the Ocean World.* New York: Thomas Y. Crowell Company, 1969, p. 235.

Page 247 Lewis Fry Richardson, quoted in Frederick V. Hunt, *Electroacoustics.* . . . Cambridge, Mass.: Harvard University Press, 1954, p. 45.

Page 248 H. J. W. Fay, "The Submarine Signal Company," *Soundings*, No. 8 (February, 1945), p. 7.

Page 251 "Report of the U.S. Hydrographic Office 17th April 1923," *Hydrographic Review*, Vol. 1, No. 2 (1924), p. 48.

Page 253 Henry B. Bigelow, *Report on the Scope, Problems and Economic Importance of Oceanography, on the Present Situation in America, and on the Handicaps to Development, with Suggested Remedies.* Report submitted to the National Academy of Sciences from the N.A.S. Committee on Oceanography, 1929, p. 111.

Page 255 Henry B. Bigelow, *Memories of a Long and Active Life.* Cambridge, Mass.: The Cosmos Press, 1964, p. 9.

Page 256 Henry B. Bigelow, "Cruise of the U.S. Fisheries Schooner 'Grampus,' in the Gulf Stream during July, 1908 . . .," *Bulletin of the Museum of Comparative Zoology*, Vol. 52, No. 12 (1909), p. 202.

Page 258 Bigelow, *Memories*, p. 23.

Page 259 Henry B. Bigelow, "Explorations in the Gulf of Maine, July and August 1912 . . .," *Bulletin of the Museum of Comparative Zoology*, Vol. 58, No. 2 (1914), p. 33.

Page 263 Bigelow, *Report on the Scope, Problems, and Economic Importance of Oceanography* . . . , p. 96.

Page 265 Edward H. Smith, "Report of Ice Observations made on the Aeroarctic Expedition with the 'Graf Zeppelin,' 1931," *International Ice Observation and Ice Patrol Service*, Bulletin No. 21, Season of 1931, p. 48.

Page 266 Andrew Carnegie, quoted in "Carnegie Institution of Washington," *Science*, Vol. 17, No. 422 (1903), p. 166.

Page 266 Andrew Carnegie, quoted in Louis M. Hacker, *The World of Andrew Carnegie.* . . . Philadelphia: Lippincott, 1968, p. 349.

Page 268 Ermund Halley, quoted in J. Harland Paul, *The Last Cruise of the Carnegie.* Baltimore: The Williams and Wilkins Co., 1932, p. 7.

Page 273 Henry B. Bigelow, "Address at the Lillie Memorial Meeting,

Woods Hole, August 11, 1948," *Biological Bulletin,* Vol. 95, No. 2 (1948), p. 158.

Page 273   F. R. Lillie, quoted in Helen Raitt and Beatrice Moulton, *Scripps Institution of Oceanography. First Fifty Years.* Los Angeles: The Ward Ritchie Press, 1967, p. 109.

Page 274   *Ibid.*

Page 274   *The Rockefeller Foundation, Annual Report for 1930,* p. 196, and "Treasurer's Report," pp. 259–338. See also: Raitt and Moulton, *op. cit.,* p. 110.

Page 275   Parker D. Trask, "Oceanography and Oil Deposits," *Bulletin of the National Research Council,* No. 61 (July, 1927), p. 235.

Page 275   Act of Incorporation, quoted in "First Annual Report for 1930," *Woods Hole Oceanographic Institution, Collected Reprints,* 1933, p. 14.

Page 275   Henry B. Bigelow, "First Annual Report for 1930," *Woods Hole Oceanographic Institution, Collected Reprints,* 1933, p. 16.

Page 279   Henry B. Bigelow, quoted in Raitt and Moulton, *op. cit.,* p. 116.

Chapter VIII

Page 282   Columbus O'D. Iselin, "Report of the Director for the Years 1943, 1944, and 1945," *Woods Hole Oceanographic Institution, Collected Reprints,* 1944–46, p. 15.

Page 287   Lieutenant C. F. Horne, Letter to Henry Bryant Bigelow, 1937. Obtainable from the archives of the Woods Hole Oceanographic Institution.

Page 294   Allyn Vine, personal communication.

Page 294   *Ibid.*

Page 296   H. U. Sverdrup, "Research Within Physical Oceanography and Submarine Geology . . . ," *Transactions of the American Geophysical Union,* Vol. 27, No. 4 (1946), p. 571.

Page 299   Quoted in Martin W. Johnson, F. Alton Everest, and Robert W. Young, "The Role of Snapping Shrimp . . . in the Production of Underwater Sound in the Sea," *Biological Bulletin,* Vol. 93, No. 2 (1947), p. 123.

Page 299   Marie Poland Fish, "The Sonic Marine Animal Problem," *Office of Naval Research, Research Reviews* (December 1954), p. 13.

Page 300   Martin W. Johnson, "Sound as a Tool in Marine Ecology . . . ," *Journal of Marine Research,* Vol. 7, No. 3 (November 1948), p. 450.

Page 301n   Johnson, Everest, and Young, "The Role of Snapping Shrimp," p. 123.

Page 305   Vaughn Cornish, *Ocean Waves and Kindred Geophysical Phenomena.* Cambridge University Press, 1934, p. 3.

Page 306   *Ibid.,* p. xiv.

Page 312   Columbus O'D. Iselin, "Report of the Director for 1946," *Woods Hole Oceanographic Institution, Collected Reprints,* 1947, p. 13.

Page 314   Roger Revelle, "Forward," in K. O. Emery, J. I. Tracey, Jr., and H. S. Ladd, *Geology of Bikini and Nearby Atolls.* U.S. Geological Survey Professional Paper 260–A. Washington, D.C.: U.S. Government Printing Office, 1954, p. iii.

Page 315   Philip Morrison, quoted in Daniel S. Greenberg, *The Politics of Pure Science.* New York: The New American Library, 1967, p. 136.

Page 315 "News and Notes," *Science*, Vol. 114, No. 2956 (August 24, 1951), p. 226.

Page 316 "Marine Sciences in the United States—1958," *Oceanography 1960 to 1970*. Washington, D.C.: National Academy of Science—National Research Council, 1959, p. 3. See also: *Marine Science Affairs— A Year of Transition*. The First Report of the President to the Congress on Marine Resources and Engineering Development. Washington, D.C.: U.S. Government Printing Office, 1967, p. 12.

Chapter IX

Page 318 Osmond Fisher, "On the Physical Cause of the Ocean Basins," *Nature*, Vol. 25, No. 636 (1882), p. 244.

Page 319 Alfred Wegener, *The Origin of Continents and Oceans*. Translated from the 3rd German edition. New York. E. P. Dutton & Co., 1924, p. 5.

Page 302n *Ibid.*, p. 10.

Page 321 *Ibid.*, p. 40.

Page 323 R. T. Chamberlin, "Some of the Objections to Wegener's Theory," *Theory of Continental Drift, a Symposium*. Tulsa, Okla.: American Association of Petroleum Geologists, 1928.

Page 323 "Birmingham Meeting, 1950," *Proceedings of the British Association for the Advancement of Science*, Vol. 7, No. 27 (1950), p. 277.

Page 323 Wegener, *op. cit.*, p. 194.

Page 324 Arthur Holmes, "Radioactivity and Earth Movements," *Transactions of the Geological Society of Glasgow*, Vol. 18, part 3 (1931), p. 579.

Page 324 Arthur Holmes, "The South Atlantic: Land Bridges or Continental Drift?" *Nature*, Vol. 171, No. 4355 (1953), p. 669.

Page 327n Thomas Gaskell, quoted in C. P. Idyll, editor, *Exploring the Ocean World*. New York: Thomas Y. Crowell Company, 1969, p. 35.

Page 329 T. Nathan Kelley, "1939–'41, Notebook," unpublished, p. 91.

Page 329 *Ibid.*

Page 329 *Ibid.*

Page 329 *Ibid.*

Page 331 Maurice Ewing, private interview, Lamont-Doherty Geological Observatory, June 1971.

Page 340n Quoted in Walter H. Munk, "Polar Wanderings: A Marathon of Errors," *Nature*, Vol. 177, No. 4508 (1956), p. 552.

Page 341 Hans Pettersson, "Exploring the Bed of the Ocean," *Nature*, Vol. 164, No. 4168 (1949), p. 469.

Page 341 Roger Revelle and Arthur E. Maxwell, "Heat Flow Through the Floor of the Eastern North Pacific," *Nature*, Vol. 170, No. 4318 (1952), p. 199.

Page 342 Henry Nottidge Moseley, *Notes by a Naturalist on the "Challenger,"* London: Macmillan and Company, 1879, p. 585.

Page 343 J. D. H. Wiseman and R. B. S. Sewell, "The Floor of the Arabian Sea," *The Geological Magazine*, Vol. 74, No. 875 (1937), p. 219.

Page 343 *Ibid.*, p. 222.

Page 343 *Ibid.*, p. 226.

Page 343 *Ibid.*, p. 227.

Page 346   Henry Menard, "Development of Median Elevations in Ocean Basins," *Bulletin of the Geological Society of America*, Vol. 69, part 2 (September 1958), p. 1180.

Page 350   H. H. Hess, "History of Ocean Basins," in A. E. J. Engel *et al.*, editors, *Petrologic Studies: A Volume in Honor of A. E. Buddington.* The Geological Society of America, 1962, p. 607.

Page 352   *Ibid.*, p. 618.

Page 352   *Ibid.*, p. 617.

Page 352   *Ibid.*, p. 615.

Page 352   *Ibid.*, p. 618.

Page 352   *Ibid.*

Page 353   R. S. Dietz, "Continent and Ocean Basin Evolution by Spreading of the Sea Floor," *Nature*, Vol. 190, No. 4779 (1961), p. 855.

Page 353   V. V. Beloussov, "Debate About the Earth. An Open Letter to J. Tuzo Wilson," *Geotimes*, Vol. 13, No. 10 (1968), p. 17.

Page 353   *Ibid.*, p. 18.

Page 354n   A. A. Meyerhoff, "Spreading Sea Floor: By Convection or Conviction," *Bulletin of the American Association of Petroleum Geologists*, Vol. 53, part 1, No. 1 (1969), p. 216.

Page 354   Francis P. Shepard, *Submarine Geology*. 2d edition. New York: Harper & Row, 1963, p. 435.

Page 355n   Quoted in John Lear, "Canada's Unappreciated Role as Scientific Innovator," *Saturday Review*, Sept. 2, 1967, p. 47.

Page 359   A. E. Maxwell *et al.*, *Initial Reports of the Deep Sea Drilling Project, Vol. 3.* Washington, D.C.: U.S. Government Printing Office, 1970, p. 8.

Page 362   "Summary of Scientific Achievements," *Ocean Industry*, Vol. 4, No. 5 (1969), p. 67.

Chapter I

"A Brief Account of the Discoveries and Results of the United States Exploring Expedition," *American Journal of Science and Arts*, Vol. 44, No. 2 (1843), pp. 393–408.

Agassiz, Alexander. "Exploration and Study of the Tropical Pacific Ocean," *Carnegie Institution of Washington Year Book No. 1*, 1902, pp. 272–74.

Aimé, Georges. "Mémoire sur un Moyen Nouveau de Sonder à la Mer," *Annales de Chimie et de Physique*, 3rd series, Vol. 7 (1843), pp. 497–505.

Allard, Dean C., Jr. "Spencer Fullerton Baird and the U.S. Fish Commission . . ." (unpublished Ph.D. dissertation, George Washington University, 1967). Diss. Abs. 28: 1750–A.

Bache, Alexander Dallas. *Report of the Superintendent of the U.S. Coast Survey for 1846.*

———. *Report of the Superintendent of the U.S. Coast Survey for 1851.*

———. "Lecture on the Gulf Stream . . .," *American Journal of Science and Arts*, 2nd series, Vol. 30, No. 90 (1860), pp. 313–29.

Bache, G. M. Letter to Alexander Dallas Bache, 1864, Appendix 4, *Report of the Superintendent of the U.S. Coast Survey for 1846*, pp. 46–51.

Bailey, Jacob W. "Microscopical Examination of Soundings Made by the U.S. Coast Survey off the Atlantic Coast of the United States," *Smithsonian Contributions to Knowledge*, Vol. 2, Article 3, 15 pp.

Carpenter, William B. "On the Gibraltar Current, the Gulf Stream, and the General Oceanic Circulation," *Proceedings of the Royal Geographical Society of London*, Vol. 15, No. 1 (1870–71), pp. 54–88.

Corbin, T. W. *The Romance of Submarine Engineering.* London: Seeley, Service and Company, Ltd., 1913.

Deacon, Margaret. "Some Early Investigations of the Currents in the Strait of Gibraltar," *Bulletin de l'Institut Océanographique de Monaco*, numéro spécial 2, Vol. 1 (1968), pp. 63–74.

Drubba, H., and H. H. Rust. "On the First Echo-Sounding Experiment," *Annals of Science*, Vol. 10, No. 1 (1954), pp. 28–32.

Dupree, A. Hunter. *Science in the Federal Government*. Cambridge: The Belknap Press of Harvard University Press, 1957.

Ferrel, William. "An Essay on the Winds and Currents of the Oceans," *Nashville Journal of Medicine and Surgery*, Vol. 12, Nos. 4 and 5, 1856.

———. "Autobiographical Sketch," *Biographical Memoirs*, National Academy of Sciences, Vol. 3 (1895), pp. 287–309.

Galtsoff, Paul S. *The Story of the Bureau of Commercial Fisheries Biological Laboratory* . . ., Circular 145, Washington, D.C., 1962.

Gerard, Robert D. "Oceanographic Measurements from Anchored and Dynamically Positioned Ships in Deep Water," preprint, p. 3 (for a volume in honor of Georg Wüst, in press. London: Gordon and Breach, Science Publications, Ltd.)

Goldsborough, Lt. L. M. Letter, 1830, in *Officers' Letters to the Secretary of the Navy*, Vol. 9, p. 53. (Files of the Navy Dept.)

Haskell, D. C. *The United States Exploring Expedition 1838–42 and Its Publications 1844–1874*. New York: New York Public Library, 1942.

Henry, Joseph. "Memoir of Alexander Dallas Bache, 1806–1867," *Biographical Memoirs*, National Academy of Sciences, Vol. 1 (1877), pp. 181–212d.

Herschel, Sir John. Letter on Ocean Currents to Dr. William B. Carpenter, 1871, in *Nature*, Vol. 4 (1871), p. 71.

Kirwan, Laurence P. *A History of Polar Exploration*. New York: W. W. Norton and Company, Inc., 1960.

Lankester, E. Ray. "An American Sea-Side Laboratory," *Nature*, Vol. 21, No. 543 (1880), pp. 497–99.

Leighly, John. "Introduction," in Matthew Fontaine Maury, *The Physical Geography of the Sea* (facsimile edition). Cambridge: The Belknap Press of Harvard University Press, 1963, pp. ix–xxx.

———. "M. F. Maury in His Time," *Bulletin de l'Institut Océanographique de Monaco*, numéro spécial 2, Vol. 1 (1968), pp. 147–61.

Lillie, Frank R. *The Woods Hole Marine Biological Laboratory*. Chicago: University of Chicago Press, 1944.

Linton, Edwin. "The Man of Science and the Public" (an appreciation of Spencer Fullerton Baird), *Science*, Vol. 48 (1918), pp. 25–34.

Manning, Thomas G. *Government in Science, The U.S. Geological Survey 1867–1894*. Lexington: University of Kentucky Press, 1967.

Maury, Matthew Fontaine. Letter to William Blackford, 1847, *Maury MSS, Letter Books*, Vol. 3 (Library of Congress, MS Division.)

———. *Annual Address Delivered Before the Maryland Institute* . . ., Oct., 1825 (pamphlet), Baltimore, 1856.

———. *The Physical Geography of the Sea*, 6th edition. New York: Harper and Bros., 1858.

"The Physical Geography of the Sea," book review, *American Journal of Science and Arts*, 2nd series, Vol. 19, No. 57 (1855), p. 449.

Pillsbury, John Elliot. "The Gulf Stream—A Description of the Methods Employed in the Investigation and the Results of the Research," Appendix 10, *Report of the Superintendent of the U.S. Coast and Geodetic Survey for 1890*, pp. 461–620.

"The Present Condition of the Coast Survey," *Science*, Vol. 8 (1886), pp. 359–60.

"Professor Agassiz's School of Natural History," *Popular Science Monthly*, Vol. 3, No. 8 (1873), pp. 123–24.

Raitt, Helen, and Beatrice Moulton. *Scripps Institute of Oceanography, First Fifty Years*, Los Angeles: The Ward Ritchie Press, 1967.

Redfield, William. "On the Prevailing Storms of the Atlantic Coast," *American Journal of Science and Arts*, Vol. 20, No. 1 (1831), pp. 17–51.

Reingold, Nathan, editor. *Science in Nineteenth-Century America, A Documentary History*. New York: Hill and Wang, 1964, p. 119.

Report on the Condition of the Sea Fisheries in 1871 and 1872. (This became the annual *Report of the Commissioner* in 1873.)

Stommel, Henry. *The Gulf Stream, A Physical and Dynamical Description.* Berkeley: University of California Press, 1958.

Thomson, C. Wyville. *The Depths of the Sea*, 2nd edition. London: Macmillan and Company, 1874.

Towle, E. L. "Science, Commerce and the Navy on the Seafaring Frontier" (unpublished Ph.D. dissertation, University of Rochester, 1966). Diss. Abs. 27: 173–A.

True, Frederick W., editor. *A History of the First Half-Century of the National Academy of Sciences, 1863–1913.* Washington, D.C.: The Lord Baltimore Press, 1913.

Weber, Gustavus A. *The Coast and Geodetic Survey. Its History, Activity, and Organization.* Institute for Government Research Monograph #16. Baltimore: Johns Hopkins Press, 1923.

———. *The Hydrographic Office. Its History, Activities, and Organization.* Institute for Government Research Monograph #42. Baltimore: Johns Hopkins Press, 1926.

Wilkes, Captain Charles. "On the Depth and Saltness of the Ocean," *American Journal of Science and Arts*, 2nd series, Vol. 5, No. 13 (1848), pp. 41–48.

Chapter II

Agassiz, Louis. "Report Upon Deep-Sea Dredging in the Gulf Stream . . . ," Appendix 10, *Report of the Superintendent of the U.S. Coast Survey for 1869*, pp. 208–19.

Anonymous. *The Ocean, A Description of the Wonders and· Important Products of the Sea.* New York: G. Lane and P. Sanford, 1844.

Carpenter, William B. In "Notes," *Nature*, Vol. 4 (1871), p. 107.

Darwin, Charles. Excerpts from *On the Origin of Species by Means of Natural Selection*, in Marston Bates and Philip Humphrey, editors, *The Darwin Reader*. New York: Charles Scribner's Sons, 1956.

Deacon, G. E. R. "Early Scientific Studies of the Antarctic Ocean," *Bulletin de l'Institut Océanographique de Monaco*, numéro spécial 2, Vol. 1 (1968), pp. 269–79.

Dollo, Louis. *Poissons*, in *Résultats du Voyage du S. Y. Belgica*, 1904, 239 pp.

Eiseley, Loren. *Darwin's Century*. Garden City, N.Y.: Doubleday and Company, 1958.

Gillispie, Charles C. *Genesis and Geology*. New York: Harper Torchbooks, 1959.

Herdman, Sir William A. *Founders of Oceanography and Their Work, An*

Introduction to the Science of the Sea. London: Edward Arnold and Company, 1923.

Huxley, T. H. "On Some Organisms Living at Great Depths in the North Atlantic Ocean," *Quarterly Journal of Microscopical Science*, New Series, Vol. 8 (1868), pp. 203–12.

Lankester, E. Ray. "Dr. Carpenter, C. B., F. R. S.," *Nature*, Vol. 33, No. 839 (1885), pp. 83–85.

Lyell, Charles. *Principles of Geology*, Vol. 1. First American edition, from 5th and last London edition. Philadelphia: James Kay, Jr., and Brothers, 1837.

Merriman, Daniel. "A Passe ad Esse," *Journal of Marine Research*, Vol. 7, No. 3 (1948), pp. 139–46.

Pourtalès, Louis François de. "Report on the Fauna of the Gulf Stream in the Strait of Florida," Appendix 16, *Report of the Superintendent of the U.S. Coast Survey for 1867*, pp. 180–82.

———. "Report on Dredging Near the Florida Reefs," Appendix 12, *Report of the Superintendent of the U.S. Coast Survey for 1868*, pp. 168–70.

Rathbun, Richard. "The American Initiative in Methods of Deep-Sea Dredging," *Nature*, Vol. 30, No. 773 (1884), pp. 399–401.

Report of the British Association for the Advancement of Science, Vol. 9 (1839).

Ritchie, G. S. "The Royal Navy's Contribution to Oceanography in the XIXth Century," *Bulletin de l'Institut Océanographique de Monaco*, numéro spécial 2, Vol. 1 (1968), pp. 121–31.

Ross, Sir James Clark. *A Voyage of Discovery and Research in the Southern and Antarctic Regions*, 2 Vols. London: John Murray, 1847.

Thomson, C. Wyville. "The Continuity of Chalk," *Nature*, Vol. 3, No. 64 (1871), pp. 225–27.

———. *The Depths of the Sea*, 2nd edition. London: Macmillan and Company, 1874.

Chapter III

Agassiz, Alexander. "On the 'Challenger' Collections," *Nature*, Vol. 15, No. 377 (1877), p. 256.

———. *Three Cruises of the U.S.C.G.S. Steamer "Blake,"* published in the *Bulletin of the Museum of Comparative Zoology*, Harvard, Vol. 14, 1888.

Agassiz, G. R. *Letters and Recollections of Alexander Agassiz.* Boston: Houghton Mifflin Company, 1913, p. 181.

Agassiz, Louis. "A Letter Concerning Deep-Sea Dredgings, Addressed to Professor Benjamin Peirce, Superintendent U.S. Coast Survey," *Bulletin of the Museum of Comparative Zoology*, Vol. 3, 1872, pp. 49–53.

Albert Ier, Prince de Monaco. *Résultats des Campagnes Scientifiques du Prince de Monaco*, No. 17, *Céphalopodes*, 1900.

———. *La Carrière d'un Navigateur.* Monaco: Éditions des Archives du Palais Princier, 1966.

Buchan, Alexander. "Report on Oceanic Circulation . . . ," in *Report on the Scientific Results of the Voyage of H.M.S. "Challenger": Summary of Scientific Results*, Vol. 2, Appendix. London: Her Majesty's Stationery Office, 1895, 33 pp.

Burstyn, Harold L. "Science and Government in the Nineteenth Century: the *Challenger* Expedition and Its Report," *Bulletin de l'Institut Océanographique de Monaco*, numéro spécial 2, Vol. 2 (1968), pp. 603–11.

Damien, Raymond. *Albert Ier, Prince Souverain de Monaco*. Paris: Institut de Valois, 1964.

Deacon, Margaret. *Scientists and the Sea, 1650–1900, A Study of Marine Science*. London: Academic Press, 1971.

Dittmar, William. "Report on Researches into the Composition of Ocean-Water Collected by H.M.S. *Challenger* . . .," in *Report on the Scientific Results of the Voyage of H.M.S. "Challenger": Physics and Chemistry*, Vol. 1, part 1. London: Her Majesty's Stationery Office, 1884, 251 pp.

Dupree, A. Hunter. *Science and the Federal Government*. Cambridge: The Belknap Press of Harvard University Press, 1957.

Herdman, Sir William A. *Founders of Oceanography and Their Work, An Introduction to the Science of the Sea*. London: Edward Arnold and Company, 1923.

Huxley, T. H. "Notes from the 'Challenger,'" *Nature*, Vol. 12, No. 303 (1875), pp. 315–16.

—— and Paul Pelseneer. "Report on Spirula," in *Report on the Scientific Results of the Voyage of H.M.S. "Challenger": Summary of Scientific Results*, Part 2. London: Her Majesty's Stationery Office, 1895, 32 pp.

Merriman, Daniel. "Speculations on Life at the Depths: a XIXth Century Prelude," *Bulletin de l'Institut Océanographique de Monaco*, numéro spécial 2, Vol. 2 (1968), pp. 377–84.

——. "Challengers of Neptune: The 'Philosophers,'" *Proceedings of the Royal Society of Edinburgh*, in press.

—— and Mary Merriman. "Sir C. Wyville Thomson's Letters to Staff-Commander Thomas H. Tizard, 1877–1881," *Journal of Marine Research*, Vol. 17 (1958), pp. 347–74.

Mill, Hugh Robert. "Mr. J. Y. Buchanan, F. R. S.," obituary, *Nature*, Vol. 116, No. 2924 (1925), pp. 719–20.

——. *An Autobiography*. London: Longmans, Green, and Company, 1951.

Moseley, Henry Nottidge. *Notes by a Naturalist on the "Challenger."* London: Macmillan and Company, 1879.

Murray, Sir John. *The Ocean, A General Account of the Science of the Sea*. New York: Henry Holt and Company [no pub. date, but about 1913].

Report on the Scientific Results of the Voyage of H.M.S. "Challenger": Narrative of the Cruise, Parts 1 and 2. London: Her Majesty's Stationery Office, 1885 and 1882.

Report on the Scientific Results of the Voyage of H.M.S. "Challenger": Summary of Scientific Results, Vols. 1 and 2. London: Her Majesty's Stationery Office, 1895.

Ritchie, G. S. *The Admiralty Chart, British Naval Hydrography in the 19th Century*. London: Hollis and Carter, 1967.

Russel, F. S., and C. M. Yonge. *The Seas, Our Knowledge of Life in the*

Sea and How It Is Gained. London: Frederick Warne and Company, 1928.

Sigsbee, Charles D. *Deep-Sea Sounding and Dredging*. Washington, D.C.: U.S. Government Printing Office, 1880.

Théodoridès, J. "Les Débuts de la Biologie Marine en France: Jean-Victor Audouin et Henri-Milne Edwards, 1826–1829," *Bulletin de l'Institut Océanographique de Monaco*, numéro spécial 2, Vol. 2 (1968), pp. 417–37.

Thomson, C. Wyville. *The Depths of the Sea*, 2nd edition. London: Macmillan and Company, 1874.

———. "The 'Challenger' Expedition," *Nature*, Vol. 14, No. 361 (1876), pp. 492–95.

———. *The Atlantic, A Preliminary Account of the General Results of the Exploring Voyage of H. M. S. "Challenger,"* 2 Vols. London: Macmillan and Company, 1877.

Chapter IV

Agassiz, Alexander. "Louis François de Pourtalès," obituary, *American Journal of Science*, 3rd Series, Vol. 20, No. 117 (1880), pp. 253–55.

———. *Three Cruises of the U.S.C.G.S. Steamer "Blake,"* published in the *Bulletin of the Museum of Comparative Zoology*, Vols. 14 and 15, 1888.

———. "On the Formation of Barrier Reefs and On the Different Types of Atolls," *Proceedings of the Royal Society of London*, Vol. 71, No. 474 (1903), pp. 412–14.

Agassiz, G. R. *Letters and Recollections of Alexander Agassiz with a Sketch of His Life and Work*. Boston: Houghton Mifflin Company, 1913.

Campbell, George Douglas, 8th Duke of Argyll, "A Conspiracy of Silence," *Nature*, Vol. 37, No. 942 (1887), pp. 53–54.

Corbin, T. W. *The Romance of Submarine Engineering*. London: Seeley, Service and Company, Ltd., 1913.

Daly, Reginald A. "Swinging Sealevel of the Ice Age," *Bulletin of the Geological Society of America*, Vol. 40 (1929), pp. 721–34.

Dana, James D. *Corals and Coral Islands*. New York: Dodd and Mead, Publishers, 1872.

Darwin, Charles. *The Structure and Distribution of Coral Reefs*. London: Smith, Elder, and Company, 1842.

———. *The Autobiography of Charles Darwin 1809–1882*. Nora Barlow, editor. London: Collins, 1958.

Darwin, Francis, editor. *Charles Darwin . . . and a Selected Series of His Published Letters*. New York: D. Appleton and Company, 1893.

Davis, William Morris. *The Coral Reef Problem*, Special Publication No. 9 of the American Geographical Society, 1928.

Donovan, Desmond T. "Henry Marc Brunel: The First Submarine Geological Survey and the Invention of the Gravity Corer," *Marine Geology*, Vol. 5 (1967), pp. 5–14.

Forbes, Henry O. "The Great Barrier Reef of Australia," *The Geographical Journal*, Vol. 2, No. 6 (1893), p. 540–45.

Herdman, William A. *Founders of Oceanography and Their Work*. London: Edward Arnold Ltd., 1923.

Huxley, T. H. "The Duke of Argyll's Charges Against Men of Science," *Nature*, Vol. 37, No. 954 (1888), p. 342.

Kirwan, Laurence P. *A History of Polar Exploration*. New York: W. W. Norton and Company, Inc., 1960.

Ladd, Harry S., and Seymour O. Schlanger. *Drilling Operation on Eniwetok Atoll*. U.S. Geological Survey Professional Paper 260-Y. Washington, D.C.: U.S. Government Printing Office, 1954, pp. 863–905.

Menard, H. W., and H. S. Ladd. "Oceanic Islands, Seamounts, and Atolls," in M. N. Hill, editor. *The Sea, Ideas and Observations*, Vol. 3. New York: Interscience, 1963.

Mitchell, Stephen W. "The 'Lost Atlantis' and the 'Challenger' Soundings," *Nature*, Vol. 15, No. 391 (1877), pp. 553–56.

Murray, John. "Preliminary Reports . . . on Work Done on Board the 'Challenger,' " *Proceedings of the Royal Society of London*, Vol. 24, No. 170 (1875), pp. 471–544.

——. "On the Structure and Origin of Coral Islands," *Proceedings of the Royal Society of Edinburgh*, Vol. 10 (1880), pp. 505–18.

——. "Alexander Agassiz—His Life and Scientific Work," *Science*, Vol. 33, No. 858 (1911), pp. 873–87.

—— and Dr. Johan Hjort. *The Depths of the Ocean*. London: Macmillan and Company, 1912, p. 144.

—— and A. F. Renard. *Report on Deep-Sea Deposits . . .*, in *Report on the Scientific Results of the Voyage of H.M.S. "Challenger."* London: Her Majesty's Stationery Office, 1891.

Pourtalès, Louis François de. "The Gulf Stream . . . ," Appendix 11, *Report of the Superintendent of the U.S. Coast Survey for 1869*, pp. 220–25.

——. "Der Boden des Golfstromes und der Atlantischen Küste Nord-Amerika's," *Petermann's Geographische Mitteilungen*, Vol. 16, 1870, pp. 393–98.

Report on the Scientific Results of the Voyage of the "Challenger": Summary of Scientific Results, Vol. 2. London: Her Majesty's Stationery Office, 1895.

Shepard, Francis P. *Submarine Geology*, 2d edition. New York: Harper & Row, 1963.

Skeats, E. W. "The Coral-Reef Problem and the Evidence of the Funafuti Borings," *American Journal of Science*, 4th series, Vol. 45, No. 266 (1918), pp. 81–90.

Sollas, W. J. "Report to the Committee of the Royal Society Appointed to Investigate the Structure of a Coral Reef by Boring," *Proceedings of the Royal Society of London*, Vol. 60, No. 367 (1896–1897), pp. 502–12.

Stevenson, Robert Louis. *In the South Seas . . .* (Facsimile reproduction). Honolulu: University of Hawaii Press, 1971.

Thomson, C. Wyville. *The Atlantic, A Preliminary Account of the General Results of the Exploring Voyage of H.M.S. "Challenger,"* 2 Vols. London: Macmillan and Company, 1877.

Viglieri, Alfredo. "La Carte Général Bathymétrique des Océans Etablie par S.A.S. le Prince Albert Ier," *Bulletin de l'Institut Océanographique de Monaco*, numéro spécial 2, Vol. 1 (1968), pp. 243–53.

Zittel, Karl Alfred. *History of Geology and Paleontology to the End of the Nineteenth Century*. London: Walter Scott, 1901.

Chapter V

Bjerknes, Vilhelm. "Ueber einen hydrodynamischen Fundamentalsatz und seine Anwendung besonders auf die Mechanik der Atmosphäre und des Weltmeeres," *Kongliga svenska Vetenskaps-Akademiens Handlingar*, Vol. 31, No. 4 (1898), pp. 1–35.

————, et al. *Dynamic Meteorology and Hydrography*. Washington, D.C.: Carnegie Institution of Washington, 1911.

Cox, Roland A. "The Salinity Problem," *Progress in Oceanography*, Vol. 1 (1963), pp. 243–61.

Defant, Albert. *Physical Oceanography*. 2 Vols. Oxford: Pergamon Press, 1961, p. 345.

Dickson, H. N. "The Movements of the Surface Waters of the North Sea," *Geographical Journal*, No. 3 (March, 1896), pp. 255–67.

————. "Surface Waters of the North Atlantic Ocean," *Philosophical Transactions of the Royal Society of London*, Vol. 196A, No. 3 (1901), pp. 61–204.

Ekman, V. W. "On Dead-Water," in F. Nansen, editor, *Scientific Results of the Norwegian North Polar Expedition, 1893–1896*, Vol. 5, No. 15 (1906), pp. 1–152.

Helland-Hansen, Bjørn. "Current-Measurements in Norwegian Fiords . . . ," *Bergens Museums Aarbog*, No. 15 (1907), pp. 3–61.

———— and F. Nansen. "The Norwegian Sea, Its Physical Oceanography Based Upon the Norwegian Researches 1900–1904," *Report on Norwegian Fishery and Marine Investigations*, Vol. 2, part 1, No. 2 (1909), pp. 1–390.

Makaroff, Vice Admiral S. O. "The 'Yermak' Ice-Breaker," *Geographical Journal*, Vol. 15, No. 1 (1900), pp. 32–46.

Merz, Alfred. "Die Atlantische Hydrographie und die Planlegung der Deutschen Atlantischen Expedition," *Sitzungberichte der Preussischen Akademie der Wissenchaften*, No. 31 (1925), pp. 562–86.

Mills, Hugh Robert. "Merz and the 'Meteor' Expedition," *Geographical Journal*, Vol. 68, No. 1 (1926), pp. 73–77.

Murray, Sir John, and Johan Hjort. *The Depths of the Ocean*. London: Macmillan and Company, 1912.

Nansen, Fridtjof. *Farthest North, Being the Record of a Voyage of Exploration of the Ship "Fram" 1893–1896*, 2 Vols. London: Harrison and Sons, 1897.

————. "Some Oceanographical Results of the Expedition with *Michael Sars* in the Summer of 1900," *Nyt Magazin for Naturvidenskaberne*, Vol. 39, No. 2 (1901), pp. 129–61.

————, editor. *Scientific Results of the Norwegian North Polar Expedition, 1893–1896*. 6 Vols. London: Longmans, Green, and Co. 1900–1906.

————. *The Waters of the North-Eastern North Atlantic*. Leipzig: Dr. Werner Klinkhardt, 1913, p. 14.

Neumann, Gerhard, and Willard J. Pierson, Jr. *Principles of Physical Oceanography*. Englewood Cliffs, N.J.: Prentice-Hall, Inc., 1966.

Pettersson, Otto. "A Review of Swedish Hydrographic Research in the Baltic and North Seas." *The Scottish Geographical Magazine*, Vol. 10 (1894), pp. 281–302.

———. "Die Wasserzirculation im Nordatlantischen Ozean," *Petermanns Geographische Mitteilungen*, Vol. 46, 1900, pp. 61–65 and 91–92.

———. "L'Atlantique, Mer Inexplorée," *La Géographie*, Vol. 13, No. 6 (1906), pp. 425–36.

Redfield, William C. "Summary Statements of Some of the Leading Facts in Meteorology," *American Journal of Science and Arts*, Vol. 25 (1834), p. 122–35.

Riley, J. P., and G. Skirrow, editors. *Chemical Oceanography*. London: Academic Press, 1965.

Rouch, J. *Traite d'Océanographie Physique*. Paris: Payot, 1943.

Sandström, Johan, and Bjørn Helland-Hansen. "Über die Berechnung von Meeresströmungen," *Report on Norwegian Fishery and Marine Investigations*, Vol. 2, part 2, No. 4 (1903), pp. 1–43.

Spiesz, F. *Die Meteor-Fahrt. Forschungen und Erlebnisse der Deutschen Atlantischen Expedition, 1925–1927*. Berlin: Dietrich Reimer, 1928.

Spiess, H. C. Fritz (same as Spiesz, F.) *Das Forschungsschiff und seine Reise Wissenschaftliche Ergebnisse Der Deutschen Atlantische, Expedition . . .*, Band 1. Berlin: Verlag von Walter de Gruyter & Co., 1932.

Stommel, Henry. *The Gulf Stream*. Berkeley: University of California Press, 1958.

Sverdrup, H. U., Martin W. Johnson, and Richard H. Fleming. *The Oceans, Their Physics, Chemistry, and General Biology*. Englewood Cliffs, N.J.: Prentice-Hall, Inc., 1942.

"The Voyage of the Meteor" (book review), *Geographical Journal*, Vol. 75, No. 2 (1930), pp. 174–76.

Thompson, D'Arcy W. "The Voyages of the *Discovery*," *Nature*, Vol. 140, No. 3543 (1937), pp. 529–32.

———. "Dr. Otto Pettersson" (obituary), *Nature*, Vol. 147, No. 3736 (1941), pp. 701–02.

Vaughan, Thomas Wayland, et al. *International Aspects of Oceanography*. Washington, D.C.: National Academy of Sciences, 1937.

Von Arx, William. *An Introduction to Physical Oceanography*. Reading, Mass.: Addison-Wesley Pub. Co., 1962.

Welander, Pierre. "Theoretical Oceanography in Sweden 1900–1910," *Bulletin de l'Institut Océanographique de Monaco*, numéro spécial 2, Vol. 1 (1968), pp. 169–73.

Wüst, Georg. "History of Investigations of the Longitudinal Deep-Sea Circulation," *Bulletin de l'Institut Océanographique de Monaco*, numéro spécial 2, Vol. 1 (1968), pp. 109–20.

Chapter VI

Barlett, J. R. "Deep Sea Soundings and Temperatures in the Gulf Stream . . .," (abstract), *Proceedings of the American Association for the Advancement of Science*, 31st meeting (Aug., 1882), pp. 349–52.

Bigelow, Henry Bryant. *Report on the Scope, Problems, and Economic Importance of Oceanography*. . . . Report submitted to the National

Academy of Sciences from the N.A.S. Committee on Oceanography, 1929.

Brandt, Karl. "On the Production and the Conditions of Production in the Sea," Appendix D, *Rapports et Proces-Verbaux*, Vol. 3 (1902–04), 12 pp.

Cleve, P. T. "Microscopic Marine Organisms in the Service of Hydrography," *Nature*, Vol. 55, No. 1413 (1896), pp. 89–90.

Dickson, H. N. "Recent Contributions to Oceanography," *Geographical Journal*, Vol. 3, No. 4 (1894), pp. 302–10.

General Secretary, The. *The International Council for the Exploration of the Sea* (Mimeographed report issued by ICES), 1970, 12 pp.

Goode, G. Brown. *The Fisheries and Fishery Industries of the United States*. Washington, D.C.: U.S. Government Printing Office, 1887.

Graham, M., editor. *Sea Fisheries, Their Investigation in the United Kingdom*. London: Edward Arnold, 1956.

Gran, H. H. "Preservation of Samples and Quantitative Determination of the Plankton," *Publications de Circonstance*, No. 62 (1912), 15 pp.

Haeckel, Ernst. "Planktonic Studies . . .," U.S. Commission of Fish and Fisheries. *Report of the Commission for 1889–91*, pp. 565–641.

Hardy, Sir Alister. *The Open Sea: Its Natural History, Part II, Fish and Fisheries*. London: Collins, 1959.

———. *Great Waters, A Voyage of Natural History to Study Whales, Plankton and the Waters of the Southern Ocean*. New York: Harper & Row, 1967.

Heincke, Friedrich. "Naturgeschichte des Herings . . .," *Abhandlungen des deutschen Seefischerei Vereins*, Vol. 2, 1898.

Hensen, Victor. "Über die Bestimmung des Planktons . . .," *Bericht der Kommissionen Wissenschaftlichen Untersuchung der deutschen Meere in Kiel*, No. 5 (1887), 109 pp.

Herdman, William A. "Stockholm International Conference on the Exploration of the Sea," *Nature*, Vol. 61, No. 1569 (1899), p. 78 and p. 177. Other articles in the same volume continue the debate on the balance between physical and biological work within ICES. See H. M. Kyle, p. 151, and E. J. Allen, p. 54 and p. 227.

Hjort, Johan. "Fluctuations in the Great Fisheries of Northern Europe," *Rapports et Proces-Verbaux*, Vol. 20 (1914), 228 pp.

——— and C. G. Joh. Petersen. "Short Review of the Results of the International Fisheries Investigations," Appendix G. *Rapports et Proces-Verbaux*, Vol. 3 (1902–04), 43 pp.

——— and Johan T. Ruud. "Deep-Sea Prawn Fisheries and Their Problems," *Hvalrådets Skrifter*, No. 17 (1938), pp. 1–144.

Johnstone, James. *Conditions of Life in the Sea: A Short Account of Quantitative Marine Biological Research*. Cambridge, England: The University Press, 1908.

McIntosh, W. C. *The Resources of the Sea. . . .* London: C. J. Clay and Sons, 1899.

Murray, Sir John, and Johan Hjort. *The Depths of the Ocean*. London: Macmillan and Company, 1912.

Petersen, C. G. Joh. "The Sea Bottom and Its Production of Fish-Food," *Report of the Danish Biological Station for 1918*, Vol. 26, 62 pp.

Pettersson, Otto, and C. F. Drechsel. "Mémoire d'une Expédition Internationale pour l'Exploration de la Mer," *Rapports et Proces-Verbaux*, Vol. 32 (1923), pp. 61–71.

Rollefsen, Gunnar. "Foreword" to "Contributions Given in Honour of Einar Koefoed . . .," *Fiskeridirektoratets Skrifter*, Vol. 13, No. 6 (1963), pp. 7–10.

———. "Norwegian Fisheries Research," *Fiskeridirektoratets Skrifter*, Vol. 14, No. 1 (1966), pp. 1–36.

Sette, Oscar E. *Biology of the Atlantic Mackerel* . . ., Fishery Bulletin No. 38 of the Fish and Wild Life Service, 1943.

Tait, John B. *Hydrography in Relation to Fisheries*. (The Buckland Lectures for 1938.) London: Edward Arnold and Company, 1952.

Thompson, John V. *Zoological Researches and Illustrations 1828–1834*. (Facsimile editions, with an introduction by Alwyne Wheeler.) London: Society for the Bibliography of Natural History, 1968.

Thompson, Ruth D'Arcy. *D'Arcy Wentworth Thompson. The Scholar-Naturalist 1860–1948*. London: Oxford University Press, 1958.

Wimpenny, R. S. *The Plankton of the Sea*. . . . London: Faber and Faber, 1966.

Wolff, Torben. *Danish Expeditions on the Seven Seas*. Copenhagen: Rhodos, 1967.

Chapter VII

Annual Report of the Board of Regents of the Smithsonian Institution, 1901.

Bigelow, Henry B. "Address at the Lillie Memorial Meeting, Woods Hole, August 11, 1948," *Biological Bulletin*, Vol. 95, No. 2 (1948), pp. 157–58.

———. "Cruise of the U.S. Fisheries Schooner 'Grampus,' in the Gulf Stream During July, 1908 . . .," *Bulletin of the Museum of Comparative Zoology*, Vol. 52, No. 12 (1909), pp. 195–210.

———. "Explorations in the Gulf of Maine, July and August 1912 . . .," *Bulletin of the Museum of Comparative Zoology*, Vol. 58, No. 2 (1914), pp. 29–147.

——— and William W. Welsh. *Fishes of the Gulf of Maine*, published in the *Bulletin of the U.S. Bureau of Fisheries*, Vol. 40, Part I, 1925.

———. *Memories of a Long and Active Life*. Cambridge, Mass: The Cosmos Press, 1964.

———. *Plankton of the Offshore Waters of the Gulf of Maine* and *Physical Oceanography of the Gulf of Maine*, published in the *Bulletin of the U.S. Bureau of Fisheries*, Vol. 40, Part 2, 1926–1927.

———. *Report on the Scope, Problems, and Economic Importance of Oceanography, on the Present Situation in America, and on the Handicaps to Development, with Suggested Remedies*, report submitted to the National Academy of Sciences from the N.A.S. Committee on Oceanography, 1929. Also published as: *Oceanography, Its Scope, Problems, and Economic Importance*. Boston: Houghton Mifflin Company, 1931.

Carnegie Institution of Washington; Year Book, Nos. 1, 2 and 3, 1902–1904.

"Carnegie Institution of Washington," *Science*, Vol. 17, No. 422 (1903), pp. 166–70.

Dupree, A. Hunter. *Science in the Federal Government: A History of Politics and Activities to 1940.* Cambridge, Mass.: The Belknap Press of Harvard University Press, 1957.

Fay, H. J. W. "The Submarine Signal Company," *Soundings,* No. 8 (Feb., 1945).

Graham, Michael. "Henry Bryant Bigelow" (obituary), *Deep-Sea Research,* Vol. 15 (April, 1968), pp. 125–32.

Hacker, Louis M. *The World of Andrew Carnegie.* . . . Philadelphia: Lippincott, 1968.

Hunt, Frederick V. *Electroacoustics, The Analysis of Transduction and Its Historical Background.* Cambridge, Mass.: Harvard University Press, 1954.

Idyll, C. P., editor. *Exploring the Ocean World.* New York: Thomas Y. Crowell Company, 1969.

"International Ice Observation and Ice Patrol," *International Ice Observation and Ice Patrol Service,* Bulletin No. 9, Season of 1921, pp. 3–6.

Lillie, Frank R. *The Woods Hole Marine Biological Laboratory.* Chicago: University of Chicago Press, 1944.

Oceanography 1960–1970. Washington, D.C.: National Academy of Science –National Research Council, 1962.

Paul, J. Harland. *The Last Cruise of the Carnegie.* Baltimore: The Williams and Wilkins Co., 1932.

Raitt, Helen, and Beatrice Moulton. *Scripps Institution of Oceanography. First Fifty Years.* Los Angeles: The Ward Ritchie Press, 1967.

"Report by the U.S. Hydrographic Office 17th April 1923," *Hydrographic Review,* Vol. 1, No. 2 (1924), pp. 43–48.

Ricketts, Nobel G., and Parker D. Trask. *The "Marion" Expedition to Davis Strait and Baffin Bay under direction of the U.S. Coast Guard, 1928. Scientific Results, Part 1,* published in *International Ice Observation and Ice Patrol Service,* Bulletin No. 19 (1932), pp. 1–81.

Scientific Results of Cruise VII of the "Carnegie" during 1928–1929 . . . , 4 Vols. Washington, D.C.: Carnegie Institution of Washington, Publication 544, 1942–1944.

"Scientific Results of the 'Nautilus' Expedition, 1931, Under the Command of Captain Sir Hubert Wilkins," *Papers in Physical Oceanography and Meteorology,* Vol. 2, Nos. 1 and 3, 1933.

Smith, Edward H. "Report of Ice Observations Made on the Aeroarctic Expedition with the 'Graf Zeppelin,' in 1931," *International Ice Observation and Ice Patrol Service,* Bulletin No. 21, Season of 1931, pp. 44–52.

Sverdrup, H. U., Martin W. Johnson, and Richard H. Fleming. *The Oceans, Their Physics, Chemistry, and General Biology.* Englewood Cliffs, N.J.: Prentice-Hall, Inc., 1942.

The Rockefeller Foundation, Annual Report for 1930.

Trask, Parker D. "Oceanography and Oil Deposits," *Bulletin of the National Research Council,* No. 61 (July, 1927), pp. 235–40.

Vaughan, T. Wayland. "Oceanography in Its Relations to Other Earth Sciences," *Journal of the Washington Academy of Sciences,* Vol. 14, No. 14 (1924), pp. 307–33.

Weber, Gustavus A. *The Coast and Geodetic Survey. Its History, Activity, and Organization.* Institute for Government Research Service Monograph #16. Baltimore: Johns Hopkins Press, 1923.

——. *The Hydrographic Office. Its History, Activities, and Organization.* Institute for Government Research Service Monograph #42. Baltimore: Johns Hopkins Press, 1926.

Whitcroft, Thomas H. "Sonic Sounding as Developed by the U.S. Navy," *U.S. Naval Institute Proceedings,* Vol. 69 (Feb., 1943), pp. 216–23.

Wolff, Torben. *Danish Expeditions on the Seven Seas.* Copenhagen: Rhodos, 1967.

Woods Hole Oceanographic Institute, Collected Reprints, 1938. (Includes Annual Reports for 1930–1933.)

Chapter VIII

Bates, Charles C. "Utilization of Wave Forecasting in the Invasions of Normandy, Burma and Japan," *Annals of the New York Academy of Sciences,* Vol. 51, Art. 3 (May, 1949), pp. 545–72.

Bigelow, Henry Bryant, and W. T. Edmundson. *Wind Waves at Sea, Breakers and Surf.* Washington, D.C.: U.S. Navy Department, Hydrographic Office, 1947.

Cornish, Vaughn. *Ocean Waves and Kindred Geophysical Phenomena.* Cambridge, England: Cambridge University Press, 1934.

Deck Log of the "Atlantis." February, 1937. Unpublished. Obtainable from the Woods Hole Oceanographic Institution.

Drury, A. T. "NRL Looks Back," *Office of Naval Research, Research Reviews* (July, 1953), pp. 14–24.

Dupree, A. Hunter. *Science in the Federal Government.* Cambridge, Mass.: Harvard University Press, 1957.

Emery, K. O., J. I. Tracey, Jr., and H. S. Ladd. *Geology of Bikini and Nearby Atolls.* Geological Survey Professional Paper 260–A. Washington, D.C.: U.S. Government Printing Office, 1954.

Fish, Marie Poland. "The Sonic Marine Animal Problem," *Office of Naval Research, Research Reviews* (Dec., 1954), pp. 13–18.

Glover, R. O. " 'Hydro' Charts a War," *U.S. Naval Institute Proceedings,* Vol. 73, No. 527 (Jan., 1947), pp. 27–37.

Greenberg, Daniel S. *The Politics of Pure Science.* New York: The New American Library, 1967.

Hersey, J. B., and R. H. Backus. "Sound Scattering by Marine Organisms," in M. N. Hill, editor, *The Sea, Vol. 1, Physical Oceanography.* New York: Interscience, 1962, pp. 498–539.

Hodgson, Sloat. "The Development of a Mechanical Bathythermograph" (unpublished article written for the Woods Hole Oceanographic Institution, 1966).

Iselin, Columbus O'D., and Maurice Ewing. *Sound Transmission in Sea Water, a Preliminary Report.* Woods Hole Oceanographic Institution for the National Defense Research Committee, 1941.

Johnson, Martin W. "Sound as a Tool in Marine Ecology . . . ," *Journal of Marine Research,* Vol. 7, No. 3 (November 1948), pp. 443–58.

——, F. Alton Everest, and Robert W. Young, "The Role of Snapping Shrimp . . . in the Production of Underwater Sound in the Sea," *Biological Bulletin,* Vol. 93, No. 2 (1947), pp. 122–38.

Knudsen, Vern O., R. S. Alford, and J. W. Emling. "Underwater Ambient Noise," *Journal of Marine Research*, Vol. 7, No. 3 (1948), pp. 410–29.

Marine Fouling and Its Prevention. Woods Hole Oceanographic Institution for the Bureau of Ships, in cooperation with the U.S. Naval Institute. Annapolis, 1952.

"News and Notes," *Science*, Vol. 114, No. 2956 (August 24, 1951), p. 226.

Oceanography in Japan. U.S. Naval Technical Mission to Japan, 1946.

Oceanography 1960 to 1970. Washington, D.C.: National Academy of Science—National Research Council, 1962.

Rossby, C. G. "On Displacements and Intensity Changes of Atmospheric Vortices, *Journal of Marine Research*, Vol. 7, No. 3 (Nov., 15, 1948), pp. 175–87.

Scripps Institution of Oceanography, Collected Reprints, 1939–1947.

Scripps Institution of Oceanography. "On Wave Heights in Straits and Sounds Where Incoming Waves Meet a Strong Tidal Current." Wave Report No. 11. Mimeographed. La Jolla, California, 1944.

Shepard, F. P., K. O. Emery, and H. R. Gould. "Distribution of Sediments on East Asiatic Continental Shelf," *Allan Hancock Foundation Publications*, Occasional Paper No. 9 (1949), pp. 1–64.

Spilhaus, Athelstan. "A Bathythermograph," *Journal of Marine Research*, Vol. 1, No. 2 (April 9, 1938), pp. 95–100.

Stommel, Henry. "The Westward Intensification of Wind-Driven Ocean Currents," *Transactions, American Geophysical Union*, Vol. 29, No. 2, (1948), pp. 202–06.

Sverdrup, H. U. "Research Within Physical Oceanography and Submarine Geology . . . ," *Transactions of the American Geophysical Union*, Vol. 27, No. 4 (1946), pp. 571–73.

———, and W. H. Munk. *Wind, Sea, and Swell: A Theory of Relations for Forecasting*. . . . Washington, D.C.: U.S. Navy Department, Hydrographic Office, 1947.

Taylor, Albert Hoyt. *The First Twenty-five Years of the Naval Research Laboratory*. Washington, D.C.: U.S. Government Printing Office, 1948.

Weyl, F. Joachim, editor. *Research in the Service of National Purpose*. Washington, D.C.: U.S. Government Printing Office, 1966.

Woods Hole Oceanographic Institution, Collected Reprints, 1939–1945.

Chapter IX

Beloussov, V. V. "Debate About the Earth. An Open Letter to J. Tuzo Wilson," *Geotimes*, Vol. 13, No. 10 (1968), pp. 17–19.

"Birmingham Meeting, 1950," *Proceedings of the British Association for the Advancement of Science*, Vol. 7, No. 27 (1950), pp. 275–78.

Blackett, P. M. S., Sir Edward Bullard, and S. K. Runcorn, editors. *A Symposium on Continental Drift*. London: The Royal Society, 1965.

Chamberlin, R. T. "Some of the Objections to Wegener's Theory," *Theory of Continental Drift, a Symposium*. Tulsa, Okla.: American Association of Petroleum Geologists, 1928, pp. 83–87.

Cox, Allen, and Richard Doell. "Geomagnetic Polarity Epochs and Pleistocene Geochronometry," *Nature*, Vol. 198, No. 4885 (1963), pp. 1049–51.

Dietz, R. S. "Continent and Ocean Basin Evolution by Spreading of the Sea Floor," *Nature*, Vol. 190, No. 4779 (1961), pp. 854–57.

Du Toit, A. L. *Our Wandering Continents, an Hypothesis of Continental Drifting.* Edinburgh: Oliver & Boyd, 1937.

Ewing, Maurice, and Bruce Heezen. "Oceanographic Research Programs of the Lamont Geological Observatory," *Geographical Review,* Vol. 46, No. 4 (1956), pp. 508–35.

———, A. P. Crary, and H. M. Rutherford. "Geophysical Investigations in the Emerged and Submerged Atlantic Coastal Plain," *Bulletin of the Geological Society of America,* Vol. 48 (June, 1937), pp. 753–802.

Fisher, Osmond. "On the Physical Cause of the Ocean Basins," *Nature,* Vol. 25, No. 636 (1882), pp. 243–44.

Heezen, B. C., M. Tharp, and Maurice Ewing. "The Floors of the Ocean. I. North Atlantic," *Geological Society of America Special Paper 65,* 1959, 122 pp.

Heirtzler, J. R., *et al.* "Marine Magnetic Anomalies, Geomagnetic Field Reversals and Motions of the Ocean Floor and Continents," *Journal of Geophysical Research,* Vol. 73 (1968), pp. 2119–36.

Hess, H. H. "Major Structural Features of the Western North Pacific, an Interpretation of H.O. 5485, Bathymetric Chart, Korea to New Guinea," *Bulletin of the Geological Society of America,* Vol. 59 (May, 1948), pp. 417–46.

———. "Evolution of Ocean Basins," manuscript prepared in December 1960 for M. N. Hill, editor, *The Sea, Ideas and Observations,* but not published in that work. It appeared instead as "History of Ocean Basins."

———. "History of Ocean Basins," in A. E. J. Engel *et al.,* editors, *Petrologic Studies: A Volume in Honor of A. E. Buddington.* The Geological Society of America, 1962, pp. 599–620.

Holmes, Arthur. "Radioactivity and Earth Movements," *Transactions of the Geological Society of Glasgow,* Vol. 18, part 3 (1931), pp. 559–606.

———. "The South Atlantic: Land Bridges or Continental Drift?" *Nature,* Vol. 171, No. 4355 (1953), pp. 669–71.

Idyll, C. P., editor. *Exploring the Ocean World.* New York: Thomas Y. Crowell Company, 1969.

JOIDES. "Deep-Sea Drilling Project," *Bulletin of the American Association of Petroleum Geologists,* Vol. 51, (part II), No. 9 (1967), pp. 1787–1802.

Kelley, T. Nathan. "1939–'41, Notebook," unpublished.

Lear, John. "Canada's Unappreciated Role as Scientific Innovator," *Saturday Review,* Sept. 2, 1967, pp. 45–50.

Mason, R. G. "Geophysical Investigations of the Sea Floor," *Liverpool and Manchester Geological Journal,* Vol. 2, part 3 (1960), pp. 389–410.

Mather, Kirtly F., editor. *Source Book in Geology.* Cambridge, Mass.: Harvard University Press, 1967.

Maxwell, A. E., *et al. Initial Reports of the Deep Sea Drilling Project, Vol. 3.* Washington, D.C.: U.S. Government Printing Office, 1970.

McKenzie, D. P. "Plate Tectonics and Continental Drift," *Endeavour,* Vol. 29, No. 106 (Jan., 1970), pp. 39–44.

Menard, Henry. "Development of Median Elevations in Ocean Basins," *Bulletin of the Geological Society of America,* Vol. 69, part 2 (Sept., 1958), pp. 1179–85.

Meyerhoff, A. A. "Arthur Holmes: Originator of Spreading Ocean Floor Hypothesis," *Journal of Geophyiscal Research*, Vol. 37, No. 20 (1968), pp. 6563–65.

———. "Spreading Sea Floor: By Convection or Conviction," *Bulletin of the American Association of Petroleum Geologists*, Vol. 53, part 1, No. 1 (1969), pp. 215–16.

Moseley, Henry Nottidge. *Notes by a Naturalist on the "Challenger."* London: Macmillan and Company, 1879, p. 585.

Munk, Walter H. "Polar Wandering: A Marathon of Errors," *Nature*, Vol. 177, No. 4508 (1956), pp. 551–54.

Pettersson, Hans. "Exploring the Bed of the Ocean," *Nature*, Vol. 164, No. 4168 (1949), pp. 468–70.

Revelle, Roger, and Arthur E. Maxwell. "Heat Flow Through the Floor of the Eastern North Pacific," *Nature*, Vol. 170, No. 4318 (1952), pp. 199–200.

Ross, David A. *Introduction to Oceanography.* New York: Appleton-Century-Crofts, 1970.

Runcorn, S. K., editor. *Continental Drift.* New York: Academic Press, 1962.

Rupke, N. A. "Continental Drift Before 1900," *Nature*, Vol. 227, No. 5255 (1970), pp. 349–50.

Scientific American, Vol. 221, No. 3 (Sept., 1969), an issue devoted to the ocean.

Shepard, Francis P. *Submarine Geology.* 2d edition. New York: Harper & Row, 1963.

"Summary of Scientific Achievements," *Ocean Industry*, Vol. 4, No. 5 (1969), p. 67.

Takeuchi, H., S. Uyeda, and H. Kanamori. *Debate About the Earth.* San Francisco: Freeman, Cooper & Co., 1967.

The Navy-Princeton Gravity Expedition to the West Indies in 1932. Washington, D.C.: U.S. Hydrographic Office, 1933.

Vine, Frederick J. "Evidence from Submarine Geology," *Proceedings of the American Philosophical Society*, Vol. 112, No. 5 (1968), pp. 325–34.

———, and D. H. Matthews. "Magnetic Anomalies Over Oceanic Ridges," *Nature*, Vol. 199, No. 4897 (1963), pp. 947–49.

Von Herzen, R. P., and M. G. Langseth. "Present Status of Oceanic Heat-flow Measurements," *Physics and Chemistry of the Earth*, Vol. 6 (1966), pp. 365–408.

Wegener, Alfred. *The Origin of Continents and Oceans.* Translated from the 3rd German edition. New York: E. P. Dutton & Co., 1924.

Wilson, J. Tuzo. "Continental Drift," *Scientific American*, Vol. 208, No. 4 (1963), pp. 86–100.

———. "Static or Mobile Earth: The Current Scientific Revolution," *Proceedings of the American Philosophical Society*, Vol. 112, No. 5 (1968), pp. 309–20.

Wiseman, J. D. H., and R. B. S. Sewell. "The Floor of the Arabian Sea," *The Geological Magazine*, Vol. 74, No. 875 (1937), pp. 219–30.

Page numbers for illustrations are in italics.